HYDRATION AND INTERMOLECULAR INTERACTION

Infrared Investigations with Polyelectrolyte Membranes

HYDRATION AND INTERMOLECULAR INTERACTION

Infrared Investigations with Polyelectrolyte Membranes

GEORG ZUNDEL

PHYSIKALISCH-CHEMISCHES
INSTITUT DER UNIVERSITÄT
MÜNCHEN, GERMANY

ACADEMIC PRESS New York and London 1969

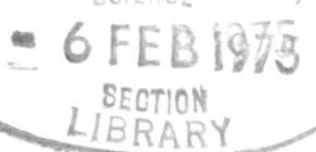

ACADEMIC PRESS, INC.
111 Fifth Avenue, New York, New York 10003

United Kingdom Edition published by
ACADEMIC PRESS, INC. (LONDON) LTD.
Berkeley Square House, London W1X 6BA

LIBRARY OF CONGRESS CATALOG CARD NUMBER: 68–23495

PRINTED IN THE UNITED STATES OF AMERICA

FOREWORD

It is with great pleasure that I offer my best wishes for the success of this work. My modest contribution to it, at best, is that I drew the attention of Dr. Zundel to the intriguing possibilities of ion-exchange resins. My interest originally came from the study of catalysis, and it is exclusively to the author's credit that he recognized that in water-containing ionized resins we have a unique model of electrolyte solutions, enabling one to study the behavior of water in the solvation sphere without interference from the bulk of solvent.

At first alone, and later with a group of able young collaborators, Dr. Zundel exhaustively studied these phenomena using infrared spectroscopy. Although normally used for the study of intramolecular structure problems, infrared spectroscopy proved especially useful for the study of intermolecular interaction. By varying the cations, fixed ions, water isotopes, or degree of swelling, it became possible, for the first time, to obtain a clear picture of the interactions of water with cations, with anions, with groups of both, and even of interactions among its own molecules. The findings on the tunneling effect of protons in hydrogen bridges within hydronium structures were a far-reaching and unforeseeable result of these studies, findings which are in agreement with current results from other research.

Beyond doubt, the material offered in this book will be of great value to chemists, electrochemists, and spectroscopists, among others. I would like to express my satisfaction that these successful investigations stemmed from this institution. I hope that this work will enhance our knowledge of ion exchangers and electrolyte solutions.

GEORG-MARIA SCHWAB
Institute of Physical Chemistry
Munich, Germany

ACKNOWLEDGMENT

Since 1958, our group at the Physical Chemistry Institute of the University at Munich has been investigating the molecular processes of hydration, particularly the hydration of polyelectrolytes, using infrared spectroscopy. This monograph gives a comprehensive survey of the results obtained to date and describes in detail the methods developed.

I most sincerely thank Professor Dr. h. c. mult. G.-M. Schwab, Director of the Institute, who has always supported my work in every possible way. Special thanks are due my co-workers Dr. A. Murr, Dr. H. Metzger, Dr. W. D. Mross, and Dr. Ilse Kampschulte-Scheuing, who have worked untiringly with me.

My collaboration with Doz. Dr. E. G. Weidemann (Institute for Theoretical Physics, University of Munich) and Doz. Dr. T. Ackermann (Institute for Physical Chemistry, University of Muenster, Westphalia) resulted in important data. I am grateful for the stimulating discussions with Professor Dr. H. Noller (Institute of Physical Chemistry), Dr. K. H. Preuss (Max Planck Institute for Theoretical Physics, Munich), Dr. M. I. Vinnik (Institute for Chemical Physics, Academy of Sciences of the USSR, Moscow). I am also grateful to Doz. Dr. T. Ackermann for having read the manuscript and for his many valuable comments.

It is with great pleasure that I thank Professor Dr. F. C. Nachod for inspiring me to write this monograph, for checking the English version of the manuscript, and for the generous hospitality he extended to me in his home in Kinderhook, New York.

I am grateful to Mr. A. Roehrich for supplying glassware and chemicals and to Mr. M. Zellner for building research apparatus. Finally, I thank Mr. K. L. Johnson, B. Sc., for translating the manuscript into English and Miss I. Straub for executing the drawings and for typing.

I also express my gratitude to the Deutsche Forschungsgemeinschaft for providing almost all the means for this work.

GEORG ZUNDEL

Haisterkirch, Germany

October, 1969

CONTENTS

Chapter V **The Acids**

Chapter VI **Preparation of the Membranes**

Chapter VII **IR Investigation Method**

HYDRATION AND INTERMOLECULAR INTERACTION

Infrared Investigations with Polyelectrolyte Membranes

CHAPTER I

INTRODUCTION

More than forty years ago, Debye and Hueckel[1] succeeded in describing the thermodynamic properties of very dilute, completely dissociated electrolyte solutions on the basis of molecular concepts. Some years later, Bjerrum[2] introduced the idea of the ion pair into the discussion of electrolyte solutions, and Bernal and Fowler,[3] in their well-known work, provided a starting point for the study of the structure of water and electrolyte solutions. At that time, however, there existed very little experimental information on the specific interactions between ions, between ions and solvent molecules, and between solvent molecules. For this reason, the models proposed at that time were based only on quite general ideas. In recent years, knowledge of the physical properties of liquids, and particularly electrolyte solutions, has been widely extended. Classical experimental methods have been supplemented by methods of molecular physics which give detailed insights into molecular processes. Among them, methods—such as, for example, infrared spectroscopy—which necessitate but little external action on the system are particularly valuable.

In spectroscopic investigations of homogeneous liquid electrolyte solutions, a serious difficulty is encountered. In addition to the water of hydration, there is also water which is barely or not at all influenced by ions. However, the absorption bands of this water overlap those of the water of hydration in a troublesome way. Spectroscopy is therefore particularly well suited for the investigation of very concentrated electrolyte solutions, especially polyelectrolytes.

The investigations described in the present monograph have been preferentially carried out with polyelectrolytes, namely organic ion exchangers. In ion

[1] P. Debye and E. Hueckel, *Z. Physik* **24**, 185 (1923).
[2] N. Bjerrum, *Kgl. Danske Videnskab. Selskab Mat. Fys. Medd.* **7**, (9), 1 (1926).
[3] J. D. Bernal and R. H. Fowler, *J. Chem. Phys.* **1**, 515 (1933).

exchangers based on macromolecular networks*[4–13a], one kind of ion is bound by chemical linkage to the polymer network, while the other is more or less freely mobile. If this system is dried, the number of water molecules per ion may be reduced to any required degree without subsequent crystallization, since one kind of ion is bound to the polymer network. Glueckauf[14] pointed this out more than ten years ago.

For the present investigations, ion exchangers were prepared in the form of membranes 5 μ thick, as described in Chapter VI. A variety of membranes containing specific anions and cations have been examined, embracing more than 30 salts of polystyrenesulfonic acid. The dependence on the degree of hydration, both with H_2O and with D_2O, was studied. The procedure is described in Chapter VII. These investigations yielded detailed information on the interactions between cations and anions and between ions and water of hydration. In addition, an exact model of the hydration structure in polyelectrolytes was developed.

In Chapter IV, the hydration of the salts is reviewed. The bands resulting from the stretching vibrations of the bonds in the anions give information on the interaction between cation and anion, and hence on the location of the cations with respect to the anions, the linking of anions by polyvalent cations, the dissociation process, and the solvation dependence of the bonds in ion pairs. At low degrees of hydration, the position of the band of the stretching vibration of the molecules of the water of hydration depends on the nature of

* In references 4–13a some monographs and textbooks giving more detailed information on the nature of ion exchangers are listed.

[4] F. C. Nachod, "Ion Exchange, Theory and Application." Academic Press, New York, 1949.

[5] F. C. Nachod and J. Schubert, "Ion Exchange Technology." Academic Press, New York, 1956.

[6] R. Griessbach, "Austauschadsorption in Theorie und Praxis." Akademie Verlag, Berlin, 1957.

[7] C. Calmon and T. R. E. Kressman, "Ion Exchangers in Organic and Biochemistry." Wiley (Interscience), New York, 1957.

[8] F. Helfferich, "Ion Exchange." McGraw-Hill, New York, 1962.

[9] "Anomalien bei Ionenaustauschvorgängen." (K. Issleib, ed.). Akademie Verlag, Berlin, 1962.

[10] K. Dorfer, "Ionenaustauscher, Eigenschaften und Anwendungen." De Gruyter, Berlin, 1963.

[11] J. A. Kitchener, "Ion Exchange Resins." Methuen, London, 1957.

[12] J. Inczédy, "Analytische Anwendungen von Ionenaustauschern." Akadémiai Kiadó, Budapest, 1964.

[13] B. N. Laskorin, N. M. Smirnova, and M. N. Gantman, "Ionenaustauschermembranen und ihre Anwendung." Akademie Verlag, Berlin, 1966.

[13a] R. Hering, "Chelatbildende Ionenaustauscher," Akademie Verlag, Berlin, 1967.

[14] E. Glueckauf, *Proc. Roy. Soc.* (*London*) **A214**, 207 (1952).

both the cation and the anion. Under these conditions, the hydrogen bridges linking the water molecules with the anions are stronger, the more strongly the hydrogen atoms of the water become positively charged by the cations, and the stronger the hydrogen-bridge acceptor property of the oxygen atoms of the anions. The cation–water and anion–water interactions are coupled at low degrees of hydration. It is apparent that a slightly disintegrated network of ions and hydration-water molecules is present. The extent to which this network is disintegrated in particular cases can be determined from the band of the free OH groups. As the degree of hydration increases, it is possible to see how the hydrogen bridges to the oxygen atoms of the anions become weaker, and how a second layer of water molecules is inserted between the cation and neighboring anions.

Acids are discussed in Chapter V. The bands obtained for the OH group of the acid in the hydrogen bridges give information on the degree of association of these groups. Further, from the stretching vibration bands of the bonds in the anions, the true degree of dissociation can be followed, since the electrons in the anions are rearranged at the approach of the acid proton, and other bands are observed. The acids studied had widely differing true degrees of dissociation, the causes of which are explained. In all these acids, the spectroscopic results give a picture of the hydrate structures and their dependence on the degree of hydration. The results relating to the dissociated hydrated excess proton are particularly interesting. The $H_5O_2^+$ and $H_9O_4^+$ groups occupy a special position within the hydrate structures. The excess proton tunnels in the hydrogen bridges of these groups. In such systems, i.e., those with tunneling protons, an intense continuous absorption is observed in the IR spectrum. This indicates that, in a noncrystalline medium of the investigated type, we have a continuous energy-level distribution for the excess protons because the tunneling protons are more or less strongly coupled by so-called proton dispersion forces. These forces are of a magnitude which is comparable with other intermolecular forces.

This brief survey shows how exhaustive and detailed are the contributions which IR spectroscopy can make to the solution of the problem of polyelectrolyte hydration.

Apart from this, our results relating to interaction and structure in polyelectrolytes hardly differ from those obtained with very concentrated electrolyte solution. The smallness of the differences is shown in the results given in Section V.7, and, in particular, by the comparison of the spectrum of saturated aqueous *p*-toluenesulfonic acid solution with that of polystyrenesulfonic acid, in Fig. 96. Thus the results obtained here, if restricted to the immediate neighborhood of the ions, can be generalized to homogeneous liquid electrolyte solutions.

Finally, the present investigations can serve as a starting point for corresponding studies of biopolymers, since ion-exchange resins, though simpler in their structure, still resemble biopolymers closely in many respects.

Before discussing the results in detail, it is necessary to consider a peculiarity of IR spectroscopic investigations of electrolyte solutions, which is as follows. The hydrate structures are frequently transformed through the thermal motions of the hydration water molecules.[15-18] Hence, IR spectroscopic investigations of electrolyte solutions give information on structures present only in dynamic equilibrium with analogous structures. They give us essentially a picture of the nature of temporary equilibrium configurations. When a hydrate structure and its properties are mentioned in the text, it must always be borne in mind that this structure exists only in dynamic equilibrium with analogous structures.

[15] T. J. Swift and R. E. Connick, *J. Chem. Phys.* **37**, 307 (1962).
[16] H. G. Hertz and M. D. Zeidler, *Ber. Bunsenges. Phys. Chem.* **67**, 774 (1963).
[17] H. G. Hertz and M. D. Zeidler, *Ber. Bunsenges. Phys. Chem.* **68**, 821 (1964).
[18] M. Eigen, *Pure Appl. Chem.* **6**, 97 (1963).

CHAPTER II

ASSIGNMENT OF THE IR BANDS

Spectral bands for our compounds can usually be assigned by comparing them with the assignments given by other workers in their investigation of similar compounds. For assignment, monographs (see footnotes 1–4) and tables (footnote 5) were consulted. In the cases of polystyrene and polystyrenesulfonic acid, comparison with the corresponding perdeuterated compounds proved quite informative.[6] The assignments are summarized in the tables in the Appendix. Readers only interested in problems related to hydration may use these band assignments and omit the remainder of this chapter. Additional IR- and Raman-spectroscopic studies with synthetic ion-exchange resins have been reported.[7-12]

II.1. Polystyrene

Figure 1(a) shows the spectrum of a polystyrene membrane, and Fig. 1(g) that of a membrane of perdeuterated polystyrene. The assignment of the bands is given in Table A.1.*

* Tables A.1–A.5 will be found in the Appendix beginning on p. 273.

[1] R. N. Jones and C. Sandorfy, in "Chemical Applications of Spectroscopy" (W. West, ed.). Wiley (Interscience), New York, 1958.

[2] L. J. Bellamy, "The Infra-Red Spectra of Complex Molecules," 2nd Ed. Methuen, London, 1958.

[3] W. Bruegel, "Einfuehrung in die Ultrarotspektroskopie," 3rd Ed. Steinkopff, Darmstadt, 1962.

[4] K. Nakamoto, "Infrared Spectra of Inorganic and Coordination Compounds," 1st Ed. Wiley, New York, 1963.

[5] W. Otting, "Spektrale Zuordnungstabelle der Infrarotabsorptionsbanden." Springer, Berlin, 1963.

[6] W. D. Mross and G. Zundel, *Spectrochim. Acta*, in press.

[7] A. Strasheim and K. Buijs, *Spectrochim. Acta* **17**, 388 (1961).

[8] K. Buijs, *J. Inorg. Nucl. Chem.* **24**, 229 (1962).

[9] J. M. Martscheskaia, O. D. Kurulenko, and S. W. Gerei, *Ukr. Khim.* **31**, 717 (1965).

[10] J. E. Gordon, *J. Phys. Chem.* **66**, 1150 (1962).

[11] S. Lapanje and St. A. Rice, *J. Am. Chem. Soc.* **83**, 496 (1961).

[12] A. Strasheim and K. Buijs, *Spectrochim. Acta* **16**, 1010 (1960).

The broad periodic humps in these spectra are due to interferences (see Section VI.1, p. 235).

The bands of the stretching vibrations of the >CH (>CD) groups in the benzene ring occur in the region 3100–3000 cm^{-1} (2300–2250 cm^{-1}).[13] As can be seen from Schmid and Langenbucher,[14] these bands cannot be assigned to individual vibrations of these groups, particularly since coupling with combination vibrations occurs.

The band at 2924 cm^{-1} (2193 cm^{-1}) is caused by the antisymmetric and that at 2851 cm^{-1} (2100 cm^{-1}) by the symmetric stretching vibration of the $—CH_2—$ ($—CD_2—$) group. They have been assigned according to Wiberley *et al.*[15] The stretching vibration of the tertiary CH group is generally not observed, according to Jones and Sandorfy (footnote 1, p. 340). Actually no band which could be assigned to this vibration is found. However, if only the α-position is deuterated or, conversely in an otherwise fully deuterated molecule only the α-position is not deuterated, then, most surprisingly, this band is encountered.[16–20]

Four weak bands are observed in the region 2000–1700 cm^{-1}. According to Whiffen,[21] they are caused by combination vibrations and by overtones of the out-of-plane bending vibrations of the >CH groups of the benzene ring.

The skeleton stretching vibrations of the benzene ring appear as two doublets[13]: at 1601 and 1583 cm^{-1} and at 1494 and 1453 cm^{-1}. The corresponding bands of the perdeuterated compound occur at 1568 and 1540 cm^{-1} and at 1376 and 1323 cm^{-1}. The fifth skeleton vibration has not been observed.

The band of the scissor vibration of the $—CH_2—$ group is masked by the skeleton vibration at 1453 cm^{-1}. In isotactic polystyrene the band is observed as a shoulder at 1444 cm^{-1}.[15] The corresponding $—CD_2—$ scissor vibration is found at 1051 cm^{-1}.

The bands of the in-plane bending vibrations of the benzene ring are also two doublets.[13] They are found at 1181 and 1154 cm^{-1} as well as at 1069 and 1028 cm^{-1}; the corresponding bands of the perdeuterated compound occur at 866 and 839 cm^{-1} and at 820 and 786 cm^{-1}.

[13] R. R. Randle and D. H. Whiffen, *in* "Molecular Spectroscopy" (G. Sell, ed.) p. 111. Inst. Petrol, London, 1955.
[14] E. D. Schmid and F. Langenbucher, *Spectrochim. Acta* **22**, 1621 (1966).
[15] S. E. Wiberley, S. C. Bunce, and W. H. Bauer, *Analyt. Chem.* **32**, 217 (1960).
[16] M. Kobayashi, *Bull. Chem. Soc. Japan* **33**, 1416 (1960).
[17] M. Kobayashi, *Bull. Chem. Soc. Japan* **34**, 560 (1961).
[18] M. Kobayashi, *Bull. Chem. Soc. Japan* **34**, 1045 (1961).
[19] H. Tadokoro, Y. Nishijama, S. Nozakura, and S. Murahashi, *Bull. Chem. Soc. Japan* **34**, 381 (1961).
[20] T. Onishi and S. Krimm, *J. Appl. Phys.* **32**, 2320 (1961).
[21] D. H. Whiffen, *Spectrochim. Acta* **7**, 253 (1955).

The weak bands at 907 and 841 cm^{-1} (759 and 655 cm^{-1}), according to Randle and Whiffen,[13] are due to out-of-plane bending vibrations of the $\gt$CH ($\gt$CD) groups of the benzene ring.

Two intense bands are characteristic for monosubstituted benzene rings. One is at 760 cm^{-1} (675 cm^{-1}, considerably weaker than for H), the other at 698 cm^{-1} (549 cm^{-1}). The band at 760 cm^{-1} must be assigned to an out-of-plane bending vibration of the five $\gt$CH groups of the benzene ring. According to an investigation by McMurry and Thornton,[22] (also see footnote 1, p. 391), in systems of the type $C_6H_5CHR_2$, this band occurs as an exception not in the region 747–737 cm^{-1}, but in the region 763–758 cm^{-1}, i.e., exactly where we have observed it. The band at 698 cm^{-1} is that of an out-of-plane skeleton bending vibration of the benzene ring.

Assignments of polystyrene, deuterated to various degrees, can be found in the literature.[16–19] Wave number values have been calculated by Schmid *et al.*[23,24] for mono- and *p*-disubstituted benzene rings. These calculations confirm our assignments in all cases. The modes of all vibrations of the benzene ring can be obtained from these papers, for monosubstituted rings,[23] and for *p*-disubstituted rings.[24]

The assignment of a considerable number of the bands of the acids and salts is not discussed in Sections II.2–II.4. It follows directly by comparison with the assignment of corresponding bands for polystyrene here reported.

II.2. Polystyrenesulfonyl Chloride, Polystyrenesulfonic Acid and Its Salts

The spectra of these substances are shown in Fig. 1(b–f, h, i). The assignments of the bands are collected in Table A.1.[25]

On incorporation of the —SO_2Cl groups, the intensities of the bands of the stretching vibrations of $\gt$CH groups of the benzene ring in the region 3100–3000 cm^{-1} diminish considerably and remain quite weak when these groups are converted into —SO_3^- ions. The same is observed on incorporation of other anions, as can be seen in Figs. 4, 5, 6, and 7. Further, it is found that the positions and particularly the relative intensities of these three bands change, depending on the anion (Tables A.1–A.5). This is understandable in

[22] H. L. McMurry and V. Thornton, *Anal. Chem.* **24**, 318 (1952).

[23] E. W. Schmid, J. Brandmueller, and G. Nonnenmacher, *Z. Elektrochem.* **64**, 726 (1960).

[24] E. W. Schmid, J. Brandmueller, and G. Nonnenmacher, *Z. Elektrochem.* **64**, 940 (1960).

[25] G. Zundel, H. Noller, and G.-M. Schwab, *Z. Naturforsch.* **16b**, 716 (1961).

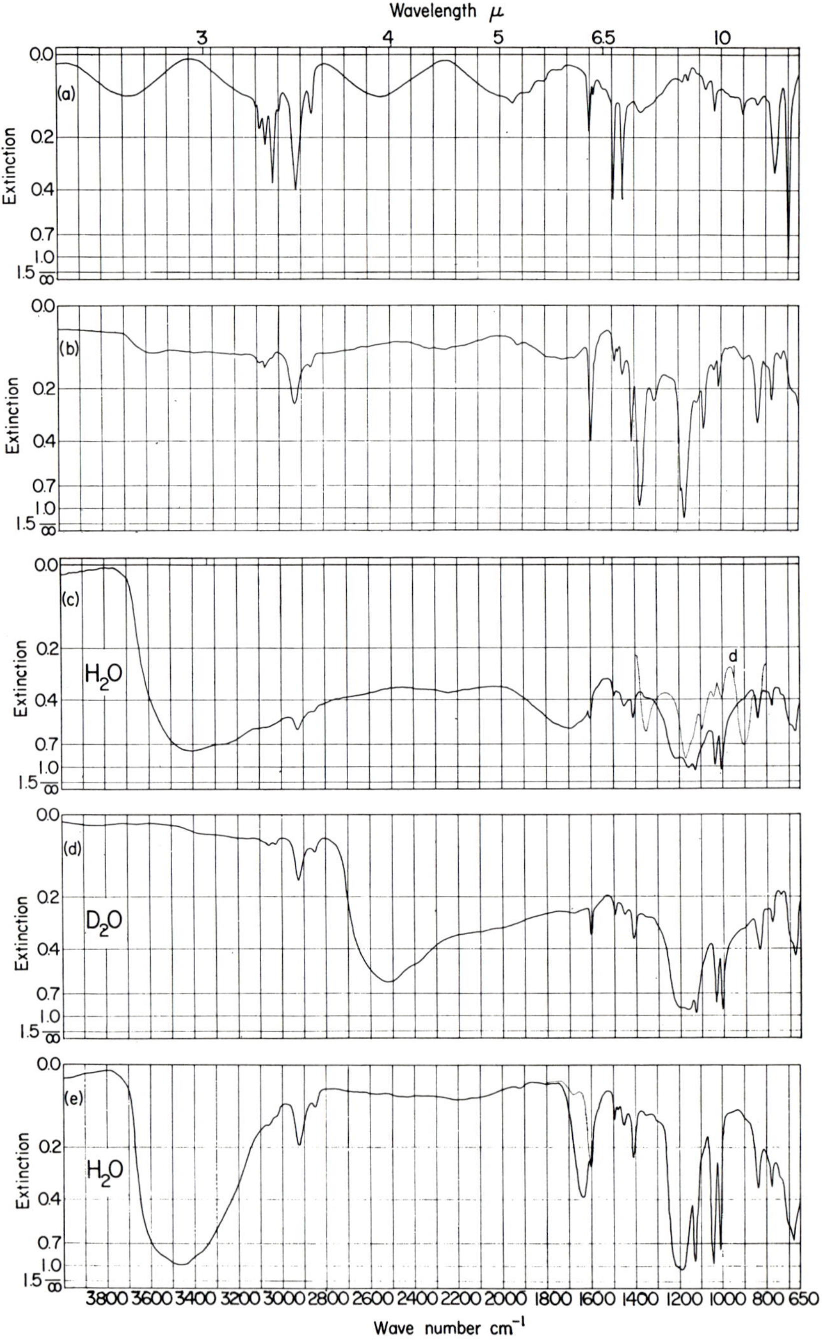

Wavelength μ
3
4
5
6.5
10
Extinction
0.0
0.2
0.4
0.7
1.0
1.5
∞
(a)
(b)
(c)
(d)
(e)
H2O
D2O
H2O
d
3800 3600 3400 3200 3000 2800 2600 2400 2200 2000 1800 1600 1400 1200 1000 800 650
Wave number cm-1

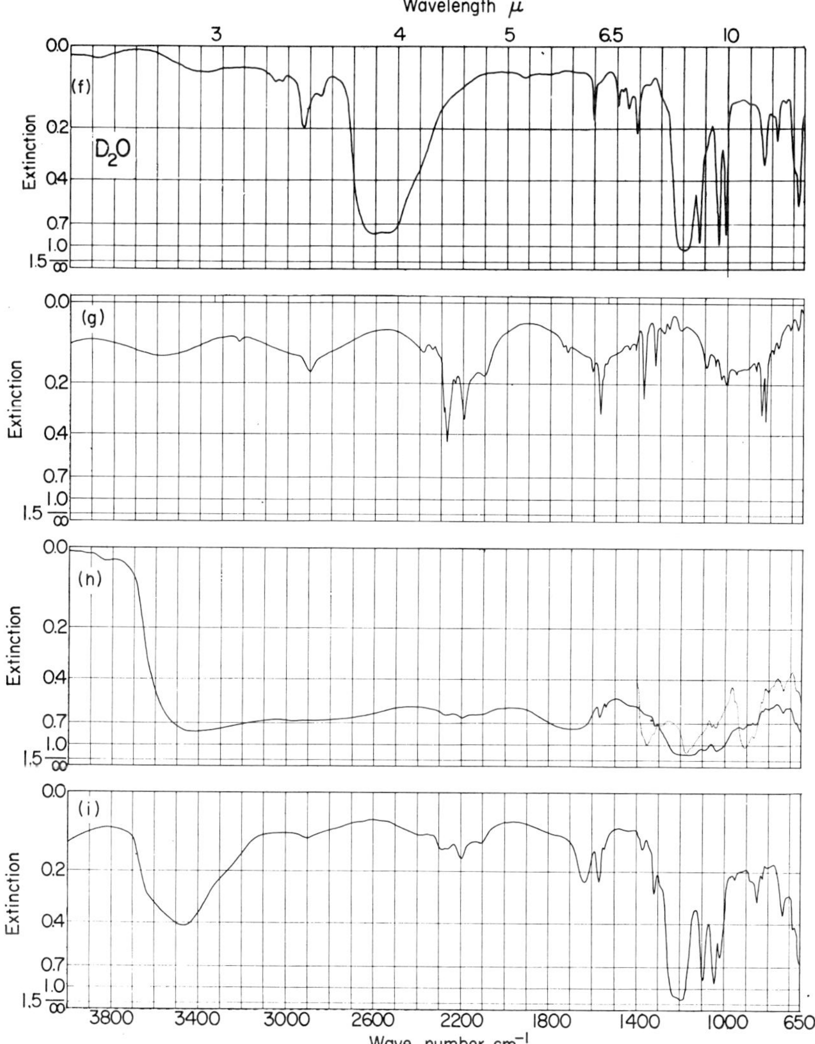

FIG. 1. IR spectra. (a) Polystyrene, 5% cross-linked (see p. 232). (b) Polystyrenesulfonyl chloride, 5% cross-linked, 3 days of sulfonation. (c) Polystyrenesulfonic acid, 5% cross-linked, 3 days sulfonation, H_2O hydrated: ——, at 98% relative atmospheric humidity (relative humidity of the atmosphere with which the membrane was in equilibrium at the time the spectrum was taken); ····, thoroughly dried membrane. (d) Polystyrenesulfonic acid, 5% cross-linked, 3 days of sulfonation, D_2O-hydrated: ——, at atmospheric humidity over a saturated solution of $BaCl_2$ in D_2O. (e) Na^+ salt of polystyrenesulfonic acid, 5% cross-linked, 3 days of sulfonation, H_2O hydrated:

Continued at foot of p. 10.

the light of results reported in the literature. Schmid and co-workers[14,26–31] have shown that the intensity and position of these bands depend on the attractive force which the substituents exert on the electrons in the benzene ring. Staab (see footnote 32, p. 575) in discussing these substituents writes: "... sulfonyl, sulfonic acid, and sulfonamide groups all belong to the —M substituents, just as do other substituents in which a S=O double bond is in the neighborhood of a C=C double bond system.[33]"

Assignment of the bands of the stretching vibrations of the SO bonds: See the works consulted here (footnotes 34–52).

[26] E. D. Schmid and J. Bellanato, *Z. Elektrochem.* **65**, 362 (1961).
[27] E. D. Schmid, *Z. Elektrochem.* **66**, 53 (1962).
[28] E. D. Schmid, V. Hoffmann, R. Joeckle, and F. Langenbucher, *Spectrochim. Acta* **22**, 1615 (1966).
[29] E. D. Schmid and V. Hoffmann, *Spectrochim. Acta* **22**, 1633 (1966).
[30] E. D. Schmid and R. Joeckle, *Spectrochim. Acta* **22**, 1645 (1966).
[31] E. D. Schmid, *Spectrochim. Acta* **22**, 1659 (1966).
[32] H. A. Staab, "Einfuehrung in die theoretische organische Chemie." Verlag Chemie, Weinheim, 1959.
[33] H. Kloosterziel and H. J. Backer, *Rec. Trav. Chim.* **71**, 295 (1952).
[34] G. Kresze, E. Ropte, and B. Schrader, *Spectrochim. Acta* **21**, 1633 (1965).
[35] K. C. Schreiber, *Anal. Chem.* **21**, 1168 (1949).
[36] N. S. Ham and A. N. Hambly, *Australian J. Chem.* **6**, 33 (1953).
[37] R. N. Haszeldine and J. M. Kidd, *J. Chem. Soc.* (*London*) p. 4228 (1954).
[38] A. Simon and H. Kriegsmann, *Z. Physik. Chem.* (*Leipzig*) **204**, 369 (1955).
[39] A. Simon and H. Kriegsmann, *Chem. Ber.* **89**, 1718 (1956).
[40] A. Simon, H. Kriegsmann, and H. Dutz, *Chem. Ber.* **89**, 1883 (1956).
[41] A. Simon, H. Kriegsmann, and H. Dutz, *Chem. Ber.* **89**, 1990 (1956).
[42] A. Simon, H. Kriegsmann, and H. Dutz, *Chem. Ber.* **89**, 2378 (1956).
[43] A. Simon and H. Kriegsmann, *Chem. Ber.* **89**, 2384 (1956).
[44] H. Siebert, *Z. Anorg. Allgem. Chem.* **289**, 15 (1957).
[45] L. J. Bellamy and R. L. Williams, *J. Chem. Soc.* (*London*) p. 861 (1957).
[46] H. Gerding and J. W. Maarsen, *Rec. Trav. Chim.* **77**, 374 (1958).
[47] T. Nortia, *Suomen Kemistilehti* **B32**, 83 (1959).
[48] G. A. Chatschkuruso, *Opt. i Spektroskopiya* **8**, 40 (1960).
[49] A. Simon and W. Schmidt, *Z. Elektrochem.* **64**, 737 (1960).
[50] E. R. Lippincott and F. E. Welsh, *Spectrochim. Acta* **17**, 123 (1961).
[51] K. Stopperka, *Z. Anorg. Allgem. Chem.* **344**, 263 (1966).
[51a] K. Stopperka, *Z. Anorg. Allgem. Chem.* **345**, 264, 277 (1966).
[52] J. H. R. Clarke and L. A. Woodward, *Trans. Faraday Soc.* **62**, 2226 (1966).

——, at 98% relative atmospheric humidity; ····, thoroughly dried membrane. (f) Na^+ salt of polystyrenesulfonic acid, 5% cross-linked, 3 days of sulfonation, D_2O-hydrated: ——, at atmospheric humidity over a saturated solution of $BaCl_2$ in D_2O. (g) Perdeuteropolystyrene, 2% cross-linked (see p. 240). (h) Perdeuteropolystyrenesulfonic acid, 2% cross-linked, sulfonation see Section VI.2.1.2, H_2O-hydrated: ——, at 98% relative atmospheric humidity; ····, thoroughly dried membrane. (i) Na^+ salt of perdeuteropolystyrenesulfonic acid 2% cross-linked, sulfonation see Section VI.2.1.2, H_2O-hydrated, at about 40% relative atmospheric humidity.

$-S(=O)(=O)-Cl$ *Group in Polystyrenesulfonyl Chloride*

The band at 1374 cm^{-1} is due to the antisymmetric and that at 1172 cm^{-1} to the symmetric stretching vibration of the two SO double bonds.[38]

$-SO_3^-$ *ion:* In the Na^+ salt of polystyrenesulfonic acid [Fig. 1(e, f)], a doublet is found at about 1200 and a further band at 1040 cm^{-1}. Corresponding bands are found with hydrated polystyrenesulfonic acid at 1200 and 1034 cm^{-1} [Fig. 1(c, d)].

Simon and Schmidt[49] showed that the HSO_3^- ion in the crystals of the Rb and Cs salts has a pyramidal structure, i.e., it has C_{3v} symmetry and all SO bonds are alike. They assigned a band at 1200 cm^{-1} to the antisymmetric and a band at 1034 cm^{-1} to the symmetric stretching vibrations of the $—SO_3^-$ ion. If we compare our bands with these results, we see a difference: In our case a doublet occurs in place of the band of the antisymmetric stretching vibration at 1200 cm^{-1}. The antisymmetric stretching vibration of a group with C_{3v} symmetry is doubly degenerate. This means that there are two modes of vibration with the same energy and hence, in the degenerate case, only one band. We shall see in Section IV.2 that the band of the antisymmetric stretching vibration is split in the present materials, since the degeneracy is removed. The band at 1040 cm^{-1} is to be assigned to the symmetric stretching vibration of the $—SO_3^-$ ion. This is confirmed by the following: If the acid membrane is dried (cf. Section V.1), the $—SO_3^-$ ion is rearranged in such a way that it can no longer execute a completely symmetric vibration. Thus it is to be expected that in the acid this band should vanish on drying. This is actually the case, as shown in Fig. 2.

$-S(=O)(=O)-OH$ *group:* Polystyrenesulfonic acid shows three other bands on thorough drying, one at 1350 cm^{-1},* one at 1172 cm^{-1}, and one at 907 cm^{-1},† Fig. 1(c), dotted curve.

Simon and Kriegsmann[43] assigned the bands of the spectrum of $CH_3-S(=O)(=O)-OH$.

If the bands of dry polystyrenesulfonic acid are compared with these results, we see that the band at 1350 cm^{-1} should be assigned to the antisymmetric

* In the $-S(=O)(=O)-OD$ group this band lies not at 1350 cm^{-1} but at 1340 cm^{-1}.

† Positions of the band at 25°C. See Result 72, p. 132.

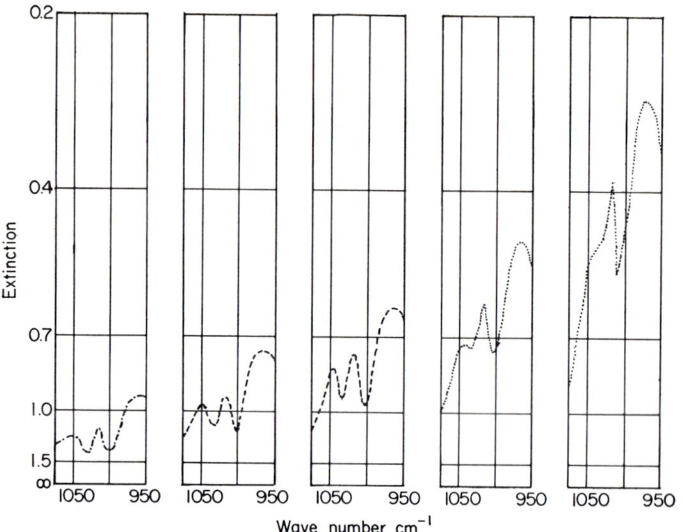

FIG. 2. Polystyrenesulfonic acid, 1% cross-linked, 4 days of sulfonation (band of the symmetric stretching vibration of the —SO_3^- ion), rapidly dried membrane. Before taking the spectra, the membrane was kept in the cell for about 6 hours at a definite atmospheric humidity. The drying stages were carried out, one after another, without interruption (atmospheric humidity decreasing from left to right, thoroughly dried membrane at the extreme right).

and the band at 1172 cm^{-1} to the symmetric stretching vibration of the two SO bonds with double-bond character; the band at 907 cm^{-1} should be assigned to the stretching vibration of the SO bond with single-bond character.

Bands Characteristic of Disubstituted Benzene Rings

The bands at 1128 and 1097 cm^{-1} can easily be assigned by comparison with those of the perdeuterated compounds, since upon deuteration they are only slightly displaced to 1090 and 1056 cm^{-1}, respectively. This shows that in-plane skeleton stretching vibrations are involved. According to Kresze *et al.*,[34] the substituent participates in these vibrations (see Section V.1.3).

The assignment of the bands at 838 and 777 cm^{-1} is decided by comparison with the spectra of *p*- and *o*-toluenesulfonic acids, respectively (Fig. 3), whereby the corresponding bands of *p*- and *o*-toluenesulfonic acids can be assigned in accordance with Randle and Whiffen.[13] The corresponding bands of the perdeuterated compounds lie at 732 and 684 cm^{-1}.

The band of *p*-toluenesulfonic acid at 817 cm^{-1} is to be ascribed to the out-of-plane bending vibration of two pairs of ⟩CH groups of the benzene ring.

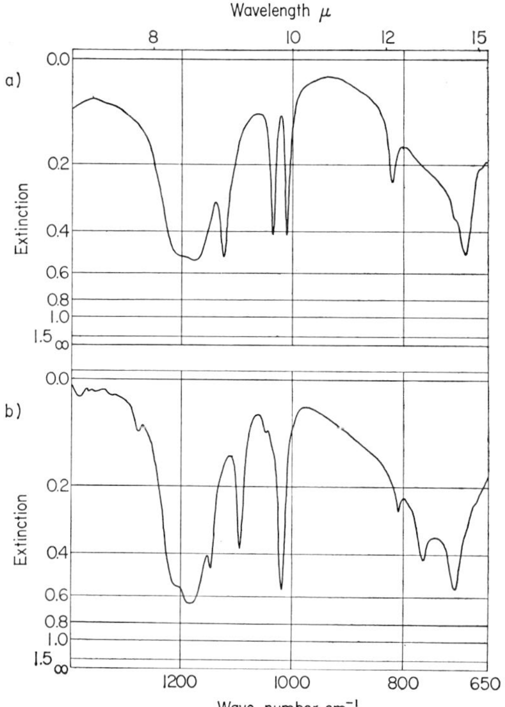

FIG. 3. *p*- And *o*-toluenesulfonic acids, saturated aqueous solution. (In taking these IR spectra, the acid solutions were placed between germanium dishes. The strong continuous absorption is compensated by weakening the reference beam.) (a) *p*-Toluenesulfonic acid. (b) *o*-Toluenesulfonic acid.

From this it follows that the band at 838 cm^{-1} for the acid and the salt, respectively, must also be ascribed to the out-of-plane bending vibrations of two pairs of CH groups of the benzene ring, i.e., to the CH groups of *p*-substituted benzene rings (see also Section VI.2.1.1).

The band of *o*-toluenesulfonic acid at 765 cm^{-1} must be ascribed to the out-of-plane bending vibration of four CH groups of the benzene ring. From this it follows, that the band at 777 cm^{-1} for the acid and the salt, respectively, must also be ascribed to the out-of-plane bending vibration of four CH groups of the benzene ring, i.e., to the CH groups of *o*-substituted benzene rings.

In the case of hydrated polystyrenesulfonic acid, a band is found at 671 cm^{-1} (649 cm^{-1} for the perdeuterated compound). This band is shifted approximately 10 cm^{-1} to smaller wave numbers on thorough drying. This band is found only in the spectra of polystyrenesulfonic acid and its salts, but not in the spectra of the membrane containing other anions (Figs. 4–7). According to Kresze *et al.*[34] this band is to be ascribed to a "ring vibration involving the substituent in the sense of a stretching vibration." This means that a CS stretching vibration is involved (also cf. footnote 2, pp. 355–357). For the hydrated Na^+ salt, it occurs at 676 cm^{-1}, for the perdeuterated compound at 649 cm^{-1}.

II.3. Polystyreneselenonic and -seleninic Acids and Their Na^+ Salts

The spectra of these substances are shown in Figs. 4(a–d) and 5(a–d). The assignments of these bands are collected in Tables A.2 and A.3.[53]

The most important bands whose assignment is still unknown are the bands of the SeO stretching vibrations. Numerous authors have investigated these bands.[54–71] Our assignments are derived from the literature cited.

Polystyreneselenonic Acid and Its Na^+ Salt

—Se(O)(O)(O)$^-$ *ion:* In the thoroughly dried Na^+ salt of polystyreneselenonic

[53] G. Zundel, *Z. Naturforsch.* **23b**, 119 (1968).
[54] S. Detoni and D. Hadži, *J. Chim. Phys.* **53**, 760 (1956).
[55] M. Falk and P. A. Giguere, *Can. J. Chem.* **36**, 1680 (1958).
[56] P. A. Giguere and M. Falk, *Spectrochim. Acta* **16**, 1 (1960).
[57] A. Simon and R. Paetzold, *Z. Anorg. Allgem. Chem.* **301**, 246 (1959).
[58] A. Simon and R. Paetzold, *Z. Anorg. Allgem. Chem.* **303**, 39 (1960).
[59] A. Simon and R. Paetzold, *Z. Elektrochem.* **64**, 209 (1960).
[60] A. Simon and R. Paetzold, *Z. Anorg. Allgem. Chem.* **303**, 46 (1960).
[61] R. Paetzold and E. Roensch, *Z. Anorg. Allgem. Chem.* **315**, 64 (1962).
[62] R. Paetzold and H. Amoulong, *Z. Anorg. Allgem. Chem.* **317**, 166 (1962).
[63] R. Paetzold and H. Amoulong, *Z. Anorg. Allgem. Chem.* **317**, 288 (1962).
[64] R. Paetzold, *Z. Anorg. Allgem. Chem.* **325**, 47 (1963).
[65] R. Paetzold and K. Aurich, *Z. Anorg. Allgem. Chem.* **335**, 281 (1965).
[66] R. Paetzold, Friedrich Schiller University, Jena, private communication on investigations and assignments of selenonates (1964).
[67] R. Paetzold, H. Amoulong, and A. Ruzicka, *Z. Anorg. Allgem. Chem.* **336**, 278 (1965).
[68] G. E. Walrafen, *J. Chem. Phys.* **36**, 90 (1962).
[69] G. E. Walrafen, *J. Chem. Phys.* **37**, 1468 (1962).
[70] K. Sathianandan, L. D. McCory, and J. L. Margrave, *Spectrochim. Acta* **20**, 957 (1964).
[71] V. Lorenzelli, F. Gesmundo, and J. Lecomte, *J. Chim. Phys.* **62**, 320 (1965).

acid two bands are found, one at 909 and one at 860 cm^{-1} [Figs. 4(c, d)*].

Paetzold and Amoulong,[62] in interpreting their results obtained with the $HOSeO_3^-$ ion, assume that the group $—SeO_3^-$ has "local" symmetry C_{3v}. If we compare with the assignment (see footnotes 62, 64, 66), it follows that the band at 909 cm^{-1} is to be ascribed to the doubly degenerate antisymmetric stretching vibration of the $—SeO_3^-$ ion; and the band at 860 cm^{-1} to its symmetric stretching vibration. This is supported by the behavior of the Sc^{3+} salt; the band of the antisymmetric stretching vibration at 909 cm^{-1} is split, i.e., the degeneracy is removed by the cation field [for further details see Section IV.2 and Fig. 16(b)].

In the spectrum of hydrated polystyreneselenonic acid two bands are observed, one at 879 and one at 847 cm^{-1} [Fig. 4(a, b)], which are correspondingly assigned.

$—Se(=O)(=O)OH$ *group:* When the acid is thoroughly dried [dotted curves in Fig. 4(a, b)], three bands are found at 965, 911, and 725 cm^{-1}. From comparison of the dependence on the degree of hydration of these bands with that of the SO bands of polystyrenesulfonic acid (Section V.1.2, Figs. 57 and 58), and with the help of the references already cited, we arrive at the following assignment: The bands at 965 and 911 cm^{-1} are to be assigned to the antisymmetric and the symmetric stretching vibrations of the two SeO bonds with double-bond character and the band at 725 cm^{-1} to the stretching vibration of the SeO bond with single-bond character.

Polystyreneseleninic Acid and Its Na^+ Salt

$—SeO_2^-$ *ion:* Spectra of the Na^+ salt of polystyreneseleninic acid are shown in Fig. 5(c, d). A broad band showing definite structure is to be seen at about 800 cm^{-1} (803 and 776 cm^{-1}). This structure depends on the degree of hydration and, especially, on whether the membrane has been hydrated with H_2O or D_2O [see also Fig. 19 and Section IV.4.(c)].

If this is contrasted with the results in the literature,[54,59] it follows that the band at 803 cm^{-1} is to be ascribed to the symmetric and the band at 776 cm^{-1} to the antisymmetric stretching vibration of the $—SeO_2^-$ ion. In this case both SeO bonds are almost identical, for no band corresponding to a stretching vibration of a SeO single bond is observed.

* Along with the $—SeO_3^-$ ions, a small number of $—SeO_2^-$ ions is to be found in the membrane. The shoulder, or weak bands, which may be observed on the side toward small wave numbers of the bands discussed here, are the bands of the SeO stretching vibrations of these $—SeO_2^-$ ions.

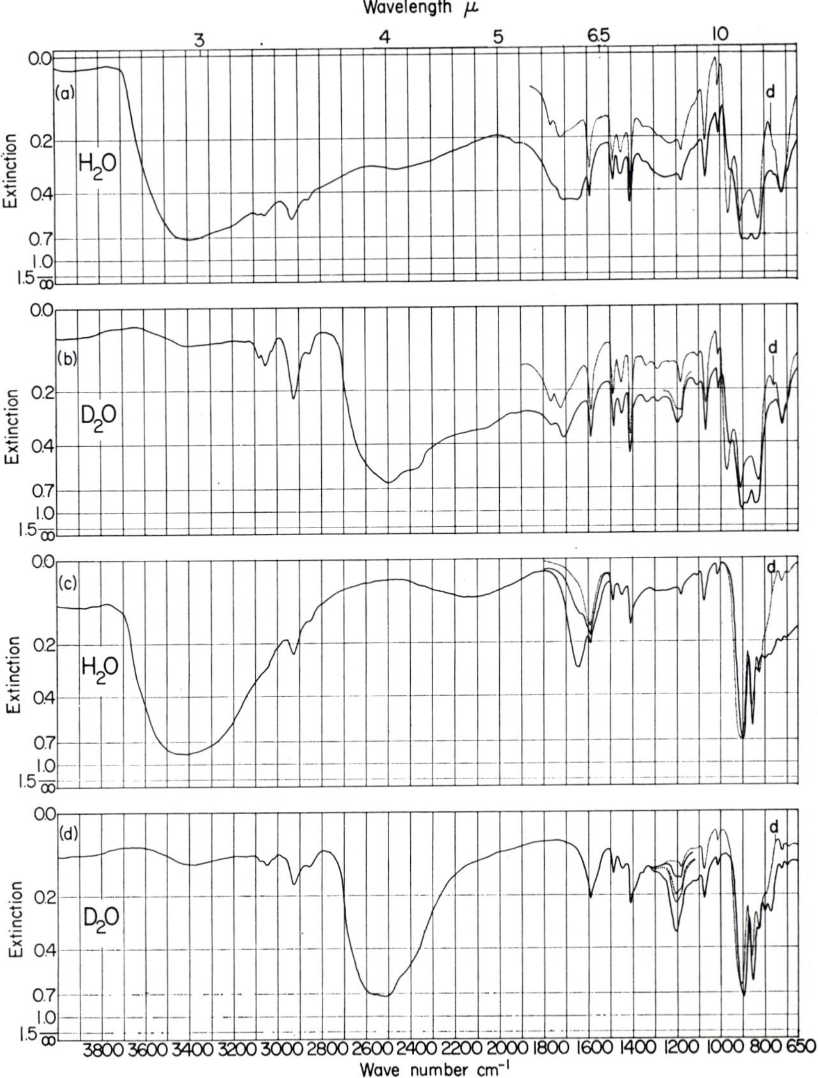

FIG. 4. IR spectra. (a) Polystyreneselenonic acid, 7% cross-linked, 3 days of selenonation (see Section VI.2.2), H_2O hydrated: ——, at 98% relative atmospheric humidity; · · · ·, thoroughly dried membrane. (b) Polystyreneselenonic acid, 7% cross-linked, 3 days of selenonation, D_2O hydrated: ——, at 100% relative atmospheric humidity; · · · ·, thoroughly dried membrane. (c) Na^+ salt of polystyreneselenonic acid, 7% cross-linked, 3 days of selenonation (particularly thin membrane), H_2O hydrated: ——, at 98% relative atmospheric humidity; · · · ·, thoroughly dried membrane. (d) Na^+ salt of polystyreneselenonic acid, 7% cross-linked, 3 days of selenonation (particularly thin membrane), D_2O hydrated: ——, at 100% relative atmospheric humidity; · · · ·, thoroughly dried membrane.

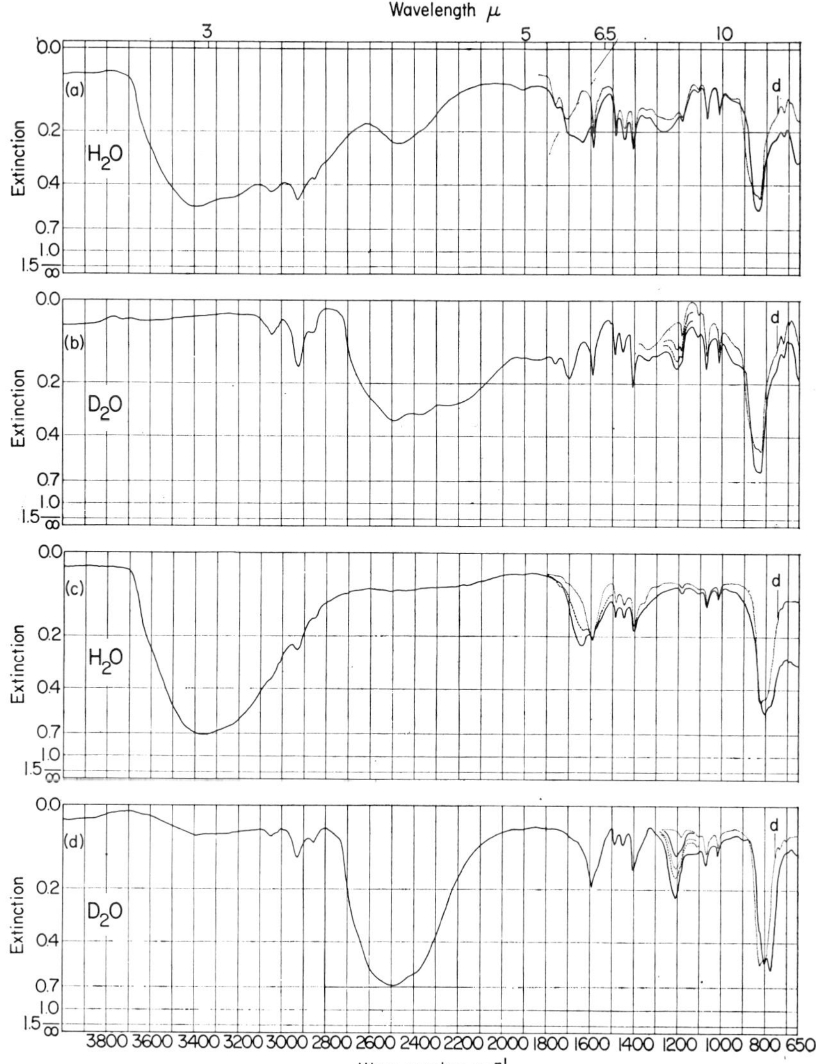

FIG. 5. IR spectra. (a) Polystyreneseleninic acid, 7% cross-linked, 3 days of selenonation (see Section VI.2.2 and 2.3), H_2O hydrated: ——, at 98% relative atmospheric humidity; · · · ·, thoroughly dried membrane. (b) Polystyreneseleninic acid, 7% cross-linked, 3 days of selenonation, H_2O hydrated: ——, at atmospheric humidity over saturated $BaCl_2$ solution in D_2O. (c) Na^+ salt of polystyreneseleninic acid, 7% cross-linked, 3 days of selenonation (particularly thin membrane, $H)_2O$ hydrated: ——, at 98% relative atmospheric humidity; · · · ·, thoroughly dried membrane. (d) Na^+ salt of polystyreneseleninic acid, 7% cross-linked, 3 days of selenonation (particularly thin membrane), D_2O hydrated: ——, at 100% relative atmospheric humidity; · · · ·, thoroughly dried membrane.

$-Se(=O)OH$ *group:* Figure 5(a) and (b), as well as Fig. 59, show spectra of polystyreneseleninic acid. At 839 cm^{-1} a broad band is found. Along with these bands, a band at 658 cm^{-1} is observed with the acid, in contrast to the Na^+ salt.

If we compare the results of Detoni and Hadži[54] with our data, it follows that the band at 839 cm^{-1} must be assigned to the SeO stretching vibration with double-bond character, and that at 658 cm^{-1} to the stretching vibration of the SeO bond with single-bond character (for further details, see Section V.1.2).

Other Bands of These Compounds

With polystyreneselenonic acid two further bands are found, one at 1766 and one at 1722 cm^{-1}. Corresponding bands are found with polystyreneseleninic acid at 1765 and at 1705 cm^{-1}. These bands were measured on thoroughly dried membranes. With an increasing degree of hydration, in the case of polystyreneselenonic acid, the first band is shifted by about 5 cm^{-1}, the second by about 10 cm^{-1}, toward smaller wave numbers. The bands of polystyreneseleninic acid show a similar behavior. These bands occur independently of whether the membranes are hydrated with H_2O or D_2O. The sharpness of these bands with the selenonic acid is somewhat affected by the rate of drying.

Concerning the assignment, the available results indicate that, since the presence of the bands is independent of whether hydration is with H_2O or D_2O, they have nothing to do with vibrations of OH groups in hydrogen bridges. The nature of the vibrations causing these bands is still unexplained. It is likely that they are due to overtones of the out-of-plane bending vibrations of the $>$CH groups of the benzene ring or to overtones of the SeO stretching vibrations.

With these compounds, only one band is found in the region 625–425 cm^{-1}, at 547 cm^{-1} with polystyreneselenonic acid, and at 551 cm^{-1} with polystyreneseleninic acid. These bands can be assigned by comparison with the results obtained with polystyrenesulfonic acid (where a band occurs at 671 cm^{-1}) and by reference to Kresze *et al.*[34] The bands must be ascribed to a "ring vibration with strong participation of the substituent in the sense of a stretching vibration."

II.4. Polystyrenephosphinic and Polystyrenethiophosphonic Acids and Their Na^+ Salts

Spectra of polystyrenephosphinic acid and its Na^+ salt are shown in Fig. 6 and of polystyrenethiophosphonic acid and its Na^+ salt in Fig. 7.* The

* Some unneutralized acid OH or OD groups are probably still present in membranes of the Na^+ salt of polystyrenethiophosphonic acid (see Section VI.3). The number of such groups, however, is very small in any case.

assignment of the bands are given in Tables A.4 and A.5.[72] In Fig. 8 the spectra of polystyrenephosphonic acid and its Na^+ salt are given; in Fig. 9, for comparison, those of aqueous solutions of *p*-benzenephosphinic and *p*-benzenephosphonic acids. Several bands have yet to be assigned: First the bands of the PH group in the case of polystyrenephosphinic acid and its Na^+ salt; second, the bands of the groups containing PO and PS bonds; third, the bands arising from the bonding of the P atom to the benzene ring.

The Bands of the PH Group

With polystyrenephosphinic acid [Fig. 6(a)] a band is found at 2365 cm^{-1}. In the Na^+ salt we find this band at 2304 cm^{-1}. This band must be ascribed to the stretching vibration of the PH group. This is supported by the following observations: First, this band vanishes on oxidation of the $-\mathrm{P}(\mathrm{H})(=\mathrm{O})(\mathrm{OH})$ groups to $-\mathrm{P}(\mathrm{OH})(=\mathrm{O})(\mathrm{OH})$ groups [cf. Fig. 6(a, c) with Fig. 8*]. Second, this band does not occur with polystyrenethiophosphonic acid and its Na^+ salt [cf. Fig. 6(a, c) with Fig. 7(a, c)]. Third, the observation that, with the acid, this band vanishes on going from H_2O to D_2O hydration, with the appearance of a band at 1722 cm^{-1}. This band is to be ascribed to the PD stretching vibration [cf. the spectrum 12 in Fig. 116(a) with that in (b)]. In the Na^+ salt, no exchange of H for D takes place at room temperature [cf. Fig. 6(c) with 6(d)]. Comparison with published result[73–83] proves the correctness of the assignment of these bands.

The PH bending vibration is found in the acid at 977 cm^{-1}. The corresponding band for the acid hydrated with D_2O, for the PD bending vibration, is

* In making this comparison, it must be borne in mind that some unoxidized groups are still present in this membrane (Section VI.2.4), i.e., the PH band (marked with an arrow in the spectrum) has not disappeared, but is considerably weakened.

[72] G. Zundel, *Kunststoff-Rundschau* **15**, 166 (1968).
[73] L. W. Daasch and D. C. Smith, *Anal. Chem.* **23**, 853 (1951).
[74] L. J. Bellamy and L. Beecher, *J. Chem. Soc.* (*London*) p. 1701 (1952).
[75] L. J. Bellamy and L. Beecher, *J. Chem. Soc.* (*London*) p. 728 (1953).
[76] D. E. C. Corbridge and E. J. Lowe, *J. Chem. Soc.* (*London*) p. 493 (1954).
[77] D. E. C. Corbridge and E. J. Lowe, *J. Chem. Soc.* (*London*) p. 4555 (1954).
[78] D. E. C. Corbridge, *J. Appl. Chem.* **6**, 456 (1956).
[79] L. C. Thomas, *Chem. Ind.* (*London*) p. 198 (1957).
[80] M. Tsuboi, *J. Am. Chem. Soc.* **79**, 1351 (1957).
[81] J. R. Ferraro and J. S. Ziomek, *J. Inorg. Nucl. Chem.* **26**, 1397 (1964).
[82] R. W. Lovejoy and E. L. Wagner, *J. Phys. Chem.* **68**, 544 (1964).
[83] R. A. Chittenden and L. C. Thomas, *Spectrochim. Acta* **21**, 861 (1965).

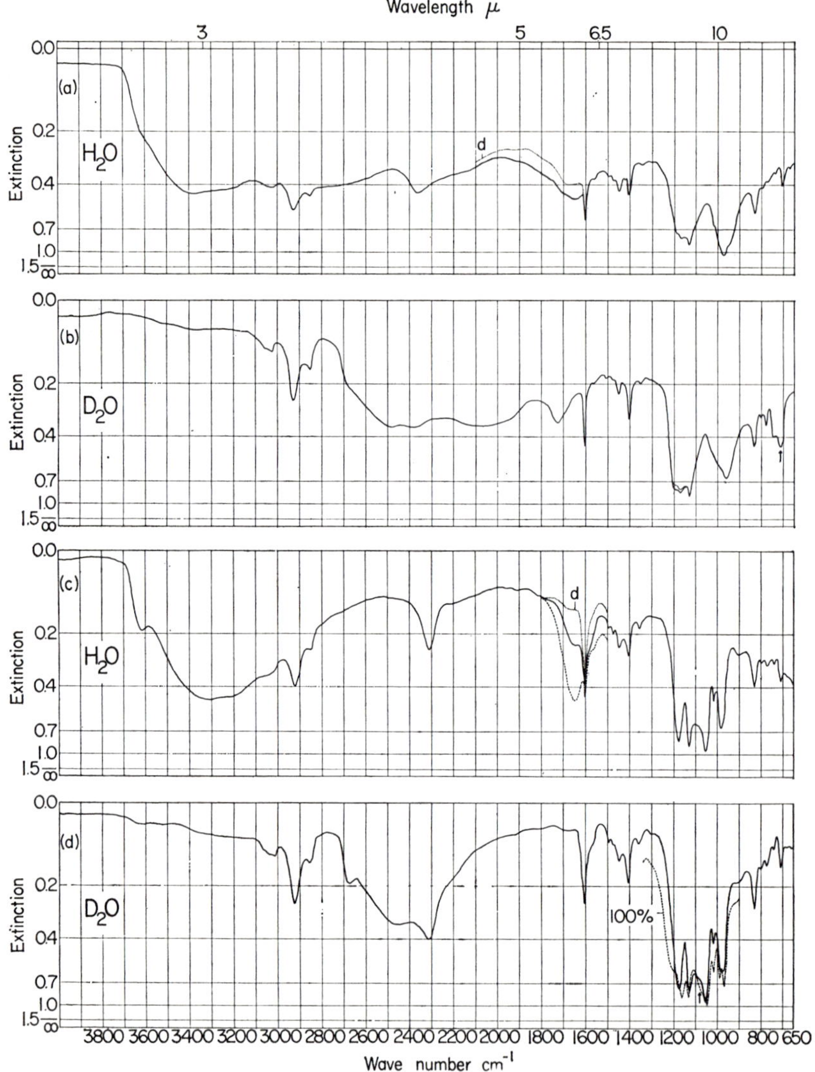

FIG. 6. IR spectra. (a) Polystyrenephosphinic acid, 7% cross-linked, 3 days phosphinated (Section VI.2.4), H_2O hydrated: ——, at 98% relative atmospheric humidity; ····, thoroughly dried membrane. (b) Polystyrenephosphinic acid, 7% cross-linked, 3 days phosphinated, D_2O hydrated: ——, at atmospheric humidity over saturated solution of $BaCl_2$ in D_2O. (c) Na^+ salt of polystyrenephosphinic acid, 7% cross-linked, 3 days phosphinated, H_2O hydrated: – – – –, at 98% relative atmospheric humidity; ——, at 11% relative atmospheric humidity; ····, thoroughly dried membrane. (d) Na^+ salt of polystyrenephosphinic acid, 7% cross-linked, 3 days phosphinated, D_2O hydrated: – – – –, at 100% relative atmospheric humidity; ——, at moderate atmospheric humidity.

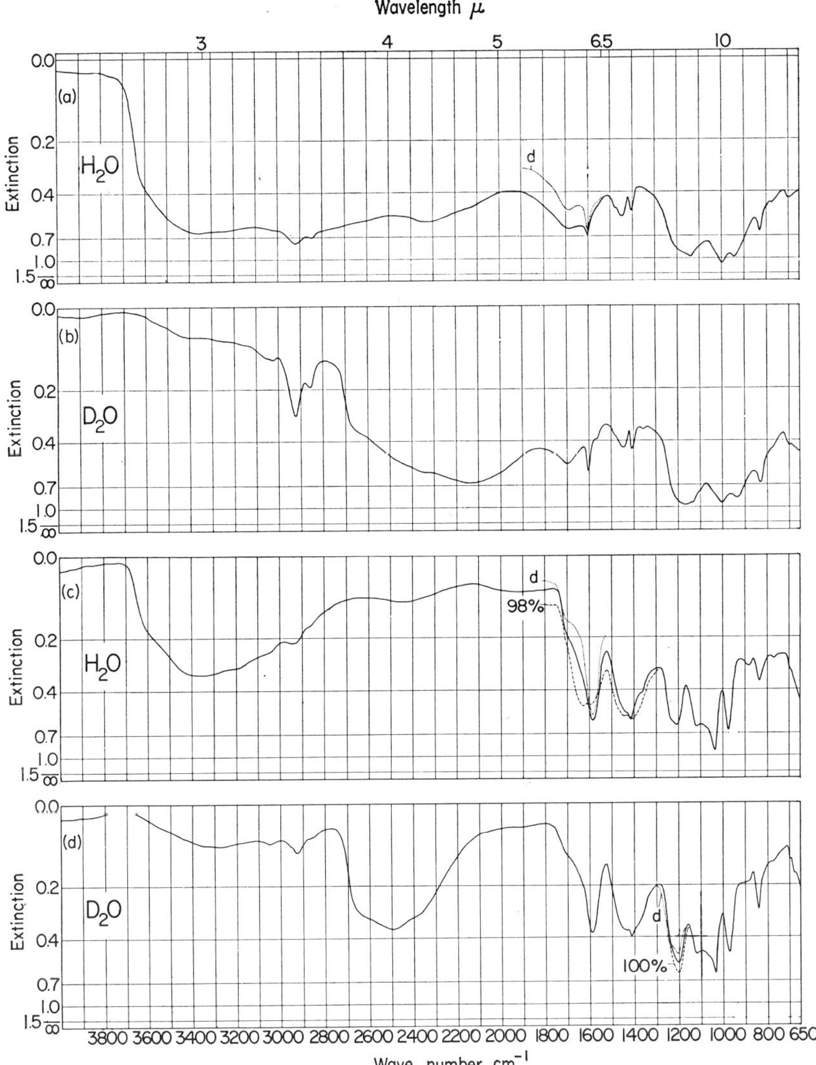

FIG. 7. IR spectra. (a) Polystyrenethiophosphonic acid, 7% cross-linked, 3 days thiophosphonated (see Section IV.2.5) (particularly thick membrane), H_2O hydrated: ——, at 98% relative atmospheric humidity; · · · ·, thoroughly dried membrane. (b) Polystyrenethiophosphonic acid, 7% cross-linked, 3 days thiophosphonated (particularly thick membrane), D_2O hydrated: ——, at 100% relative atmospheric humidity. (c) Na^+ salt of polystyrenethiophosphonic acid, 7% cross-linked, 3 days thiophosphonated, –·–·, at 98% relative atmospheric humidity; H_2O hydrated: ——, at 11% relative atmospheric humidity; · · · ·, thoroughly dried membrane. (d) Na^+ salt of polystyrenethiophosphonic acid, 7% cross-linked, 3 days thiophosphonated, –·–·, at 100% relative atmospheric humidity; D_2O hydrated: ——, at moderate atmospheric humidity; · · · ·, thoroughly dried membrane.

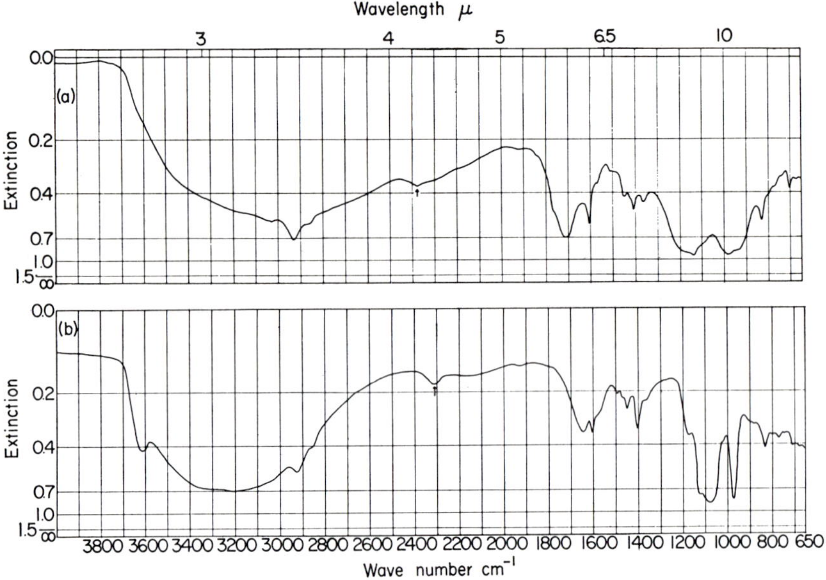

FIG. 8. IR spectra. (a) Polystyrenephosphonic acid (some —P(H)(=O)(OH) groups present), 7% cross-linked, 3 days phosphinated (Section VI.2.4), H_2O hydrated, at about 40% relative atmospheric humidity. (b) Na^+ salt of polystyrenephosphonic acid (still contains some —P(H)(⋯O)(⋯O)$^-$ ions), 7% cross-linked, 3 days phosphinated, H_2O hydrated, at about 40% relative atmospheric humidity.

found at 716 cm^{-1} [cf. Fig. 6(a), with (b)].* In the Na^+ salt, the circumstances relating to these bands are very obscure [See Section IV.4(d)].

The Bands of the Stretching Vibration of the PO and PS Bonds

The assignments of these bands have been studied by numerous workers.[73–101]

* Note: With the acid hydrated with D_2O another very weak band is found at 775 cm^{-1} and a shoulder at 740 cm^{-1}. These bands are faint, or are not found at all, with the acid hydrated with H_2O.

[84] E. D. Bergmann, U. Z. Littauer, and S. Pinchas, *J. Chem. Soc.* (*London*) p. 847 (1952).

[85] H. J. Emeleus, R. N. Haszeldine, and R. C. Paul, *J. Chem. Soc.* (*London*) p. 563 (1955).

[86] C. Duval and J. Lecomte, *Compt. Rend.* **240**, 66 (1955).

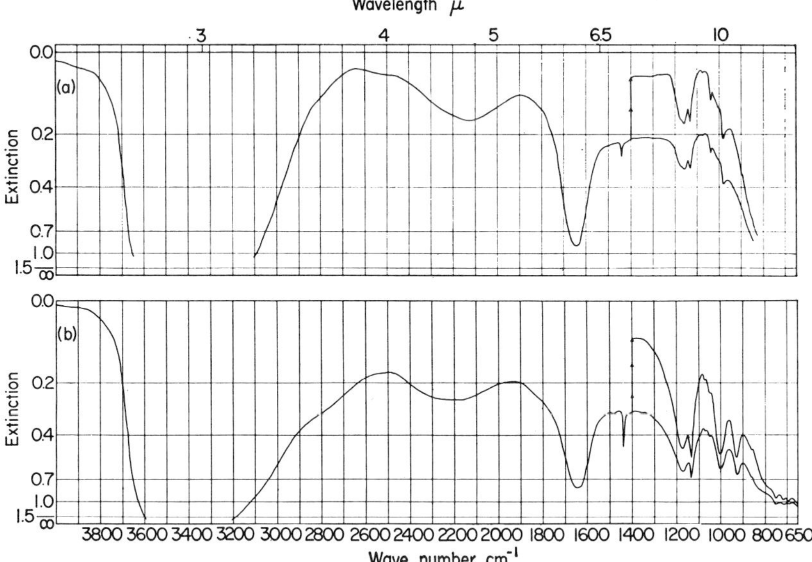

FIG. 9. IR spectra; (acid solutions were placed between germanium dishes). (a) *p*-Benzene phosphinic acid, saturated aqueous solution. (b) *p*-Benzenephosphonic acid, saturated aqueous solution.

The Na⁺ Salts

In the Na^+ salt of polystyrenephosphinic acid [(Fig. 6(c, d)] we find bands at 1183 and 1056 cm^{-1}. In the Na^+ salt of polystyrenethiophosphonic acid

[87] F. A. Miller and C. H. Wilkins, *Anal. Chem.* **24**, 1253 (1952).
[88] C. Duval and J. Lecomte, *Mikrochim. Acta* p. 454 (1956).
[89] D. F. Peppard, J. R. Ferraro, and G. W. Mason, *J. Inorg. Nucl. Chem.* **7**, 231 (1958).
[90] D. F. Peppard, J. R. Ferraro, and G. W. Mason, *J. Inorg. Nucl. Chem.* **12**, 60 (1959).
[91] D. F. Peppard, J. R. Ferraro, and G. W. Mason, *J. Inorg. Nucl. Chem.* **16**, 246 (1961).
[92] L. C. Thomas, R. A. Chittenden, and H. E. R. Hartley, *Nature* **192**, 1283 (1961).
[93] E. Steger and K. Herzog, *Z. Anorg. Allgem. Chem.* **331**, 169 (1964).
[94] A. C. Chapman and L. E. Thirlwell, *Spectrochim. Acta* **20**, 937 (1964).
[95] L. C. Thomas and R. A. Chittenden, *Spectrochim. Acta* **20**, 467 (1964).
[96] L. C. Thomas and R. A. Chittenden, *Spectrochim. Acta* **20**, 489 (1964).
[97] R. A. Chittenden and L. C. Thomas, *Spectrochim. Acta* **20**, 1679 (1964).
[98] M. I. Kabaznik, T. A. Mastrjukowa, E. I. Matrosow, and B. Fischer, *Zh. Strukt. Khim.* **6**, 691 (1965).
[99] M. Tsuboi, *J. Am. Chem. Soc.* **79**, 1351 (1957).
[100] J. S. Ziomek, J. R. Ferraro, and D. F. Peppard, *J. Mol. Spectr.* **8**, 212 (1962).
[101] T. Shimanouchi, M. Tsuboi, and Y. Kyogoku, *in* "The Structure and Properties of Biomolecules and Biological Systems" (J. Duchesne, ed.), p. 467 ff. Wiley (Interscience) New York, 1964.

[Fig. 7(c, d)] a doublet* is found for the thoroughly dried membrane at 1218 cm^{-1} and a band at 1037 cm^{-1}. With an increasing degree of hydration, the splitting of the band at 1218 cm^{-1} disappears, and the bands at 1218 cm^{-1} and 1037 cm^{-1} are shifted by about 5 cm^{-1} toward smaller wave numbers.

Tsuboi,[80] Ferraro and Ziomek,[81] and Lovejoy and Wagner[82] have studied the $H_2PO_2^-$ ion. With a variety of salts, they always found a band at about 1190 cm^{-1} and another at about 1060 cm^{-1}. Assuming that the PO_2^- group has the "local" symmetry C_{2v}, they assign these two bands to the antisymmetric and symmetric PO stretching vibrations. A normal coordinate treatment was carried out by Shimanouchi *et al.*[101]

Contrasting our results with the conclusions above, we ascribe the bands at 1183 and 1056 cm^{-1} for the Na^+ salt of polystyrenephosphinic acid to the antisymmetric and the symmetric PO stretching vibration bands of the $\left(-\mathrm{P}\begin{matrix}\diagup \mathrm{H} \\ \cdots \mathrm{O} \\ \diagdown \mathrm{O}\end{matrix}\right)^-$ ion, in which this ion has two almost identical PO bonds. The bands at 1218 and 1037 cm^{-1} for the Na^+ salt of polystyrenethiophosphonic acid are to be ascribed to the antisymmetric and symmetric PO stretching vibrations of the $\left(-\mathrm{P}\begin{matrix}\diagup \mathrm{S} \\ \cdots \mathrm{O} \\ \diagdown \mathrm{O}\end{matrix}\right)^{2-}$ ion, in which this ion also has two almost identical PO bonds.

With the Na^+ salt of polystyrenephosphonic acid [Fig. 8(b)] a very intense band is found at approximately 1075 cm^{-1}, as well as a band at approximately 970 cm^{-1}.

Tsuboi[80] studied an aqueous solution of K_2HPO_3 and found bands at 1085 and 979 cm^{-1}. Assuming that the HPO_3^- ion possesses C_{3v} symmetry, he assigned these to the degenerate antisymmetric and the symmetric PO stretching vibrations of this group. A normal coordinate treatment is given by Shimanouchi *et al.*[101]

Contrasting our results with this, we ascribe the bands found at 1075 and 970 cm^{-1}, with the Na^+ salt of polystyrenephosphonic acid, to the degenerate antisymmetric and the symmetric PO stretching vibrations of the $\left(-\mathrm{P}\begin{matrix}\diagup \mathrm{O} \\ \cdots \mathrm{O} \\ \diagdown \mathrm{O}\end{matrix}\right)^{2-}$ ion, which possesses C_{3v} "local" symmetry.

* A similar doublet is also found for polystyrenephosphinic acid [Fig. 6(a, b)] as well as benzenephosphinic acid [Fig. 9(à)]. Similar results are reported in the literature. For example Corbridge,[78] discussed this band and noted: "... sometimes a doublet." Various causes for this splitting have been discussed. For this point, see Thomas and Chittenden.[95,102]

[102] L. C. Thomas and R. A. Chittenden, *Spectrochim. Acta* **21**, 1905 (1965).

The Acids

In polystyrenephosphinic acid [Fig. 6(a, b)], an intense doublet (see footnote on p. 24) is found at about 1170 cm^{-1} and a band at 965 cm^{-1}. The latter is observed with the D_2O-hydrated membrane, since in this case no band caused by a PH bending vibration is superimposed.

In polystyrenethiophosphonic acid (Fig. 7(a, b)] a broad and very intense band is found at about 1170 cm^{-1} and a band at 995 cm^{-1}* with a shoulder at 940 cm^{-1}. It is at first surprising that the spectrum of polystyrenethiophosphonic acid does not differ in this region from that of polystyrenephosphonic acid [cf. Figs. 7(a) and 8(a)].

Lovejoy and Wagner,[82] working on H_2POOH also found a very intense band at 1184 cm^{-1} and one at 950 cm^{-1}. The band at 950 cm^{-1} is shifted toward larger wave numbers upon deuteration, to 977 cm^{-1}. The first band they ascribe to the stretching vibration of a P═O double bond, the second to the stretching vibration of a P—O single bond. Thomas and Chittenden,[95,96] on the basis of studies of thirty compounds of the type $R—P(═O)(OH)_2$ make the following assignment: one band in the region 1220–1150 cm^{-1} arises from the stretching vibration of the P═O double bond, while two bands, one in the region 1030–972 cm^{-1} and the other in the region 950–917 cm^{-1}, arise from the stretching vibrations of the single bonds of the $P(OH)_2$ portion of this group.

Comparing our bands with these results, we can ascribe the band for polystyrenephosphinic acid, at about 1170 cm^{-1}, to the stretching vibration of the PO bond with double-bond character, and that at 965 cm^{-1} to the stretching vibration of the PO bond with single-bond character, in the —P(H)(═O)(OH) group.

The spectrum is more complicated in the case of polystyrenethiophosphonic acid. The very intense band at about 1170 cm^{-1} can also be ascribed to the stretching vibration of a PO bond with double-bond character. This band of the P═O double bond shows that most of these groups are present as —P(SH)(═O)(OH) groups. Comparison with the spectrum of polystyrenephosphinic acid shows that the band of the single bond in this group may be expected

* If the membrane was hydrated with D_2O before drying, this band is found at 1005 cm^{-1}.

to occur at about 965 cm^{-1}. In fact, however, two bands are found, one at 995 cm^{-1} and a shoulder at 940 cm^{-1}. Our result, that the spectrum of polystyrenethiophosphonic acid does not differ essentially from that of polystyrenephosphonic acid in this region, would indicate that some $-P(=S)(OH)_2$ groups are present in tautomeric equilibrium. In this case the two bands are assigned to the antisymmetric and symmetric PO stretching vibrations of the $P(OH)_2$ portion of this group, these two bands being superimposed on the band of the stretching vibration of the PO bond with single-bond character in the $-P(=O)(SH)(OH)$ group. In Section V.14 we shall obtain a result which further supports the presence of a tautomeric equilibrium between $-P(=O)(SH)(OH)$ and $-P(=S)(OH)_2$. A similar tautomeric equilibrium in similar compounds was found by Semljanski and Klimowskaja.[103]

Bands Due to the P-Phenyl Group

In the literature, for example, Corbridge[78] and Bellamy (see footnote 2, p. 320) give two bands, one in the region 1450–1435 cm^{-1} and the other in the region 1005–995 cm^{-1}, as characteristic for the P-phenyl group (see also footnote 102). However, the mode of these vibrations is not known.

Region 1000–900 *cm*$^{-1}$

With the Na^+ salt of polystyrenephosphinic acid, a band is found at 985 cm^{-1}. With the Na^+ salt of polystyrenethiophosphonic acid, a band is observed at 979 cm^{-1} after thorough drying, which is shifted to 970 cm^{-1} with increasing degree of hydration. Neither Tsuboi[80] nor Lovejoy and Wagner[82] found such a band in the spectra of the hypophosphite or the phosphite ion. Hence this band is obviously a characteristic of the P-phenyl group.

In the corresponding acids, the bands of the stretching vibrations of the PO bonds with single-bond character lie in this region. The band observed in the spectra of the salts, which is characteristic of the P-phenyl group, presumably is then masked by the band of the stretching vibration of the PO single bond. In the polystyrenephosphinic acid, we see [Fig. 6(b)] that the band at 965 cm^{-1} is definitely not bell-shaped. In this case the two bands are actually superimposed.

[103] N. I. Semljanski and L. K. Klimowskaja, *Zh. Obshch. Khim.* **30**, 4056 (1960).

Region 1450–1435 *cm*$^{-1}$

In the Na^+ salt of polystyrenethiophosphonic acid [Fig. 7(c, d)], a band is found at 1435 cm^{-1} which can be assigned to the vibration of the P-phenyl group. The intensity of this band, however, varies from one membrane to another. Its intensity is also somewhat dependent on the hydration. The reasons for this are not known as yet (see also Section VI.2.5).

The Bands in the Region 1700–1650 *cm*$^{-1}$

In polystyrenephosphinic acid (see, in particular, Fig. 116) a strongly marked band is found at 1655 cm^{-1} for the H_2O-hydrated membrane, even after complete drying. It is not possible to determine conclusively whether such a band is also present with the D_2O-hydrated membrane, since the band of the PD stretching vibration masks this region. No such band is found for the Na^+ salt of polystyrenesphosphinic acid (see, in particular, Fig. 26).

In polystyrenethiophosphonic acid (see also Fig. 120), a clearly marked band is found at 1690 cm^{-1}, independently of whether the membrane has been hydrated with H_2O or D_2O. With the Na^+ salt of polystyrenethiophosphonic acid [Fig. 7(c, d)], a faint shoulder is found.

Braunholtz and co-workers[104] have collected together the bands characteristic for the $X{\langle}^{\,O}_{\,OH}$ (X with =O and —OH) group. Along with other bands, they give one at about 1650 cm^{-1}. Some authors have supposed that this band could be connected with the OH bending vibration of the hydrogen bridges which these groups form on association. If this were the case, the band should disappear, with polystyrenethiophosphonic acid, on passing from H_2O to D_2O hydration. This does not occur and the assignment cannot therefore be correct. This is in agreement with other published results: Ryskin and Stavitskaya[105] have shown that the bending vibration in question should occur at about 1250 cm^{-1} but not at 1650 cm^{-1}. The same result is reached by other workers[90,92,106] (see also Section V.2.2, Result 76). Thomas and co-workers,[92,96] therefore, assign the band at about 1650 cm^{-1} tentatively to a combination vibration, in particular to the combination of the stretching vibration of the P=O double bond with a vibration whose band lies in the range 400–500 cm^{-1}. Such an assignment would also be possible in our case.* Nevertheless it must be borne in mind that a corresponding band is given in the summary by Braunholtz and co-workers[104] for different compounds. A conclusive assignment of the band in this region cannot be made at the present time on the basis either of published results or of our own findings.

* In the Na^+ salt of polystyrenethiophosphonic acid, the faint shoulder would then be explained by the presence of some OH-acid groups (see p. 255).

[104] J. T. Braunholtz, G. E. Hall, F. G. Mann, and N. Sheppard, *J. Chem. Soc.* (*London*) p. 868 (1959).

[105] Y. I. Ryskin and G. P. Stavitskaya, *Opt. Spectry.* 7, (USSR), (English transl.) 488 (1959).

[106] S. Detoni and D. Hadži, *Spectrochim. Acta* **20**, 949 (1964).

CHAPTER III

PURE WATER

III.1. The Free Water Molecule

The water molecule has three normal modes of vibration, a symmetrical stretching mode $\bar{\nu}_1$, a bending mode (scissor mode) $\bar{\nu}_2$, and an antisymmetric stretching mode $\bar{\nu}_3$. These three normal modes are shown in Fig. 10. The arrows show the directions of displacement of the nuclei and the lengths of the arrows indicate their relative amplitudes of vibration. The water molecule possesses C_{2v} symmetry. Hence none of these three modes is degenerate. A change of dipole moment is associated with each of the three modes, so that each must be IR active. The positions of the bands of these normal modes are given in Table 1 for the free water molecule. Further information on the modes of vibration of the water molecule is to be found in the literature

TABLE 1
THE BANDS OF THE FREE WATER MOLECULE

Molecule	$\bar{\nu}_1$ (cm^{-1})	$\bar{\nu}_2$ (cm^{-1})	$\bar{\nu}_3$ (cm^{-1})	Reference
H_2O^{16}	3657	1595	3756	Ref. 1, p. 1164, Table XI
H_2O^{18}	3647	1586	3744	theoretical values according to Ref. 2
D_2O^{16}	2671	1178	2788	Ref. 1, p. 1164, Table XI,
D_2O^{18}	$2657^a \pm 2$	1169 ± 2	2764 ± 5	Ref. 3
	$\bar{\nu}_{OD}$ (cm^{-1})	$\bar{\nu}_{2\ HDO}$ (cm^{-1})	$\bar{\nu}_{OH}$ (cm^{-1})	—
HDO^{16}	2727	1402	3707	Ref. 1, p. 1164, Table XI, and Ref. 4

[a] Measured in the Raman spectrum.

[1] W. S. Benedict, N. Gailar, and K. Plyler, *J. Chem. Phys.* **24**, 1139 (1956).
[2] G. Gompertz and W. J. Orville-Thomas, *J. Phys. Chem.* **63**, 1331 (1959).
[3] S. Pinchas and M. Halmann, *J. Chem. Phys.* **31**, 1692 (1959).
[4] R. D. Waldron, *J. Chem. Phys.* **26**, 809 (1957).

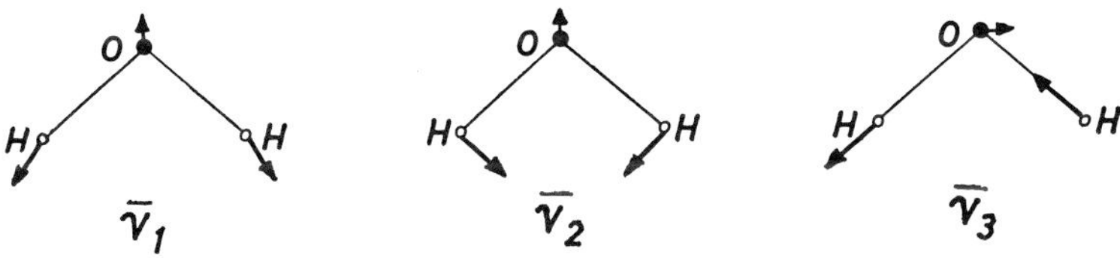

FIG. 10. The normal modes of the vibrations of the water molecule.

(see footnote 5, p. 280 ff., footnote 6, p. 506 ff., and footnote 7, p. 15 ff.). Table 2 contains the most important data of the free H_2O molecule.

TABLE 2
DATA FOR THE FREE H_2O MOLECULE

Property	Value	Reference
Bond length r_{OH}	0.965 Å	8, p. 96
Valence angle 2 γ[a]	104.5°	8, p. 96
Force constant in the direction of the bond	8.256×10^5 dyne/cm	9[b]
Dipole moment μ	1.84 [ESU cm $\times 10^{-18}$]	11, p. 86
Polarizability α_W	1.48×10^{-24} cm^3	6, p. 295
Ionization energy I_W	12.56 eV	12, p. 2649

[a] To avoid confusion with the polarizability, the valence angle of the water molecule is here and subsequently termed 2 γ.

[b] This value agrees almost exactly with that of 8.23×10^5 dyne/cm previously given by Bonner.[10]

III.2 The Bound Water Molecule

III.2.1 INTERMOLECULAR INTERACTIONS AND THE NORMAL MODES OF VIBRATION

If the water molecule is bound to its surroundings by hydrogen bridges, the bands $\bar{\nu}_1$ and $\bar{\nu}_3$ of the stretching vibrations of the water molecule are

[5] G. Herzberg, "Molecular Spectra and Molecular Structure." Vol. 2. Van Nostrand, Princeton, New Jersey 1962.

[6] M. W. Wolkenstein, "Struktur und physikalische Eigenschaften der Moleküle." Teubner, Leipzig, 1960.

[7] K. Nakamoto, "Infrared Spectra of Inorganic and Coordination Compounds." Wiley, New York, 1963.

[8] L. Pauling, "Die Natur der chemischen Bindung." Verlag Chemie, Weinheim, 1962.

[9] G. Strey, Ph.D. Thesis, Munich University, 1964 (under Prof. J. Brandmueller).

[10] L. G. Bonner, *Phys. Rev.* **45**, 458 (1934).

[11] J. W. Smith, "Electric Dipole Moments." Butterworths, London and Washington, D.C., 1955.

[12] C. D. Hodgman, ed. "Handbook of Chemistry and Physics." 44th Ed. Chem. Rubber Publ. Co., Cleveland, Ohio, 1961.

shifted toward smaller wave numbers, and the band $\bar{\nu}_2$ of the scissor vibration is shifted toward larger wave numbers, compared with that of free water (cf. Table 1 and Table 3).

The behavior of the two bands of the stretching vibrations should be considered in greater detail. Greinacher *et al.*[13] investigated water dissolved in organic solvents, which interact with the water molecule, in varying strengths. Ackerman[14] studied the temperature and concentration dependence of the bands of the stretching vibrations of the water molecule under the same conditions. These authors showed that the band of the antisymmetric stretching vibration is shifted toward smaller wave numbers more strongly than those of the symmetric, if interaction of the water with its surroundings becomes stronger. This interaction causes the bands of these two vibrations to draw together and finally to coincide. The same effect is discussed from another point of view in Section IV.5.1.

Liquid Water

The positions of the bands of liquid water are summarized in Table 3. In the few cases in which IR data are not available, the Raman data are given. These values are identified with the letter R. All these bands are very broad, and the results given by different authors often differ considerably. The positions of these bands are somewhat dependent on temperature, as already shown by Fox and Martin[17] and, more recently, in the elegant experiment of Fishman and Saumagne.[18] A corresponding finding is reported in the Raman-spectroscopic investigations of Cross *et al.*[31] and of Weston.[24] If the

[13] E. Greinacher, W. Luettke and R. Mecke, *Z. Elektrochem.* **59**, 23 (1955).
[14] T. Ackermann, *Z. Physik. Chem.* (*Frankfurt*) **42**, 119 (1964).
[15] E. K. Plyler and D. Williams, *J. Chem. Phys.* **4**, 157 (1936).
[16] C. H. Cartwright, *Phys. Rev.* **49**, 470 (1936).
[17] J. J. Fox and A. E. Martin, *Proc. Roy. Soc.* (*London*) **A174**, 234 (1940).
[18] E. Fishman and P. Saumagne, *J. Phys. Chem.* **69**, 3671 (1965).
[19] J. Lecomte, *J. Chim. Phys.* **50**, C53 (1953).
[20] P. A. Giguere and K. B. Harvey, *Can. J. Chem.* **34**, 798 (1956).
[21] M. O. Bulanin, *Opt. i Spektroskopiya* **2**, 557 (1957).
[22] M. Falk and P. A. Giguere, *Can. J. Chem.* **35**, 1195 (1957).
[23] G. E. Walrafen, *J. Chem. Phys.* **36**, 1035 (1962).
[24] R. E. Weston, Jr., *Spectrochim. Acta* **18**, 1257 (1962).
[25] J. W. Schultz and D. F. Hornig, *J. Phys. Chem.* **65**, 2131 (1961).
[26] G. E. Walrafen, *J. Chem. Phys.* **40**, 3249 (1964).
[27] G. E. Walrafen, *J. Chem. Phys.* **44**, 1546 (1966).
[28] W. A. Senior and W. K. Thompson, *Nature* **205**, 170 (1965).
[29] W. K. Thompson, *Trans. Faraday Soc.* **61**, 2635 (1965).
[30] T. T. Wall and D. F. Hornig, *J. Chem. Phys.* **43**, 2079 (1965).
[31] P. C. Cross, J. Burnham, and P. A. Leighton, *J. Am. Chem. Soc.* **59**, 1134 (1937).

temperature at which the authors carried out their measurements is known, it is stated in Table 3 in brackets, following the wave number value.

In studying the bands of the HDO molecule, the procedure is as follows: H and D are rapidly exchanged in a mixture of H_2O and D_2O. Hence the OH bands of HDO can be observed if a little H_2O is added to D_2O; the OD bands of HDO can be observed if a little D_2O is added to H_2O.*

From Fig. 11 it is seen that in the spectrum of HDO, in the region from 3700 to 3000 cm^{-1}, a band is found at about 3410 cm^{-1} with an extremely weak shoulder at about 3615 cm^{-1}. In the spectrum of H_2O, a broad band is found, with a maximum at about 3420 cm^{-1}. A very weak shoulder on this band can also be observed at about 3615 cm^{-1}. In contrast to the findings with HDO, however, a clearly marked shoulder is also found at about 3250 cm^{-1}. Further, the band for H_2O is also broader on the side toward larger wave numbers than in the case with HDO.

Table 3, columns 2–4, shows that all these observations agree well with

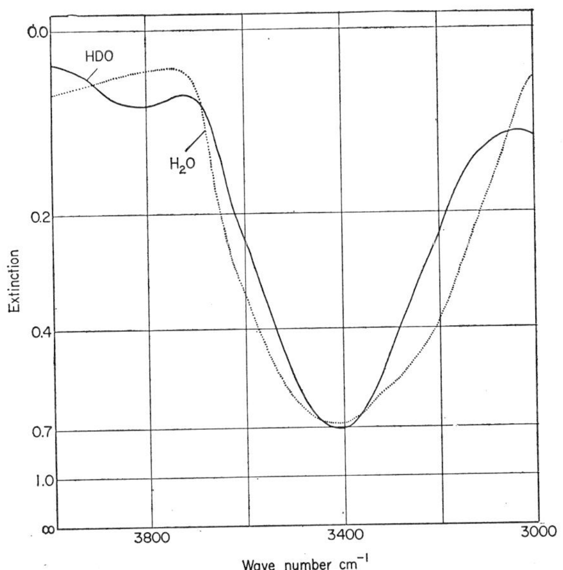

FIG. 11. The spectra of liquid water at 25°C: · · · ·, H_2O; ——, HDO (mixture of 10 mole % H_2O molecules and 90 mole % D_2O molecules).

* On the IR-spectroscopic analysis of a mixture of H_2O and D_2O (see footnotes 32 and 32a).

TABLE 3

THE BANDS OF LIQUID WATER[a]

Molecule	Stretching vibrations (cm^{-1}): free OH or OD groups		Stretching vibrations (cm^{-1}): OH or OD groups in the hydrogen bridges		Overtones of the scissor vibration $2\bar{\nu}_2$ (cm^{-1})		Scissor vibration + torsion vibration or, alternatively, $3\bar{\nu}_\beta$ (for further details see text, p. 35) (cm^{-1})	
1	2		3		4		5	
H_2O	3650 R	(23)	3382 (3°C)		3290	(21)	2146 (3°C)	
	3610 R[c]	(24)	3395 (19°C)	(17)	3225 (25°C)	(23)	2128 (21°C)	(17)
	3613 R	(25)	3426 (60°C)		3225	(27) (28)	2092 (70°C)	
	3615 (always a weak shoulder only)	Fig. 11	3428	(13)			2128	(19)
			3425	(21)			2110	(13)
			3400	(22)			2120	(22)
			3450 (25°C)	(23)			2115 (25°C)	(23)
	3590 (200°C)		3450	(27)				
	3625 (300°C)	(18)[d]		(28)				
	3650 (374°C)		3420 (30°C)					
			3450 (100°C)					
			3480 (200°C)	(18)[d]				
			3520 (300°C)					
			3545 (374°C)					
D_2O	2680 R[c]	(24)	2500	(15)	2435	(21)	1620	(22)
	2617 R	(25)	2550	(21)				
	(always a weak shoulder only)		2500	(22)				
HDO	at about 3615 (only a weak shoulder)	Fig. 11	$\bar{\nu}$OH					
			3405	(4)				
			3420	(21)				
			$\bar{\nu}$OD					
			2520	(4)				
			2510	(21)				

published results. Comparison of the spectra of H_2O and HDO has already led numerous authors[21,23,24,33] to the following conclusions:

If, as occasionally supposed in earlier works, the two bands of the H_2O molecule (3420 and 3250 cm^{-1}) are to be ascribed to H_2O molecules bound in

[32] J. Oostens and J. J. Fripiat, *Beckman Rept.* ed. Beckman Instruments, GmbH München Heft **4** p.8 (1959).

[32a] E. G. Mahadevan, *Analyst*, **92**, 717 (1967).

[33] D. F. Hornig, *J. Chem. Phys.* **40**, 3119 (1964).

Scissor vibration $\bar{\nu}_2$ (cm^{-1})		Torsion vibrations $\bar{\nu}_\beta$ (cm^{-1})		(see text, p. 34) (cm^{-1})		Translational vibration $\bar{\nu}_\sigma$ (cm^{-1})		$\bar{\nu}$ (For further details see text p. 35) (cm^{-1})	
6		7		8		9		10	
1649 (3°C)	(17)	710 (10°C)	(20)	500 [b]	(16)	160	(16)	60 R	(26)
1646 (20°C)	(17)	710	(22)	450 ± 5 R	(27)			60 R	(27)
1642 (70°C)	(17)	705	(26)						
1637	(13)	~780 R	(27)						
1640	(22)								
1645 (25°C)	(23)								
1640	(27)								
1640	(28)								
1220	(15)	530 (10°C)	(20)	360 [b]	(16)	160	(16)	60 R	(26)
1205	(22)	530	(21)			167 ± 5	(27)	60 R	(27)
		525	(26)						
		550 R	(26)						
1455	(4)								

[a] Wave numbers of the bands of IR spectra in older publications are given by Giguere and Harvey,[20] 'ables II and III.

Corresponding summaries of the bands observed in Raman spectra are given by Giguere and Iarvey,[20] Weston,[24] Walrafen,[26] and Wall and Hornig.[30]

[b] Found by other workers only in Raman-spectroscopic investigations.

[c] Further details on the Raman investigations of this shoulder are given in footnote 24, Table 3.

[d] In (18) the bands in columns 2 and 3 are assigned as an alternative to $\bar{\nu}_1$ and $\bar{\nu}_3$ This assignment ay be correct at high temperatures if the hydrogen bridges are weakened. However, this will not be e case at 25°C (see Sect. IV.5.1).

two different ways in the hydrate structure, the band at about 3250 cm^{-1} should also be observed with HDO. This is not the case. The band therefore is due to the overtone of the scissor vibration of the H_2O molecule $2\bar{\nu}_2$. Fermi-resonance between $2\bar{\nu}_2$ and $\bar{\nu}_1$ is also possible, as was pointed out by Hornig.[33]

In liquid H_2O, the band of the OH-stretching vibration of the hydrogen bridges is found at about 3420 cm^{-1}. Under the present conditions, the bands of the antisymmetric $\bar{\nu}_3$, and symmetric $\bar{\nu}_1$, stretching vibrations almost coincide. This is a consequence of the interactions, in this case the interactions of the water molecules with one another, as previously discussed.

The weak shoulder observed at about 3615 cm^{-1}, however, cannot be associated with the antisymmetric stretching vibration of H_2O molecules as is maintained in some publications, since this shoulder is found in spectra of both H_2O and HDO. From the position of this shoulder, the band may be ascribed to the stretching vibration of free OH groups—i.e., of groups not bound by hydrogen bridges. This is discussed in detail in Sections IV.5 and IV.12.

Therefore, with the exception of the weak shoulder at about 3615 cm^{-1}, the observed complex of bands with pure liquid water is caused by different vibrations of water molecules all bound in the same way in the hydrate structure. However, this need not be the case in electrolyte solutions, as we shall see in Section IV.18.

This leads to the following procedure in the investigation of water of hydration: If several bands are observed in the region 3800–3000 cm^{-1}, it must always be shown whether they are due to differently attached water molecules, or to different modes of vibration of water molecules all attached in the same way. Doubtful cases may be elucidated in a study of spectra of samples hydrated with HDO.

III.2.2. Restricted Torsion and Translation Vibrations

If the water molecule is not free, but bound by hydrogen bridges, bands of restricted torsions and restricted translations of the water molecules are encountered. The first is denoted by $\bar{\nu}_\beta$, the second by $\bar{\nu}_\sigma$, following the nomenclature of Pimentel and McClellan (see footnote 34, p. 132). These bands have been observed by numerous authors, in investigations of both liquid water,[16,20*,22,26,27*,31] and ice.[20*,31,35,36,37*,38*,39–41] However, not all these bands can be assigned completely unambiguously.

The bound water molecule can carry out three torsion vibrations. In liquid water, two bands are found, one at about 710 cm^{-1} and one at about 500

* These contain many references to more literature.

cm^{-1} (Table 3, columns 7 and 8), which may be due to vibrations $\bar{\nu}_\beta$.*[26,27] However, not all authors agree on how to interpret the smaller wave number band. For example, Hornig *et al.*[37] ascribe it to the combination vibration $\bar{\nu}_\beta - \bar{\nu}_\sigma$ (restricted torsion less the restricted translation, see below). However, this assignment cannot be correct, since these bands are observed in inelastic neutron scattering spectra with energy gain.[42,43] Weston[24] gives yet another assignment for the band at about 500 cm^{-1}.

Cross *et al.*[31] have already assigned the band found in investigations of liquid H_2O at about 170 cm^{-1},† (Table 3, column 9) to the restricted translation vibration of the water molecules, i.e., to the bridge vibration $\bar{\nu}_\sigma$. All other authors have accepted this assignment. Walrafen[27] investigated the temperature dependence of the intensities of these bands. He found a decrease in intensity with increasing temperature, as would be expected from this assignment. He also obtained the same result for the band at 60 cm^{-1}.

For the assignment of the band found at 60 cm^{-1} in Raman investigations of liquid H_2O and D_2O (Table 3, column 10), different hypotheses are advanced by various authors (cf. footnote 38, and in particular footnotes 26 and 27).

Finally, many authors observe a band at about 2110 cm^{-1} in investigations on liquid H_2O (see Table 3, column 4 and Fig. 90). This band is almost always assigned to a combination vibration $\bar{\nu}_2 + \bar{\nu}_\beta$ (scissor vibration plus torsion vibration). On the other hand, Hornig *et al.*[37] conjecture that it is to be assigned to the second overtone of the torsion vibration, i.e., $3\bar{\nu}_\beta$.

* Whalley *et al.*[39-41] have further investigated the fine structure of these bands in the various crystal modifications of ice, giving as their explanation that intermolecular coupling occurs in the crystal.

† These authors give as the result of their Raman-spectroscopic investigation of liquid H_2O at 40°C a wave number of 200 cm^{-1}, or for liquid D_2O at 50°C, 190 cm^{-1}; nevertheless they remark: "... no definite maximum, frequencies approximate."

[34] G. C. Pimentel and A. L. McClellan, "The Hydrogen Bond." Freeman, San Francisco, California, 1960.

[35] E. F. Gross and W. I. Walkow, *Dokl. Akad. Nauk. SSSR* **74**, 453 (1950).

[36] E. F. Gross and W. I. Walkow, *Dokl. Akad. Nauk SSSR* **81**, 761 (1951).

[37] D. F. Hornig, H. F. White, and F. P. Reding, *Spectrochim. Acta* **12**, 338 (1958).

[38] N. Ockmann, *Advan. Phys.* **7**, 199 (1958).

[39] J. E. Bertie and E. Whalley, *J. Chem. Phys.* **40**, 1637 (1964).

[40] J. E. Bertie and E. Whalley, *J. Chem. Phys.* **40**, 1646 (1964).

[41] M. J. Taylor and E. Whalley, *J. Chem. Phys.* **40**, 1660 (1964).

[42] K. E. Larsson and U. Dahlborg, *Reactor Sci. Tech.* **16**, 81 (1962).

[43] D. J. Hughes, H. Palevsky, W. Kley, and E. Tunkelo, *Phys. Rev.* **119**, 872 (1960).

CHAPTER IV

HYDRATION OF THE SALTS

Before considering the bands of water of hydration, we should consider the bands of the vibrations of the anions.[1-3] In the following, the term "anion" is used to mean a negatively charged functional group of the resin network, such as an $—SO_3^-$ ion.

IV.1. The Electron Structure of the Anions in the Salts

The assignment of the bands of the stretching vibrations which are summarized for all these anions in Table 11 (column 6) gives:

Result 1: In salts with anions $\left(—S\begin{smallmatrix}O^-\\ O\\ O\end{smallmatrix}\right)$, $\left(—Se\begin{smallmatrix}O^-\\ O\\ O\end{smallmatrix}\right)$, and $\left(—P\begin{smallmatrix}O^{2-}\\ O\\ O\end{smallmatrix}\right)$ three almost identical bonds are present, and in those with ions $\left(—Se\begin{smallmatrix}O^-\\ O\end{smallmatrix}\right)$, $\left(—P\begin{smallmatrix}S^{2-}\\ O\\ O\end{smallmatrix}\right)$, and $\left(—P\begin{smallmatrix}H^-\\ O\\ O\end{smallmatrix}\right)$ two almost identical SeO or PO bonds.

According to this, the charge is almost uniformly distributed among the XO bonds, so that the mesomerism is very strong.[3] These XO bonds are, however, exactly identical only if these anions are isolated from their surroundings. When cations are attached to these anions, the latter are somewhat polarized, as will be shown in the next section. The PH and PS bonds also take part in mesomeric bond resonance to some extent, as shall be seen in the results given in Section V.1.1.

[1] G. Zundel, H. Noller, and G.-M. Schwab, *Z. Elektrochem.* **66**, 122 (1962).
[2] G. Zundel and A. Murr, *Z. Naturforsch.* **21a**, 1640 (1966).
[3] G. Zundel, *Z. Naturforsch.* **22a**, 199 (1967).

IV.2. Removal of Antisymmetric Stretching Vibration Degeneracy of Ions $\left(-S\begin{array}{l}O^-\\ O\\ O\end{array}\right)$ and $\left(-Se\begin{array}{l}O^-\\ O\\ O\end{array}\right)$ by Cation Interaction

Experimental Findings

In Table 4 the positions of the bands of the stretching vibrations of the $\left(-S\begin{array}{l}O^-\\ O\\ O\end{array}\right)$ ion in the various salts of polystyrenesulfonic acid are collected together (spectra in Figs. 12–15). The experimental maximum errors are ± 3 cm^{-1}, in the case of the bands of the symmetrical stretching vibration $\bar{\nu}_s$, or ± 5 cm^{-1} in the case of the two bands of the doublet. In order to obtain a basis for discussing the band positions, the relative strengths of the Coulombic field of the cation on its $-SO_3^-$ ions have been calculated. These E_K values are given in the last column of the table; $(1 - A)E_K$ is the strength of the Coulombic field of the cation at the location of the nucleus of the oxygen atom, in the $\left(-S\begin{array}{l}O^-\\ O\\ O\end{array}\right)$ ion, to which this cation is attached [see Fig. 17(a)]. We shall see, in Section IV. 3, that the cation is always attached to one particular oxygen atom of the $\left(-S\begin{array}{l}O^-\\ O\\ O\end{array}\right)$ ion. The factor $(1 - A)$ takes into account the electronic shielding, and it is constant for a given anion. The E_K values have been calculated by Coulomb's formula. The distance of the cation from the oxygen nucleus is taken as the sum of the cation radius and the covalent radius of the oxygen atom.

From the data given in Figs. 12–15 and Table 4 we obtain the following results:

Result 2: The position of the symmetric stretching vibration of the $-SO_3^-$ ion, $\bar{\nu}_s$, at about 1040 cm^{-1}, is not appreciably influenced by the cation (Table 4, column 2).

Result 3: A doublet* is always observed in place of the band of the antisymmetric stretching vibration $\bar{\nu}_{as}$ of the $-SO_3^-$ ion at about 1200 cm^{-1}. The splitting increases considerably with progressive drying of the membranes.

Result 4: Comparison of columns 4 and 6 (Table 4) shows that the splitting of the band near 1200 cm^{-1} is greater, the stronger the field of the cation at the anion. This effect is found with all groups of cations. Furthermore, ions of the transition elements produce an additional effect.

* The band of the in-plane skeleton vibration of the benzene ring at 1128 cm^{-1} (Table A.1) is always superimposed on the component with the smaller wave number.

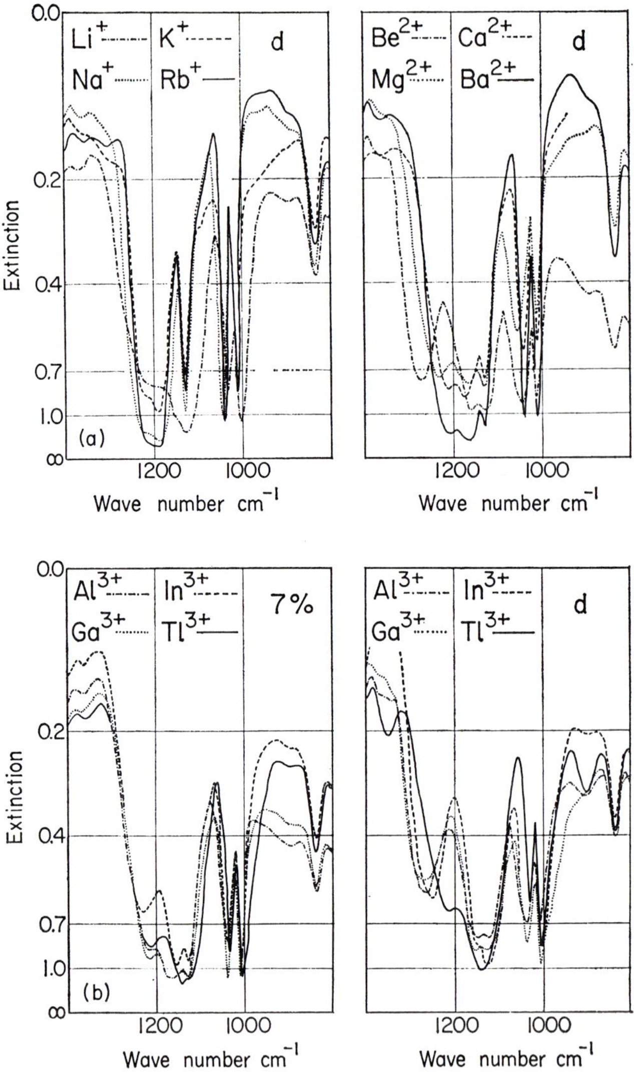

FIG. 12. Salts of polystyrenesulfonic acid, 5% cross-linked, 3 days of sulfonation, (a) Li^+, Na^+, K^+, Rb^+; Be^{2+}, Mg^{2+}, Ca^{2+}, and Ba^{2+} salts, thoroughly dried membranes (b) Al^{3+}, Ga^{3+}, In^{3+}, Tl^{3+} [$Tl(OH)_2^+$ also present (see Section IV.20)] salts, thoroughly dried membranes (d), at 7% relative atmospheric humidity (7%).

Result 5: Although E_K, the electrostatic field of the ions Cu^{2+} and Zn^{2+} at the anion, is smaller than that of Mg^{2+} (column 6), the splitting of the bands near 1200 cm^{-1} with Cu^{2+} and Zn^{2+} is much greater [cf. Fig. 12(a) with Fig. 14].

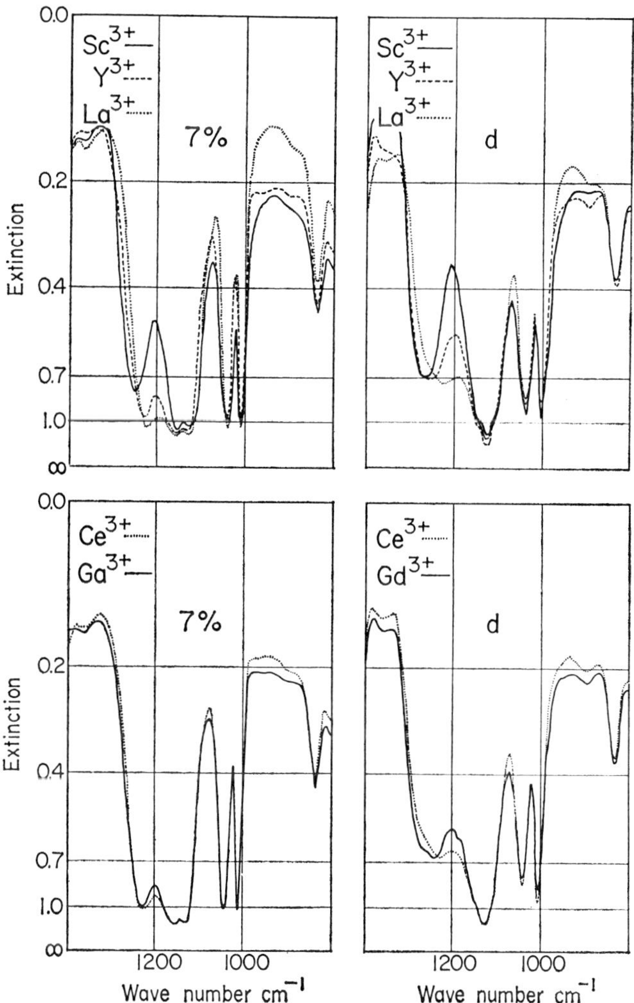

FIG. 13. Sc^{3+}, Y^{3+}, La^{3+}, Ce^{3+}, and Gd^{3+} salts of polystyrenesulfonic acid, 5% cross-linked, 3 days of sulfonation, thoroughly dried membranes (d), at 7% relative atmospheric humidity (7%).

Result 6: When a cation possessing a weak field, such as Rb^+, Fig. 12(a), is present, the band is split only slightly. This is particularly the case with a strongly hydrated salt, as is shown for the hydrated Na^+ salt in Fig. 1(e,f). Under these conditions, the band lies at 1200 cm^{-1}. After splitting has taken place, the band with the higher wave number always lies above 1200 cm^{-1}, and that with the lower wave number below that wave number.

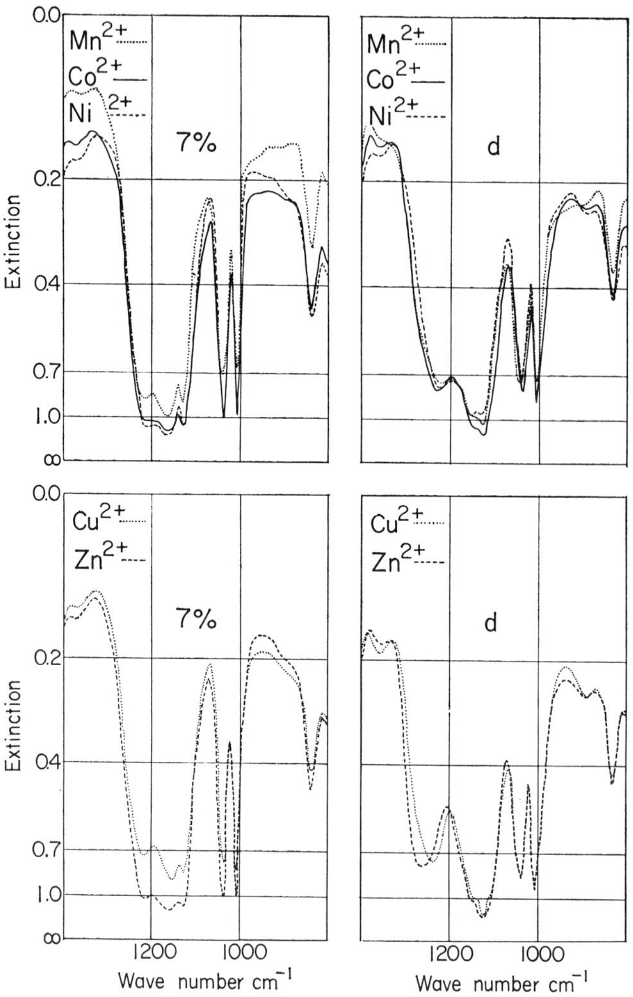

FIG. 14. Mn^{2+}, Co^{2+}, Ni^{2+}, Cu^{2+}, and Zn^{2+} salts of polystyrenesulfonic acid, 5% cross-linked, 3 days of sulfonation, thoroughly dried membranes (d), at 7% relative atmospheric humidity (7%).

From Fig. 16, we obtain:

Result 7: The antisymmetric stretching vibration $\bar{\nu}_{as}$ of the $—SeO_3^-$ ion at 909 cm^{-1} is not perceptibly split in the case of the Na^+ salt. Considerable splitting of this band is, however, observed in the case of the thoroughly dried Sc^{3+} salt.

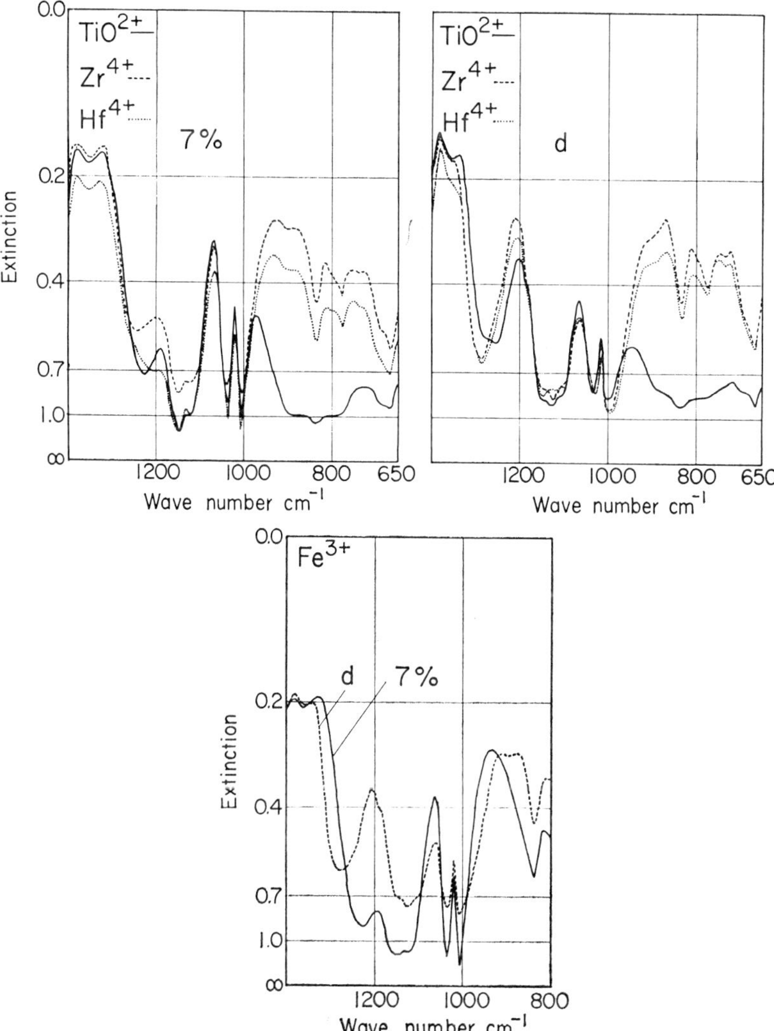

FIG. 15. Zr^{4+}, Hf^{4+}, TiO^{2+}; Fe^{3+} salts of polystyrenesulfonic acid, 5% cross-linked, 3 days of sulfonation, thoroughly dried membranes (d), at 7% relative atmospheric humidity (7%).

Interpretation of the Band Splitting

The structure of the ions $—S\left(\begin{smallmatrix}O^- \\ O \\ O\end{smallmatrix}\right)$ and $—Se\left(\begin{smallmatrix}O^- \\ O \\ O\end{smallmatrix}\right)$:

TABLE 4

THE POSITIONS OF THE BANDS OF THE —SO_3^- ION STRETCHING VIBRATIONS AS A FUNCTION OF THE NATURE OF THE CATION PRESENT

Cation	Stretching vibrations of the —SO_3^- ion (cm^{-1}) Symmetric	Antisymmetric[a]		Splitting due to removal of degeneracy	Ionic radius (Å) (after footnote 4)	E_K (ESU · cm^{-2} × 10^6)
1	2	3		4	5	6
Li^+	1030	1134	1203	69	0.68	3.18
Na^+	1042	1188	1226	38	0.97	2.08
K^+	1037	1191	1223	32	1.33	1.36
Rb^+	1036	1194	1219	25	1.47	1.18
Cs^+	1031	1194	1219	25	1.67	0.98
Be^{2+}	1044	1160	1269	109	0.35	11.9
Mg^{2+}	1052	1170	1229	59	0.66	6.58
Ca^{2+}	1040	1169	1223	54	0.99	4.05
Sr^{2+}	1044	1179	1227	48	1.12	3.45
Ba^{2+}	1039	1169	1204	35	1.34	2.70
Al^{3+}	1039	1151	1269	118	0.51	12.8
Ga^{3+}	1032	1148	1261	113	0.62	10.6
In^{3+}	1035	1141	1251	110	0.81	7.78
Tl^{3+} [b]	(1035)	(1145)	(1216)	(71)	0.95	6.40
Sc^{3+}	1040	1137	1268	131	0.81	7.78
Y^{3+}	1041	1138	1257	119	0.92	6.60
La^{3+}	1039	1145	1228	83	1.14	5.04
Ce^{3+}	1040	1143	1228	85	1.07	5.50
Gd^{3+}	1041	1143	1242	99	0.94 [c]	6.48
Zr^{4+}	1032	1138	1287	149	0.79	10.7
Hf^{4+}	1031	1150	1285	135	0.78	10.8
Mn^{2+}	1042	1149	1223	74	0.80	5.27
Co^{2+}	1039	1150	1227	77	0.72	5.96
Ni^{2+}	1039	1150	1222	72	0.69	6.23
Cu^{2+}	1038	1148	1239	91	0.72	5.96
Zn^{2+}	1037	1140	1262	122	0.74	5.78
Fe^{3+}	1036	1143	1273	130	0.64	10.2

[a] Position of the two bands arising on removing the degeneracy of the antisymmetric stretching vibration.

[b] The values are given in parentheses because $Tl(OH)_2^+$ is present in the membrane together with Tl^{3+} (on this point see Section IV.20).

[c] Personal communication from R. C. Weast, editor-in-chief, "Handbook of Chemistry and Physics," Chemical Rubber Publishing Co., Cleveland, Ohio.

[4] C. D. Hodgman, ed. "Handbook of Chemistry and Physics." 44th Ed. Chem. Rubber Publ. Co., Cleveland, Ohio, 1961.

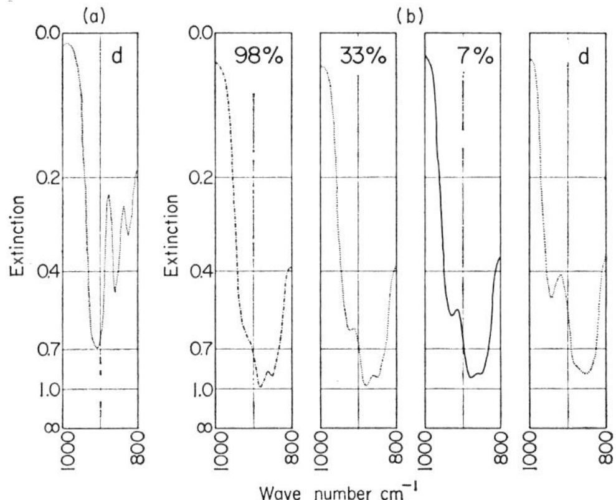

FIG. 16. Salts of polystyreneselenonic acid, 7% cross-linked, 3 days of selenonation. (a) Na^+ salt, thoroughly dried membrane, (b) Sc^{3+} salt, various atmospheric humidities.

Simon and Kriegsmann[5] have investigated the ion $CH_3SO_3^-$, Siebert[6] the ions $HOSO_3^-$ and FSO_3^-, and Simon and Schmidt[7] the ion HSO_3^-. Siebert[6] and Simon and Schmidt[7] found that under prevailing conditions of their investigations, ions with three identical SO bonds were present. Hence, in these cases, the $—SO_3^-$ ion has a pyramidal structure; i.e., it possesses C_{3v} "local" symmetry. In $CH_3OSeO_3^-$ the $—Se(O^-)(O)(O)$ group also has C_{3v} "local" symmetry.[8]

The antisymmetric stretching vibration of a group with C_{3v} "local" symmetry is doubly degenerate. If the C_{3v} symmetry is disturbed, the degeneracy of the antisymmetric stretching vibration is removed, i.e., instead of one band, two should be found for this vibration (on this point see footnote 9, p. 87).

Such a removal of the degeneracy can arise from linkage of the $—SO_3^-$ ion to another group. An example of this has been shown by Siebert[6] in his studies of $NH_2SO_3^-$ and $CH_3OSO_3^-$ ions. He found here that the band of the antisymmetric stretching vibration of the $—SO_3^-$ ion is split.

The splitting of the bands by removal of the degeneracy arises here, to a very subordinate degree only, from that disturbance of the symmetry produced by linkage of the anions to the benzene ring. This is shown by results 3, 4, and, in particular, 6. If linkage of the

[5] A. Simon and H. Kriegsmann, *Chem. Ber.* **89**, 1718 (1956).
[6] H. Siebert, *Z. Anorg. Allgem. Chem.* **289**, 15 (1957).
[7] A. Simon and W. Schmidt, *Z. Elektrochem.* **64**, 737 (1960).
[8] R. Paetzold, *Z. Anorg. Allgem. Chem.* **325**, 47 (1963).
[9] K. Nakamoto, "Infrared Spectra of Inorganic and Coordination Compounds." Wiley, New York, 1963.

anion to the benzene ring were the cause for the removal of degeneracy, then the splitting should be independent of the degree of hydration and the nature of the cation present. Furthermore it should, particularly, still be just as strong with cations which possess a weak field.

This indicates that the cation removes the degeneracy. This can be caused, on the one hand, by motion of the mass of the cation in the vibration of the $—SO_3^-$ ion. On the other hand, the interaction between the cation and the anion, and, in particular, the effect of the electrostatic field of the cation on the anion, may disturb the C_{3v} "local" symmetry and thus remove the degeneracy.

Columns 4 and 6 in Table 4 show results for the Zr^{4+} and Hf^{4+} salts.* The electrostatic field of the cations is almost equally strong for these two ions. The band splitting differs only by about 10%. The mass of the Hf^{4+} however, is approximately double that of the Zr^{4+} ion. Hence the mass of the cation must play only a subordinate part in removing the degeneracy.

It follows that the degeneracy of the antisymmetric stretching vibrations of the $—SO_3^-$ and $—SeO_3^-$ ions are removed by the interaction between cation and anion. In this, the effect of the electrostatic cation field on the anion is by far the most important cause for the removal of the degeneracy. This is proved by Result 4, the band splitting depending on the strength of the electrostatic field of the cation.

Comparison of Figs. 12(a) and 13 with Fig. 16 leads to Result 8.

Result 8: The splitting for the Na^+ salt of polystyrenesulfonic acid is 38 cm^{-1}, but with the Na^+ salt of polystyreneselenonic acid the band is not split. For the Sc^{3+} salt of polystyrenesulfonic acid the splitting is 131 cm^{-1}, for that of polystyreneselenonic acid it is only 80 cm^{-1} under comparable conditions. Thus the splitting of the bands of the antisymmetric stretching vibration is much greater in the salts of polystyrenesulfonic acid than in salts of polystyreneselenonic acid.

This result is understandable when one considers how the degeneracy is removed. Anions are polarized by the fields of their cations, as is known, for example, from the refraction experiments of Fajans and Joos.[12] If now the

* When $TiCl_4$ is dissolved in water, titanyl ions are produced. These ions are actually found in the spectrum of a membrane treated with such a solution. In such membranes a broad, very intense band is observed at about 700 cm^{-1} (Fig. 15). According to Barraclough *et al.*[10] and Dehnike and Weidlein,[11] this band is to be ascribed to the stretching vibration of $Ti{=}O^{2+}$. To determine whether a similar behavior occurs in solutions of Zr^{4+} and Hf^{4+} salts, we investigated the spectra of these membranes down to 400 cm^{-1}. No corresponding band was found. For this reason we are certain we are dealing with Zr^{4+} and Hf^{4+} ions.

[10] C. G. Barraclough, J. Lewis, and R. S. Nyholm, *J. Chem. Soc.* (*London*), p. 3552 (1959).

[11] K. Dehnike and J. Weidlein, *Angew. Chem.* **78**, 1065 (1966).

[12] K. Fajans and G. Joos, *Z. Physik* **23**, 1 (1924).

cation is unsymmetrically bound to the $—SO_3^-$ or $—SeO_3^-$ ion (on this point, see Section IV.3 and Fig. 17), the cation field—which polarizes the anion unsymmetrically—disturbs the mesomeric bond resonance in these anions, and C_{3v} "local" symmetry changes to C_s symmetry. This, in turn, removes the degeneracy of the antisymmetric stretching vibration.

This explains the dependence of the splitting on the strength of the cation field, for the stronger the field, the more strongly is the anion polarized. However, if this mechanism holds, the splitting must be greater, the more polarizable the anion. If we assume that the $—SO_3^-$ ion is more polarizable than the $—SeO_3^-$ ion, Result 8 becomes understandable.

From this point of view, the removal of the degeneracy is closely connected with the ion-induced dipole interaction between cation and anion, particularly with the interaction of the cation with the dipole which it induces in the anion (see Section IV.3). The special features observed with Cu^{2+} and Zn^{2+} salts (Result 5), however, show that other kinds of cation–anion interaction are also important for the removal of degeneracy. The ion–ion interaction, however, is not of appreciable significance, since it does not alter the structure of the anion.

Cu^{2+} and Zn^{2+} are ions of the transition elements. The *d* electrons of these ions cause a stronger interaction (covalent) of these ions with the $—SO_3^-$ ions. The nature of this interaction will be discussed later on when considering the interaction between the ion and water of hydration in Section IV.10. The effects of this covalent interaction between cation and anion on the structure of the anion obviously also lead to a splitting of the bands of the antisymmetric stretching vibration, which explains Result 5.

If these irregularities be set aside, it appears at first sight possible to formulate quantitatively the relationship between the band splitting, the electrostatic field, and the polarizability of the anion. Decius[13] has formulated a theory of such a band splitting for the sulfate ions.

Nevertheless, in considering the interaction, it must be borne in mind that the polyvalent cations cross-link the anions (Fig. 18). This means that the field of one anion weakens the cation field at the other anion. This is generally known from results obtained with ionic crystals (the Madelung factor). The same feature is shown by the comparison of band splitting and E_K values. If, for example, we compare the monovalent and divalent cations in Table 4, we see that the ratio of band splitting to E_K value is greater for the monovalent cations. This is understandable, since with the monovalent cations no second anion is close enough to exert any great effect on the corresponding anion of the cation. If we further compare, for example, band splitting and the E_K value in the presence of Be^{2+} and Al^{3+} with those in the presence of In^{3+},

[13] J. C. Decius, *Spectrochim. Acta* **21**, 15 (1965).

Sc^{3+}, Zr^{4+}, and Hf^{4+}, we see that the ratio of band splitting to E_K value is smaller for Be^{2+} and Al^{3+}. This is understandable, since, because of the small radii of these ions, the weakening of the cation field at one of its anions by the fields of its other anions is relatively large. Thus, in the presence of polyvalent cations the influence of the anions on one another plays a significant role.

Finally, there is yet another difficulty to be overcome before attempting a theoretical treatment. A small amount of water often is attached to the cation, even after vigorous drying. As we shall see in Section IV.15, this affects the splitting markedly. Hence we cannot at present undertake a quantitative formulation. However, this removal of degeneracy will lead to interesting conclusions in our further work.

In summary, we conclude that the degeneracy of the band of the antisymmetric stretching vibration of the $—SO_3^-$ and $—SeO_3^-$ ions is removed by the electrostatic field of the cation. The observed band splitting is the greater, the stronger the electrostatic field of the cation, and the more polarizable the anion. The removal of degeneracy arises from unsymmetrical polarization of the anion, i.e., from a disturbance of the mesomerism of these anions. It is therefore closely connected with the ion-induced dipole interaction between cation and anion. Other types of cation–anion interaction lead to an increase of band splitting, insofar as they alter the structure of the anion, as in covalent interaction.

IV.3. Location of the Cation with Respect to the Oxygen Atoms of the $—SO_3^-$ and $—SeO_3^-$ Ions

In the foregoing section, we have already mentioned that the fields of the cations only remove the degeneracy of the antisymmetric stretching vibration if they polarize the anion unsymmetrically. Only under these circumstances is the C_{3v} "local" symmetry changed to C_s symmetry. From the removal of the degeneracy it thus follows that the cation is not symmetrically bound to the anion.

There are two possible unsymmetrical positions of attachment, as is shown in Fig. 17(a, b). The cation is located either on one of the oxygen atoms of the $—SO_3^-$ or $—SeO_3^-$ ion or between two such atoms. The following considerations enable us to make a decision between these two possibilities. If the cation were located not at a single oxygen atom of the $—SO_3^-$ or $—SeO_3^-$ ion but between two oxygen atoms, it would be necessary to assume that the position midway between the three oxygen atoms would be preferred over one midway between two such atoms. The midway position between the three oxygen atoms, however, is just the symmetrical position which, as we know, does not occur. Consequently the cation must be located at a single oxygen atom, as shown in Fig. 17(a).

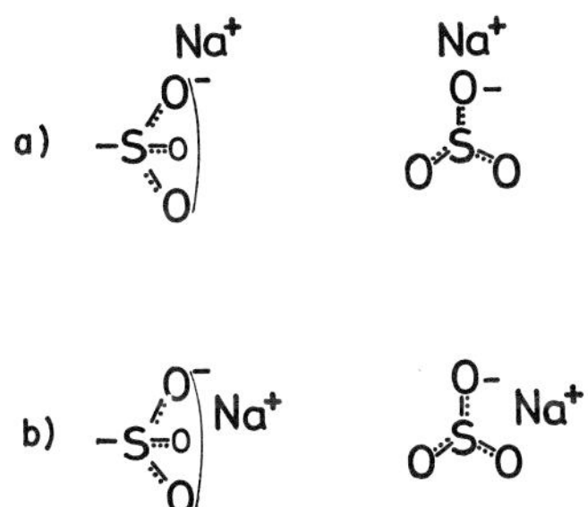

FIG. 17. The location of the cation relative to the oxygen atoms in the —SO_3^- ion.

The attachment of two cations to a single —SO_0^- or —SeO_0^- ion would also lead to a disturbance of the C_{3v} local symmetry. Such an attachment of two cations to a single —SO_3^- or —SeO_3^- ion is, however, ruled out on grounds of electrical neutrality, since the —SO_3^- ion is monovalent. It is nevertheless to be supposed that the —$SO_3^-M^+$ are turned toward one another with monovalent cations, particularly if no water is present. It is not possible to demonstrate this from the experimental results in the case of monovalent ions. The results which follow do, however, show a cross-linkage of —SO_3^- ions via the polyvalent cations.

Considered from the standpoint of the bond energy between cation and anion, it appears, at first sight, very strange that the cation should be attached unsymmetrically to the —SO_3^- or —SeO_3^- ion. Nevertheless, this can be understood on the assumption that the cation polarizes the anions. In addition to the ion–ion interaction, the interaction of the cation with the dipole induced by it in the anion also takes place. This produces a further energy decrease in the attachment of the cation. The polarizability of a bond is by far the greatest in the direction of the bond. If the effect of the cation field on a single SO or SeO bond is considered, the bond energy increase between cation and anion is greatest when the cation is attached to the oxygen atom in the direction of the SO bond. Under these conditions the bond energy increase is then greater than it would be if the cation field affected all three SO or SeO bonds without polarizing them very strongly. Consideration of the bond energy between cation and anion also is an argument in favor of the observed unsymmetrical attachment.

In all cases in which the splitting of the bands of the antisymmetric stretching vibration is large (Figs. 12–15) we note:

Result 9: Apart from the doublet arising on the removal of degeneracy, no unsplit band is present at low degree of hydration at 1200 cm^{-1}. This is particularly true with the tetravalent Zr^{4+} and Hf^{4+} ions (Fig. 15). Thus there is no band of an antisymmetric stretching vibration whose degeneracy has not been removed.

It then follows: At low degrees of hydration a cation is present at each —SO_3^- ion, i.e., at low degrees of hydration every —SO_3^- ion is turned toward a cation; for if this is possible for the tetravalent cations, despite all steric hindrance, it must also be the case for cations of lower valence. This gives a picture as shown in Fig. 18. The anions are cross-linked by polyvalent cations.

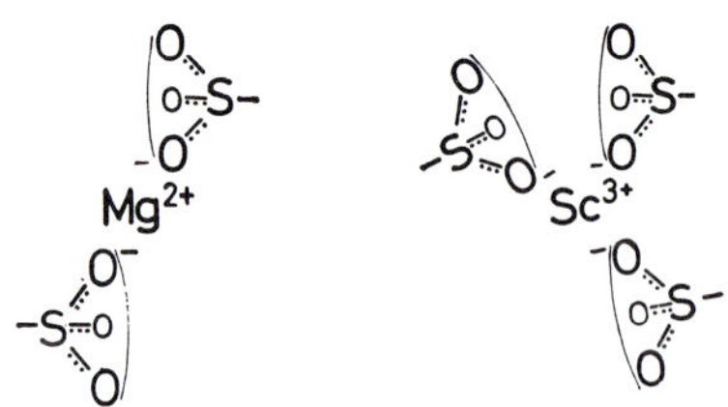

FIG. 18. Cross-linkage of the anions through the cations.

The cross-linkage of anions by polyvalent cations, observed in the present systems, may play an important role in some biological systems. For example, it is known that Ca^{2+} ions reduce the permeability of biological membranes.[14-17] Presumably, the divalent Ca^{2+} ions cross-link the functional groups of these membranes. The formation of supermolecules by Ca^{2+} ions[18] may be understood in the same manner.

In summary, we conclude that the cation is unsymmetrically attached to the —SO_3^- or —SeO_3^- ion. It is located on one of the three oxygen atoms. Each —SO_3^- ion is turned toward a cation, i.e., the polyvalent cations cross-link the anions together. Such a cross-linkage of functional groups by polyvalent cations, particularly Ca^{2+}, may play a part in biological systems.

IV.4. Anion Band Changes as a Function of Degree of Hydration

In Sections IV.1–IV.3 we have seen the important information that the bands of the anions can yield. Sections IV.2 and IV.3 were, in particular, concerned with the removal of the degeneracy of the antisymmetric stretching vibration of the —SO_3^- and —SeO_3^- ions, respectively; Table 5 gives a comprehensive survey.

[14] F. Brink, Jr., *Pharmacol. Rev.* **6**, 243 (1954).
[15] A. Fleckenstein, "Der Kalium-Natrium-Austausch als Energieprinzip in Muskel und Nerv." Springer, Berlin, 1955.
[16] H. Schmidt and R. Staempfli, *Helv. Physiol. Pharmacol. Acta* **15**, 200 (1957).
[17] H. Schmidt, *Pflüger's Arch. ges. Physiol.* **271**, 634 (1960).
[18] R. Boennimann, "Das Riesenmolekül," p. 35 ff. Huber, Bern, 1958.

TABLE 5

CHANGES IN ANION BANDS DEPENDENT ON THE DEGREE OF HYDRATION

Anion	See figure below	Section where discussed	Bands when the membranes are hydrated at relative atmospheric humidity > 90%	Bands when the membranes are thoroughly dried
1	2	3	4	5
O^- —S····O O	12–15	IV.1–3 IV.15 IV.4(a)	at about 1140 cm^{-1} symmetric SO stretching vibration at about 1200 cm^{-1} doublet[a] ⇒ further split antisymmetric SO stretching vibration (degeneracy removed)	at about 1140 cm^{-1}
	15 Zr^{4+} and Hf^{4+} salt	IV.4(e)	no band in-plane skeleton vibration of the benzene ring with strong participation of the substituents	1097 cm^{-1}
O^- —Se····O O	4(c, d) 16	IV.4(b) IV.1–3 IV.4(a)	855 cm^{-1} ←[a] 860 cm^{-1} symmetric SeO stretching vibration at about 899 cm^{-1} ← 909 cm^{-1} as a doublet with the Sc^{3+} salt ⇒ further split antisymmetric SeO stretching vibration (degeneracy removed with the Sc^{3+} salt)	
S^{2-} —P····O O	21	IV.4(b)	1212 cm^{-1} ← 1218 cm^{-1} doublet symmetric PO stretching vibration 1032 cm^{-1} ← 1037 cm^{-1} antisymmetric PO stretching vibration	
	21	IV. 4(e)	at about 1070 cm^{-1} shoulder vanishes on drying in-plane skeleton vibration of the benzene ring with strong participation of the substituents	1123 cm^{-1} appears on drying
O^- —Se O	19	IV. (4)c	D_2O hydration: 803 cm^{-1}; H_2O hydration: 803 cm^{-1} symmetric D_2O hydration: 776 cm^{-1}; H_2O hydration: 776 cm^{-1} only a weak shoulder antisymmetric SeO stretching vibration	only a broad band at about 803 cm^{-1}
H^- —P····O O	6(d)	IV. 4(b)	1045 cm^{-1} ← 1056 cm^{-1} symmetric PO stretching vibration 1162 cm^{-1} ← 1183 cm^{-1} antisymmetric PO stretching vibration	
	26	—	2313 cm^{-1} ← 2304 cm^{-1} PH stretching vibration	
	20	IV. 4(d)	D_2O hydration: 969 cm^{-1}; H_2O hydration: 969 cm^{-1} only a weak shoulder possibly PH bending vibration	this band vanishes with progressive drying

[a] ⇒ shows a change caused by the cation approaching the anion; ← shows a change caused by water in hydrogen bridge formation with the oxygen atoms of the anion.

(a) Removal of the degeneracy of vibrations.

Changes in spectra with the removal of the degeneracy of vibrations, dependent on the degree of hydration, are an important source of information when anions or cations with degenerate vibrations are present. Doubly degenerate vibrations arise when an ion has an axis of symmetry higher than C_2, triply degenerate vibrations when an ion has more than one axis of C_3 symmetry (see footnote 9, p. 22).

(b) From Table 5 (or the figures listed in column 2) we obtain Result 10.

Result 10: The SeO stretching vibrations of the $-Se(O^-)(O)(O)$ ion and the PO stretching vibrations of the $-P(S^{2-})(O)(O)$ and $-P(H^-)(O)(O)$ ions are slightly shifted toward smaller wave numbers with an increasing degree of hydration. Szymanski and Povinelli[19] report similar shifts of the bands of carbonate ions dependent on the degree of hydration.

Explanation: The formation of hydrogen bridges leads, as is known, to broadening and to shifts toward smaller wave numbers of the bands of the stretching vibrations of hydrogen-bridge acceptor groups. Hence, the bands of the SeO and PO stretching vibrations must be shifted toward smaller wave numbers when these anions are hydrated. Investigation of the bands of these anions with different cations must show to what extent, in all cases, the loosening of the cation from the anion also influences the positions of these bands.

(c) Figure 19 shows the bands of the SeO stretching vibrations of the $-Se(O^-)(O)$ ion as a function of the degree of hydration. In considering the figure it must be borne in mind that the band of the out-of-plane bending vibration of the $>CH$ groups of the benzene ring at 819 cm^{-1} is superimposed on these bands, but that, at a high degree of hydration, the former is seen only as a shoulder. From Fig. 19 we obtain:

Result 11: On hydration with D_2O, when the membrane is exposed to the atmospheric humidity over a saturated solution of $BaCl_2$ in D_2O, two strongly marked bands are found at 803 and 776 cm^{-1}. On hydration with H_2O in an atmosphere with a relative humidity of 98%, the band at 776 cm^{-1} is found only as a shoulder. After drying, after either D_2O or H_2O hydration, only a single broad band is found. This indicates, first, at a high degree of hydration, two bands of the stretching vibrations of the $-Se(O^-)(O)$ ion are observed, while after thorough drying only one is observed. Second, the splitting at a high degree of hydration is much more pronounced when the membrane is hydrated with D_2O.

An explanation of this result will only be possible when we know the dependence of the bands of this anion on the degree of hydration in the presence of other cations.

(d) The spectra in Fig. 20 show:

Result 12: If the Na^+ salt of polystyrenephosphinic acid is hydrated with D_2O, a strongly marked band is observed at 969 cm^{-1}. This vanishes with prolonged drying. If the membrane is hydrated with H_2O, a weak shoulder is observed, which also vanishes with prolonged drying.

[19] H. A. Szymanski and R. Povinelli, *Nature* **191**, 64 (1961).

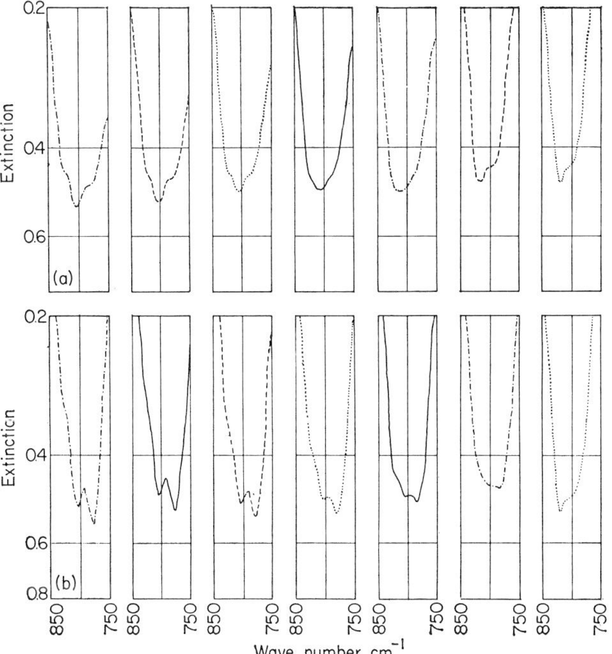

FIG. 19. Na^+ salt of polystyreneseleninic acid, 7% cross-linked, 3 days of selenonation. (a) H_2O hydrated: from left to right at 98, 53, 33, 22, and 11% relative atmospheric humidity, thoroughly dried membrane. (b) D_2O hydrated: on the left at 100% relative atmospheric humidity, from left to right decreasing atmospheric humidity, on the right a thoroughly dried membrane.

With the acid the PH bending vibration at 977 cm^{-1} is equivalent to this wave number. With the Na^+ salt, a PH bending vibration should be observable, whether the salt is hydrated with H_2O or with D_2O. The band of the PH stretching vibration in Fig. 26 shows us that the H is not exchanged for D in the Na^+ salt. It could be postulated that the IR activity of the PH bending vibration changes with the degree of hydration. However, a definite explanation of the difference between H_2O and D_2O hydration, in particular, will only be possible after further studies.

(e) Changes in the bands of the benzene ring, depending on the degree of hydration.

Randle and Whiffen,[20] like Kresze *et al.*,[21] point out that the substituents take part in some vibrations of the benzene ring. It would be expected that these bands also depend on the degree of hydration.

[20] R. R. Randle and D. H. Whiffen, *in* "Molecular Spectroscopy" (G. Sell, ed.), p. 111 Inst. Petrol., London, 1955.

[21] G. Kresze, E. Ropte, and B. Schrader, *Spectrochim. Acta* **21**, 1633 (1965).

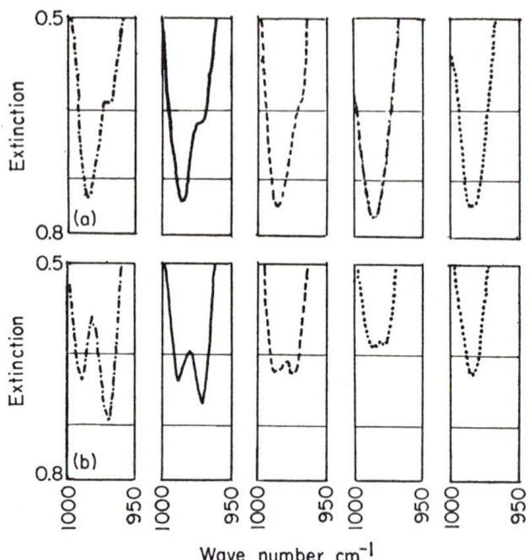

FIG. 20. Na^+ salt of polystyrenephosphinic acid, 7% cross-linked, 3 days phosphinated. (a) H_2O hydrated: beginning at the left with the membrane at 98, 71, 53, and 11% relative atmospheric humidity, on the right a thoroughly dried membrane. (b) D_2O hydrated: on the left at 100% relative atmospheric humidity, from left to right decreasing atmospheric humidity, on the right thoroughly dried membrane.

From Fig. 15, we obtain:

Result 13: In Zr^{4+} and Hf^{4+} salts of polystyrenesulfonic acid (see footnote p. 44), a band appears at 1097 cm^{-1} on thorough drying.

In Section V.1.3 we shall see that, in the case of the acid, the in-plane skeleton vibration of the benzene ring at 1097 cm^{-1} is only observed when the anions, rearranged by the protons, are attached as $-S(=O)(=O)OH$ groups to the benzene rings (see Result 61). Here, a corresponding explanation is probable, since the Hf^{4+} and Zr^{4+} ions strongly disturb the mesomeric bond resonance in the anion (Section IV.2).

The spectra in Fig. 21 show:

Result 14: In the spectrum of the Na^+ salt of polystyrenethiophosphonic acid, a shoulder is found at about 1070 cm^{-1}, if this substance is hydrated at 98% relative atmospheric humidity. The shoulder disappears with progressive drying, and a new band appears at 1123 cm^{-1}.

The bands of the benzene ring, in particular, also may show characteristic changes, dependent on the degree of hydration, which presumably are the result of changes in the anion structure dependent on the degree of hydration.

In summary, we see that various changes, dependent on the degree of hydration, are observed in the bands of the anions and in some bands of the benzene ring.

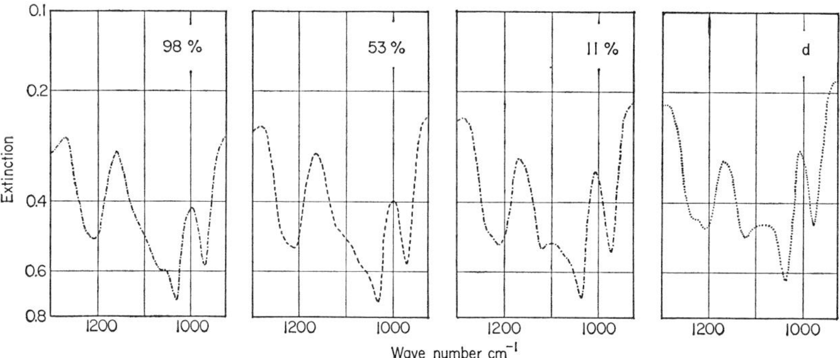

FIG. 21. Na^+ salt of polystyrenethiophosphonic acid, 7% cross-linked, 3 days thiophosphonated, H_2O hydrated at 98, 53, and 11% relative atmospheric humidity and thoroughly dried membrane (d).

IV.5. Hydration Water Bands in the Region 4000–2000 cm^{-1} at Low Degrees of Hydration

Figures 22–26 show spectra of different Na^+ salts. As a measure of the degree of hydration the relative atmospheric humidity with which the membrane was in equilibrium during the run is given (see Section VII.1). All the spectra of a single series have been taken on one and the same membrane.

The spectra of membranes hydrated with H_2O are compared with those of membranes hydrated with D_2O. Further, we will consider the dependence of the bands on the relative atmospheric humidity surrounding the membrane; finally, we take into account the dependence of the position of these bands on the nature of the ions present (see following sections). This leads to the following:

Result 15: The complex of bands in the region 3700–3000 cm^{-1} is ascribed to H_2O hydration molecules, that in the region 2750–2200 cm^{-1} to D_2O hydration molecules. In these regions, apart from the hydration bands, there are only the weak bands of the stretching vibrations of the $\rangle$CH groups of the benzene ring observed at about 3050 cm^{-1}, and, in the Na^+ salt of polystyrenephosphinic acid, the band of the PH stretching vibration at 2310 cm^{-1}.

In Figs. 22–26 and in the spectra of various salts of polystyrenesulfonic acid (see Figs. 29–33 and Figs. 37–39), we consider in each case the spectra of the membrane with the bands of water of hydration of lower intensity. Under these conditions, the following bands of water of hydration are observed:

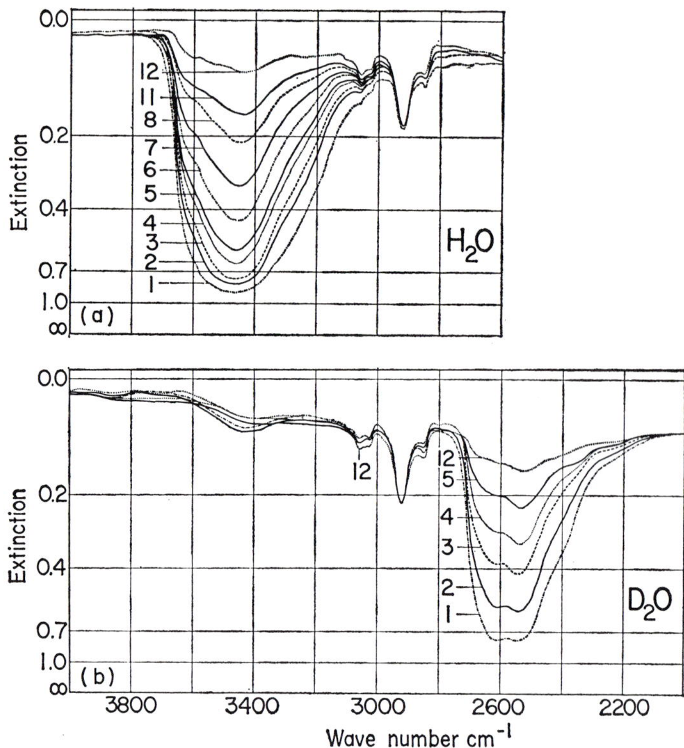

FIG. 22. Na^+ salt of polystyrenesulfonic acid, 7% cross-linked, 3 days of sulfonation. (a) H_2O hydrated: 1, 98%; 2, 71%; 3, 53%; 4, 33%; 5, 22%; 6, 11%; 7, 5%; 8, 1% relative atmospheric humidity; 11, incompletely dried; 12, thoroughly dried membrane. (b) D_2O hydrated: 1, atmospheric humidity over a saturated solution of $BaCl_2$ in D_2O; 2 → 5, decreasing atmospheric humidity; 12, thoroughly dried membrane.

Result 16:

Bands on Hydration with H_2O (Low Degree of Hydration)

A broad band is found. Its maximum lies in the region 3470–3190 cm^{-1}. On the side of this band toward smaller wave numbers a more or less pronounced shoulder is observed at about 3250 cm^{-1}. In addition there is sometimes found, e.g., as seen in Fig. 26, a particularly clear and fairly sharp band at 3615 cm^{-1}. This band, however, occurs most of the time as a shoulder only. In a very few cases, e.g., as seen in Fig. 23, a fairly sharp but very weak band is observed at 3530 cm^{-1}.

Bands on Hydration with D_2O (Low Degree of Hydration)

The intense broad hydration band sometimes becomes a doublet on hydration with D_2O, e.g., with some salts of polystyrenesulfonic acid (see Figs. 22

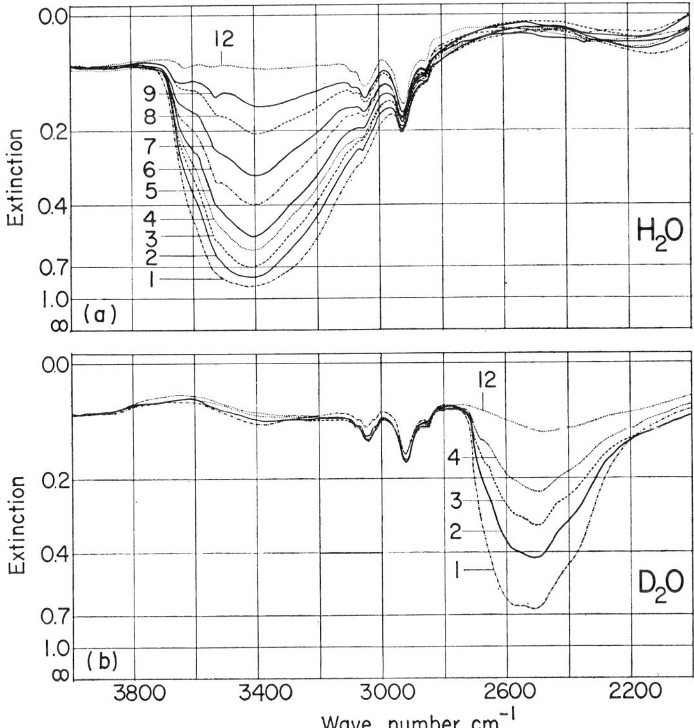

FIG. 23. Na^+ salt of polystyreneselenonic acid, 7% cross-linked, 3 days of selenonation (particularly thin membrane). (a) H_2O hydrated: 1, 98% 2, 71% 3, 53%; 4, 33%, 5, 22%; 6, 11%; 7, 5%; 8, 1% relative atmospheric humidity; 9, incompletely dried; 12, thoroughly dried membrane. (b) D_2O hydrated: 1, 100% relative atmospheric humidity; 2 → 4, decreasing atmospheric humidity; 12, thoroughly dried membrane.

and 28) and with the Na^+ salt of polystyreneselenonic acid (Fig. 23). The more or less pronounced shoulder on the side of these bands toward small wave numbers is found at 2375 cm^{-1}, e.g., in Figs. 22–25 and 28. In addition, for example in Fig. 26, the fairly sharp band on the side toward larger wave numbers is again clearly to be seen at 2670 cm^{-1}. In the other cases generally only a shoulder is observed.

Assignment of These Bands

The shoulder at about 3250 cm^{-1} (H_2O) or about 2375 cm^{-1} (D_2O):

We have seen in Section III.2.1 that a band in this region may be assignable to $2\bar{\nu}_2$, the overtone of the scissor vibration of the H_2O or D_2O molecule. Furthermore, we have seen there that a decision on this point can be reached through the investigation of samples hydrated with HDO.

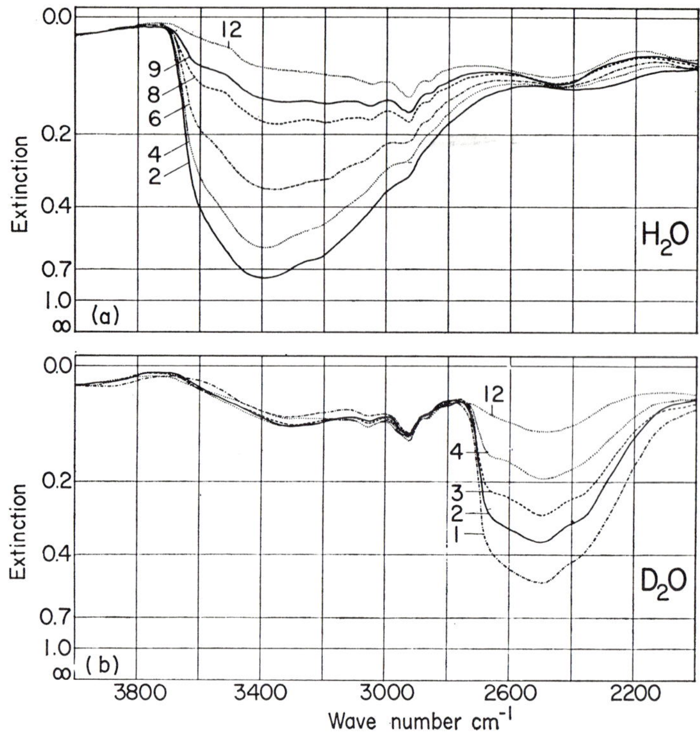

FIG. 24. Na^+ salt of polystyrenethiophosphonic acid, 7% cross-linked, 3 days thiophosphonated (membrane 3 months old). (a) H_2O hydrated: 2, 71%; 4, 33%; 6, 11%; 8, at 1% relative atmospheric humidity; 9, incompletely dried; 12, thoroughly dried membrane. (b) D_2O hydrated: 1, 100% relative atmospheric humidity; 2 → 4, decreasing atmospheric humidity; 12, thoroughly dried membrane.

From Fig. 27 we obtain Result 17.

Result 17: The shoulder at about 3250 cm^{-1} is observed under these conditions only on hydration with H_2O, and not with HDO.

It follows that this shoulder is the band of the overtone $2\bar{\nu}_2$ of the scissor vibration of the water molecule. However, at a high degree of hydration, this shoulder may arise from another cause. We shall return to this point in Section IV.17 and IV.18.

Result 18: The broad intense band of water of hydration and, in particular, the weak band at 3615 cm^{-1} is observed when the membrane has been hydrated with either H_2O or HDO.

It follows that, if these two bands (3470–3190 cm^{-1} and 3615 cm^{-1}) were to be associated with the symmetric ($\bar{\nu}_1$) and antisymmetric ($\bar{\nu}_3$) stretching vibrations of H_2O, only one band of the OH stretching vibration should be observed on hydration with HDO. The fact that both bands are still found

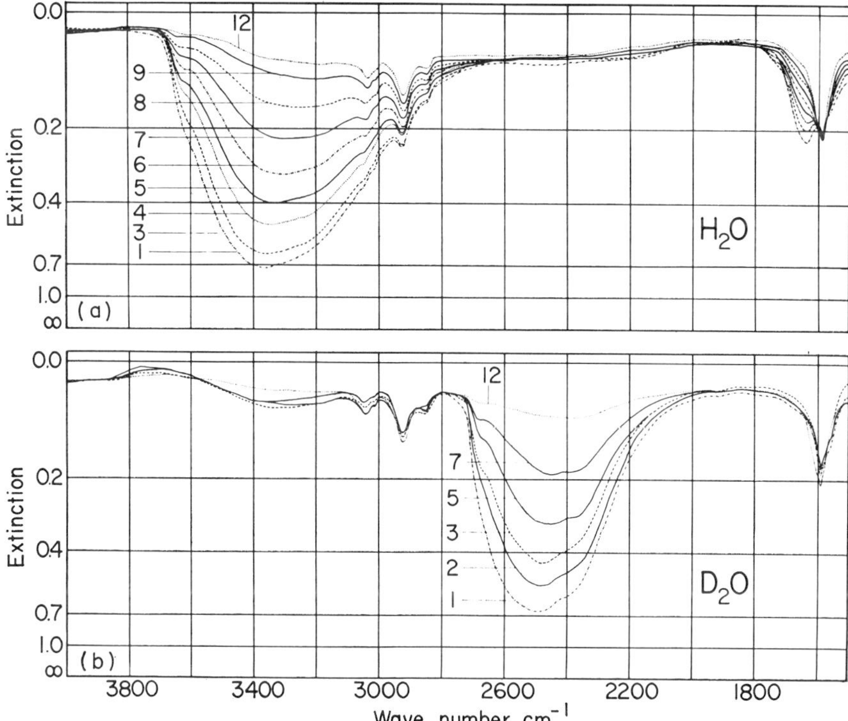

FIG. 25. Na^+ salt of polystyreneseleninic acid, 7% cross-linked, 3 days selenonated (particularly thin membrane). (a) H_2O hydrated: 1, 98%; 3, 53%; 4, 33%; 5, 22%; 6, 11%; 7, 5%; 8, 1% relative atmospheric humidity; 9, incompletely dried; 12, thoroughly dried membrane. (b) D_2O hydrated: 1, atmospheric humidity over a saturated solution of $BaCl_2$ in D_2O; 2 → 7, decreasing atmospheric humidity; 12, thoroughly dried membrane.

with HDO hydration, shows (a) that, as in pure liquid water, the symmetric ($\bar{\nu}_1$) and antisymmetric ($\bar{\nu}_3$) stretching vibrations of the H_2O coincide, and (b) that both of the two bands are associated with stretching vibrations of OH or OD groups in the molecules of water of hydration, which exhibit various bonds within the hydrate structure. With regard to the positions of these bands, it is necessary to assume that the broad band is the stretching vibration of OH or OD groups in hydrogen bridges, but that the band at 3615 cm^{-1} (OH) or 2670 cm^{-1} (OD) is the OH or OD stretching vibration of free groups.

IV.5.1. Explanation of the Doublet Structure with D_2O

While the broad intense hydration band for H_2O shows no doublet structure such a structure is found in the corresponding band for D_2O. In Section III.2.1

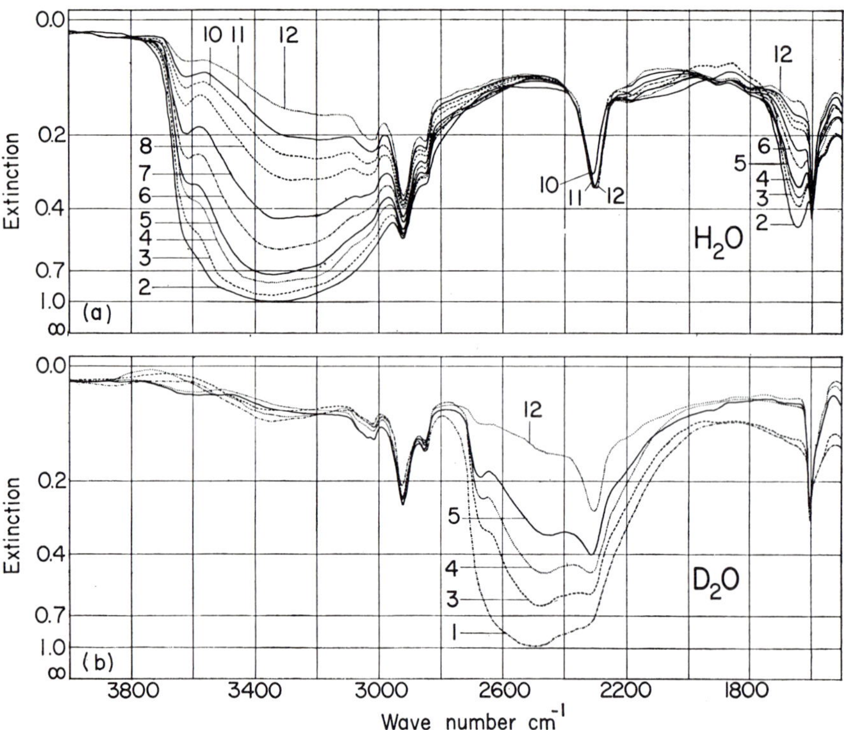

FIG. 26. Na^+ salt of polystyrenephosphinic acid, 7% cross-linked, 3 days phosphinated. (a) H_2O hydrated: 2, 71%; 3, 53%; 4, 33%; 5, 22%; 6, 11%; 7, 5%; 8, 1% relative atmospheric humidity; 10, 11, incompletely dried; 12, thoroughly dried membrane. (b) D_2O hydrated: 1, atmospheric humidity over a saturated solution of $BaCl_2$ in D_2O; 3 → 5, decreasing atmospheric humidity; 12, thoroughly dried membrane.

we have seen (from footnotes 22, 23) that the bands of the symmetric $\bar{\nu}_1$ and antisymmetric $\bar{\nu}_3$ stretching vibrations of the water molecule move closer together as the interaction of this water molecule with its surroundings increases. This was confirmed by Mohr *et al.*[47]

We attain a better understanding of these observations if we correlate them with the findings to be discussed in the following sections; first we should consider another experimental result. If the bands move together, they at first occupy the same position. If the interaction increases, it would be expected that they should split again, as $\bar{\nu}_3$ "overtake" $\bar{\nu}_1$. This should be the case with hydrated salts with particularly strong interaction between the cation and water. Let us now consider the upper family of curves in the Be^{2+},

[22] E. Greinacher, W. Luettke, and R. Mecke, *Z. Elektrochem.* **59**, 23 (1955).
[23] T. Ackermann, *Z. Physik. Chem.* (*Frankfurt*) **42**, 119 (1964).

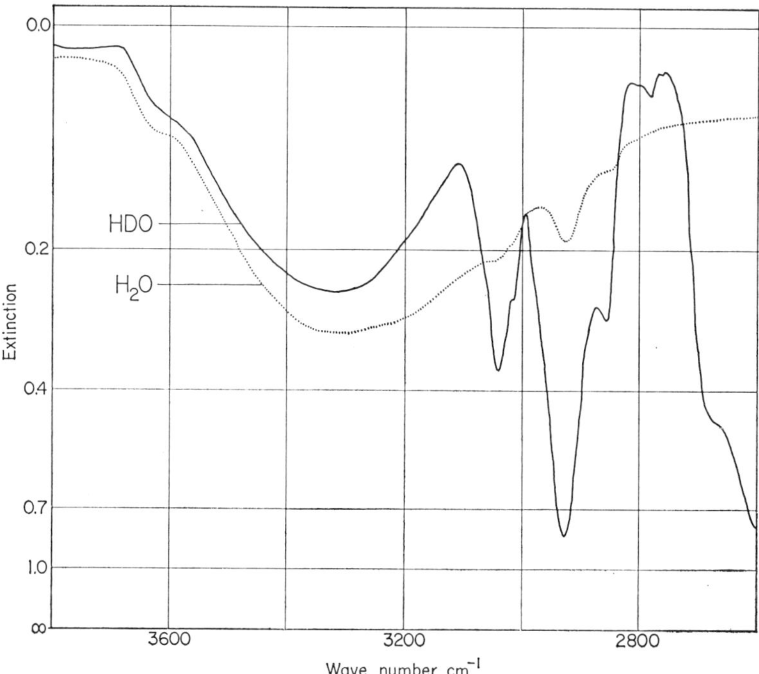

FIG. 27. Na^+ salt of polystyreneseleninic acid, 7% cross-linked, 3 days of selenonation. ——, HDO hydrated: atmospheric humidity over a saturated solution of LiCl in H_2O–D_2O (mixture of about 7 mole % H_2O and 93 mole % D_2O); · · · ·, H_2O-hydrated: 11% relative atmospheric humidity.

Al^{3+}, Ga^{3+}, In^{3+}, Zr^{4+}, and Hf^{4+} salts of polystyrenesulfonic acid in Figs. 30(a), 33, and 55. Here, even when the cations interact very strongly with water, the broad band of the OH stretching vibration of the molecules of hydration water shows no doublet structure. This can be understood only if we assume that with increasing interaction the stretching vibrations of the two OH groups of an H_2O molecule are increasingly uncoupled.

How does this uncoupling come about? In the following sections we shall see that the molecules of hydration water are bound to their surroundings by hydrogen bridges (Section IV.7). In Section IV.13 we shall discuss the fact that these hydrogen bridges are sterically heavily stressed. The two OH groups of a single water molecule are affected to differing extent by this steric constraint. As a result, the two hydrogen bridges of the H_2O molecule differ somewhat in strength. The difference is, on the average, greater as the strength of the hydrogen bridges increases. If the two hydrogen bridges do differ in strength, an uncoupling of the vibration of the two OH groups occurs; in place of the two bands $\bar{\nu}_1$ and $\bar{\nu}_3$ of the whole molecule only the one broad band of the vibration of the single OH groups of the molecule of water of hydration is found.

We can now understand why the bands of D_2O show doublet structure. The shift of the OD bands through hydrogen bridge bonding is less than that of the OH bands. Hence the difference between the two OD groups of one D_2O molecule is also, on the average, smaller, and these two vibrations can still be coupled when those of H_2O are already uncoupled. Consequently, a symmetric and an antisymmetric vibration are still found in D_2O.

The spectra in Fig. 28 confirm this explanation.

Result 19: The bands of the symmetric and antisymmetric stretching vibration of the D_2O molecule move progressively closer together in passing along the series of alkaline earth salts of polystyrenesulfonic acid with D_2O hydration, from the Ba^{2+} to the Mg^{2+} salt. For the last named, they almost coincide.

Now that we understand the difference between H_2O and D_2O hydration, the variation of the bands of the D_2O stretching vibration in passing along

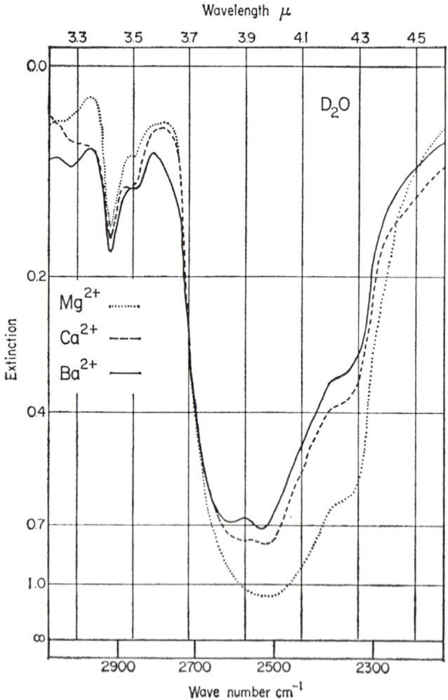

FIG. 28. Mg^{2+}, Ca^{2+}, Ba^{2+} salts of polystyrenesulfonic acid, 6% cross-linked, 4 days of sulfonation, D_2O hydrated: atmospheric humidity over a saturated solution of $MgCl_2$ in D_2O (taken with model 21, LiF prism, linear in wavelengths).

the series from Ba^{2+} to Mg^{2+} is also understandable. The hydrogen bridges forming the water hydration molecules become stronger in passing along this series (Section IV.9), so that the difference between the two OD groups also becomes, on the average, more pronounced, i.e., their vibrations become progressively uncoupled.*

In summary, in the present systems we see that the stretching vibrations of the two OH groups of H_2O are uncoupled. Only a single broad band is found. With the D_2O molecule, this uncoupling with increasing interaction occurs later than with the H_2O molecule. Hence a doublet structure is observed in the bands of hydration water with D_2O; this structure vanishes on increasing interaction of the D_2O molecule with its surroundings.

IV.6. Position of the Hydration Water Band

IV.6.1. Dependence of the Position of the Intense Hydration Water Band on Cation Nature

Spectra of hydrated salts of polystyrenesulfonic acid are shown in Figs. 29–33, 37–39, and 55. The membranes used for the salts of a single group of cations were of equal thickness so that the intensities of the bands could be compared. In these figures two spectra of the same salt are given, but we consider first the spectrum of the membrane at a smaller degree of hydration, i.e., that with the less intense OH band.†

From these figures and from Table 7, we obtain:

Result 20: The band of the OH stretching vibration of the hydration water is shifted increasingly toward smaller wave numbers as the radius of the cation in the individual groups decreases, e.g., from Ba^{2+} to Be^{2+}, from In^{3+} to Al^{3+}, from La^{3+} to Sc^{3+} (see also Section IV.14).

* The same doublet structure is found in the spectra of D_2O hydrated salts of poly-(*p*-trimethylammonium)styrene [—$N(CH_3)_3^+$ groups]. This splitting decreases in going from the I^-, Br^-, Cl^- to the F^- salt. In the latter, no doublet structure for the band having the stretching vibration of the D_2O molecule is found.[23a] This result is explained in an analogous manner.

† We have plotted water adsorption isotherms only for the acids. However, a rough estimate of the number of water molecules present per cation at a given relative humidity permits comparison with the adsorption isotherms reported in the literature.[24-29] In making this comparison it is important that the degree of cross-linking of the membranes be approximately the same.

[23a] T. Ackermann, G. Zundel, and K. Zwernemann, *Ber. Bunsenges. Physik. Chem.* (1969), in press.

[24] E. Glueckauf and G. P. Kitt, *Proc. Roy. Soc.* (*London*) **A228**, 322 (1955).

[25] G. Dickel, *Chem. Tech.* (*Berlin*), **10**, 449 (1958).

[26] G. Dickel, H. Degenhart, K. Haas, and J. W. Hartmann, *Z. Physik. Chem.* (*Frankfurt*) **20**, 121 (1959).

[27] G. Dickel and J. W. Hartmann, *Z. Physik. Chem.* (*Frankfurt*) **23**, 1 (1960).

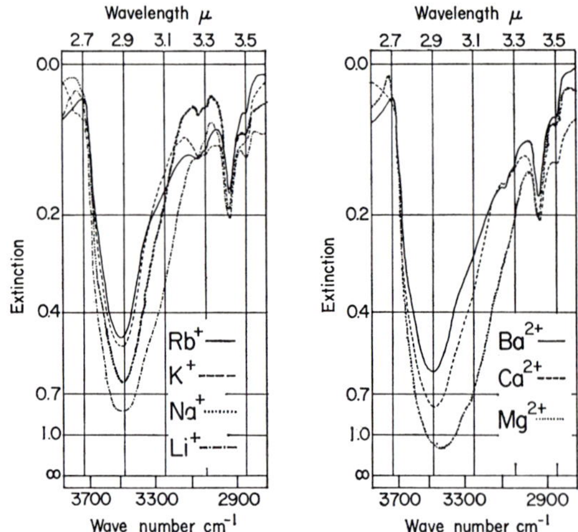

FIG. 29. Li^+, Na^+, K^+, Rb^+, Mg^{2+}, Ca^{2+}, and Ba^{2+} salts of polystyrenesulfonic acid, 6% cross-linked, 4 days of sulfonation, H_2O hydrated: 33% relative atmospheric humidity (taken with model 21, LiF prism, linear in wavelengths).

Result 21: The higher the valence of the cations the larger are the shifts. From Figs. 29 and 40, we obtain:

Result 22: The intensity of these OH stretching vibration bands of the hydration water increases progressively in the series of the alkali metal ions from Rb^+ to Li^+ and also in the series of the alkaline earth metal ions from Ba^{2+} to Mg^{2+}, if the membranes are hydrated at the same relative humidity. The same holds true for the band, observed at about 1640 cm^{-1}, of the scissor vibration of this hydration water (Fig. 40 and Result 33). Thus the quantity of water of hydration under comparable conditions increases in both series with decreasing radius of the cations.* This is, however, precisely what is

* Since the ionic field also alters the extinction coefficients of the vibrations of the molecules of hydration water, caution must be exercised in drawing conclusions from the intensities of the bands of hydration water. It is always necessary to take into account the intensities both of the stretching and of the scissor vibrations. For, according to Greinacher *et al.*,[22] interaction with the surroundings changes the extinction coefficients of the stretching and scissor vibrations in opposite directions. If the intensities of stretching and scissor vibrations change in the same direction, the changes in the extinction coefficients can only play a subordinate role. However, we shall see in Section V.10.1 that, in the case of a very strong interaction between the water molecule and its surroundings, i.e., in acids, the extinction coefficients no longer change in opposite directions.

[28] D. Dolar, S. Lapanje, and S. Paljk, *Z. Physik. Chem.* (*Frankfurt*) **34**, 360 (1962).

[29] G. Dickel and K. Bunzl, *Z. Physik. Chem.* (*Frankfurt*) **39**, 198 (1963).

[30] R. Griessbach, "Austauschadsorption in Theorie und Praxis." Akademie Verlag, Berlin, 1957.

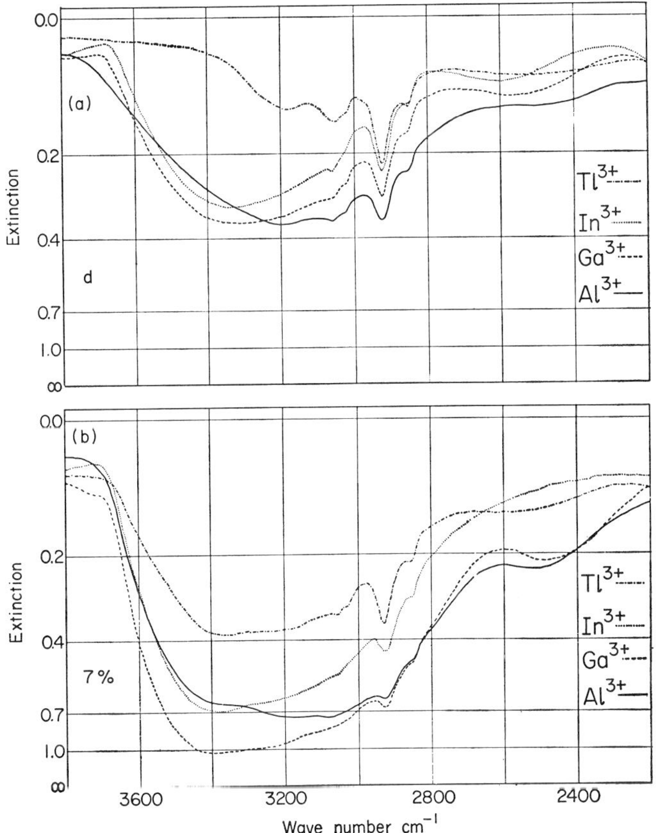

FIG. 30. Al^{3+}, Ga^{3+}, In^{3+}, Tl^{3+} [$Tl(OH)^{2+}$ also present (see Section IV.20)] salts of polystyrenesulfonic acid, 5% cross-linked, 3 days of sulfonation. H_2O hydrated: (a) thoroughly dried membranes; (b) at 7% relative atmospheric humidity.

expected with water of hydration interacting with the cations (e.g. see the summary in footnote 30, p. 253).

From Results 20, 21, and 22, as well as from 19, it follows that the molecule of hydration water which causes the intense band of the OH stretching vibration interacts with the cation.

A summary of publications on homogeneous liquid electrolyte solutions follows. As early as 1933, Suhrmann and Breyer[31] investigated the shifts of the bands of the first overtone in homogeneous liquid electrolyte solutions under the influence of Li^+, Mg^{2+}, and Ca^{2+} ions. This work was extended by Meerlender.[32] Further investigations of

[31] R. Suhrmann and F. Breyer, *Z. Physik. Chem. (Leipzig)* **B20**, 17 (1933).

[32] G. Meerlender, Ph.D. Thesis under R. Suhrmann, Technische Hochschule, Braunschweig, 1959.

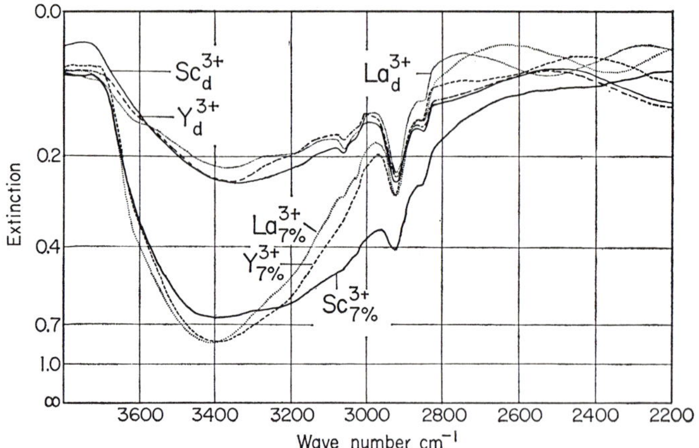

FIG. 31. Sc^{3+}, Y^{3+}, and La^{3+} salts of polystyrenesulfonic acid, 5% cross-linked, 3 days of sulfonation, H_2O hydrated: 7% relative atmospheric humidity (7%), thoroughly dried membrane (d).

hydration in the overtone region are reported (see footnotes 33–38). In the region of the fundamentals, the authors (footnotes 39–48) investigated homogeneous liquid electrolyte solutions. Of these, two publications (footnotes 44, 47) deal with the interaction between ions and water in other solvents. Raman spectra of homogeneous liquid electrolyte solutions are given (footnotes 49–62).

[33] W. J. Biermann and J. B. Gilmour, *Can. J. Chem.* **37**, 1249 (1959).
[34] G. R. Choppin and K. Buijs, *J. Chem. Phys.* **39**, 2042 (1963).
[35] H. Yamatera, B. Fitzpatrick, and G. Gordon, *J. Mol. Spectry.* **14**, 268 (1964).
[36] V. M. Vdovenko, D. N. Suglobov, and A. P. Taranov, *Radiokhimiya* **6**, 559 (1964).
[37] W. Luck, *Ber. Bunsenges. Physik. Chem.* **69**, 69 (1965).
[38] K. W. Bunzl, *J. Phys. Chem.* **71**, 1358 (1967).
[39] R. D. Waldron, *J. Chem. Phys.* **26**, 809 (1957).
[40] M. Falk and P. A. Giguere, *Can. J. Chem.* **35**, 1195 (1957).
[41] J. D. S. Goulden, *Spectrochim. Acta* **15**, 657 (1959).
[42] J. D. S. Goulden, *Spectrochim. Acta* **16**, 715 (1960).
[43] J. D. S. Goulden, *Chem. Ind. (London)*, p. 721 (1960).
[44] L. D. Shcherba and A. M. Sukhotin, *Russ. J. Phys. Chem. (English Transl.)* **33**, 448 (1959).
[45] T. Ackermann, *Z. Physik. Chem. (Frankfurt)* **27**, 253 (1961).
[46] J. W. Moisejew and M. I. Vinnik, *Zh. Strukt. Khim.* **4**, 336 (1963).
[47] S. C. Mohr, W. D. Wilk, and G. M. Barrow, *J. Am. Chem. Soc.* **87**, 3048 (1965).
[48] W. K. Thompson, *Trans. Faraday Soc.* **62**, 2667 (1966).
[49] W. R. Busing and D. F. Hornig, *J. Phys. Chem.* **65**, 284 (1961).
[50] J. W. Schultz and D. F. Hornig, *J. Phys. Chem.* **65**, 2131 (1961).
[51] R. E. Weston, Jr., *Spectrochim. Acta* **18**, 1257 (1962).
[52] H. A. Lauwers and G. P. van der Kelen, *Bull. Soc. Chim. Belges* **72**, 477 (1963).
[53] P. M. Vollmar, *J. Chim. Phys.* **39**, 2236 (1963).
[54] Z. Keçki, *Roczniki Chem.* **38**, 329 (1964).
[55] G. E. Walrafen, *J. Chem. Phys.* **36**, 1035 (1962).

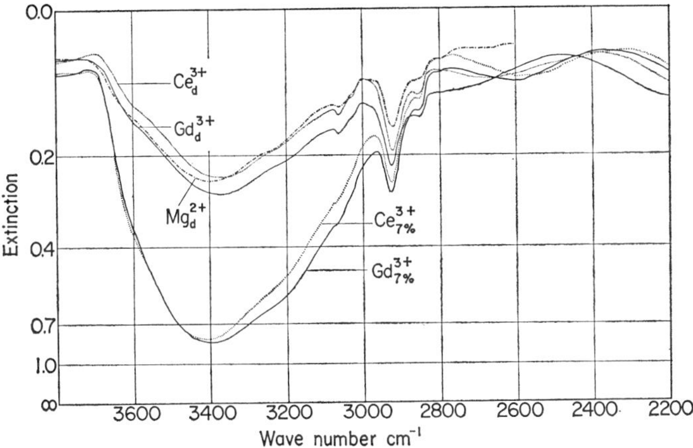

FIG. 32. Gd^{3+}, Ce^{3+} and Mg^{2+} salts of polystyrenesulfonic acid, 5% cross-linked, 3 days of sulfonation. H_2O hydrated: at 7% relative atmospheric humidity (7%), thoroughly dried membrane (d).

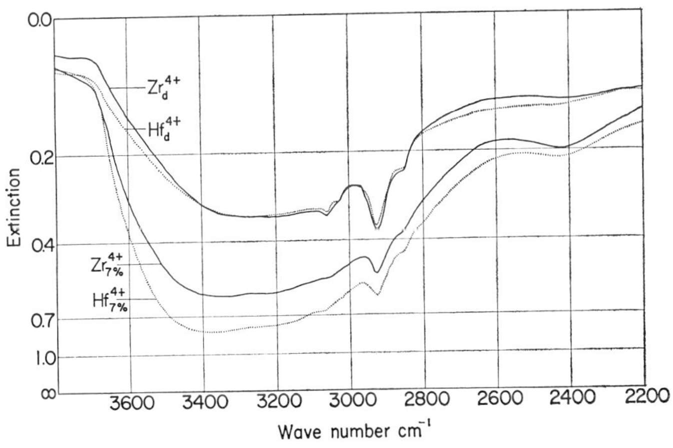

FIG. 33. Zr^{4+} and Hf^{4+} salts of polystyrenesulfonic acid, 5% cross-linked, 3 days of sulfonation, H_2O hydrated: at 7% relative atmospheric humidity (7%), thoroughly dried membrane (d) (the membrane Zr^{4+} examined at 7% relative atmospheric humidity is not the same as the thoroughly dried membrane; the former is thinner).

[56] R. E. Hester, R. A. Plane, and G. E. Walrafen, *J. Chem. Phys.* **38**, 249 (1963).
[57] G. E. Walrafen, *J. Chem. Phys.* **40**, 3249 (1964).
[58] O. Bonner, K. W. Bunzl, and G. B. Woolsey, *Spectrochim. Acta* **22**, 1125 (1966).
[59] G. E. Walrafen, *J. Chem. Phys.* **44**, 1546 (1966).
[60] G. E. Walrafen, *J. Chem. Phys.* **44**, 3726 (1966).
[61] J. H. R. Clarke and L. A. Woodward, *Trans. Faraday Soc.* **62**, 2226 (1966).
[62] D. E. Irish and G. E. Walrafen, *J. Chem. Phys.* **46**, 378 (1967).

The authors who have investigated homogeneous liquid electrolyte solutions frequently come to the conclusion that the shift of the band of the OH stretching vibration under the influence of the cations is insignificant. This is in contradiction with our observations. An explanation will be given in Section IV.18.

IV.6.2. Dependence of the Position of the Hydration Water Band on Anion Nature

The question arose whether the position of the intense band of the OH stretching vibration of the hydration water is dependent on the properties of the anion. To solve this, we compared the spectra of the Na^+ salts of the series with the anions

$$\left(-\mathrm{S}{\genfrac{}{}{0pt}{}{\mathrm{O}^-}{\genfrac{}{}{0pt}{}{\cdots\mathrm{O}}{\mathrm{O}}}}\right),\quad \left(-\mathrm{Se}{\genfrac{}{}{0pt}{}{\mathrm{O}^-}{\genfrac{}{}{0pt}{}{\cdots\mathrm{O}}{\mathrm{O}}}}\right),\quad \left(-\mathrm{P}{\genfrac{}{}{0pt}{}{\mathrm{S}^{2-}}{\genfrac{}{}{0pt}{}{\cdots\mathrm{O}}{\mathrm{O}}}}\right),\quad \left(-\mathrm{Se}{\genfrac{}{}{0pt}{}{\mathrm{O}^-}{\mathrm{O}}}\right),\quad \text{and}\quad \left(-\mathrm{P}{\genfrac{}{}{0pt}{}{\mathrm{H}^-}{\genfrac{}{}{0pt}{}{\cdots\mathrm{O}}{\mathrm{O}}}}\right)$$

i.e., the spectra in Figs. 22–26. A selection of the spectra is summarized in Fig. 34. The data are collected in Table 6. The wave numbers of the maxima* have been measured on membranes in which one or two water molecules per ion pair were present.† From these figures‡ and Table 6 we obtain:

Result 23: The position of the intense band of the OH stretching vibration of the hydration water depends on the nature of the anions present. It is shifted progressively toward smaller wave numbers from $\left(-\mathrm{S}{\genfrac{}{}{0pt}{}{\mathrm{O}^-}{\genfrac{}{}{0pt}{}{\cdots\mathrm{O}}{\mathrm{O}}}}\right)$ to $\left(-\mathrm{P}{\genfrac{}{}{0pt}{}{\mathrm{H}^-}{\genfrac{}{}{0pt}{}{\cdots\mathrm{O}}{\mathrm{O}}}}\right)$.

It follows that, under the existing conditions, the molecules of hydration water are bound to the anions naturally by means of hydrogen bridges. The changes in the bands of the stretching vibrations of the bonds of the anions depending upon the degree of hydration (see Result 10 and its explanation) have supported the same conclusion.

IV.7. Attachment of Water Molecules at Low Degrees of Hydration

The results given in Section IV.6.1 indicate that the intense band of the OH stretching vibration of the water of hydration depends on the nature of the cation present. We have seen that the position of this band depends on

* On this point see Section VII.4.

† See footnote † on p. 61.

‡ In assessing Figs. 25, 26, and 34, it must be borne in mind that the membrane of the Na^+ salt of polystyreneseleninic acid was particularly thin, while that of the Na^+ salt of polystyrenephosphinic acid was particularly thick.

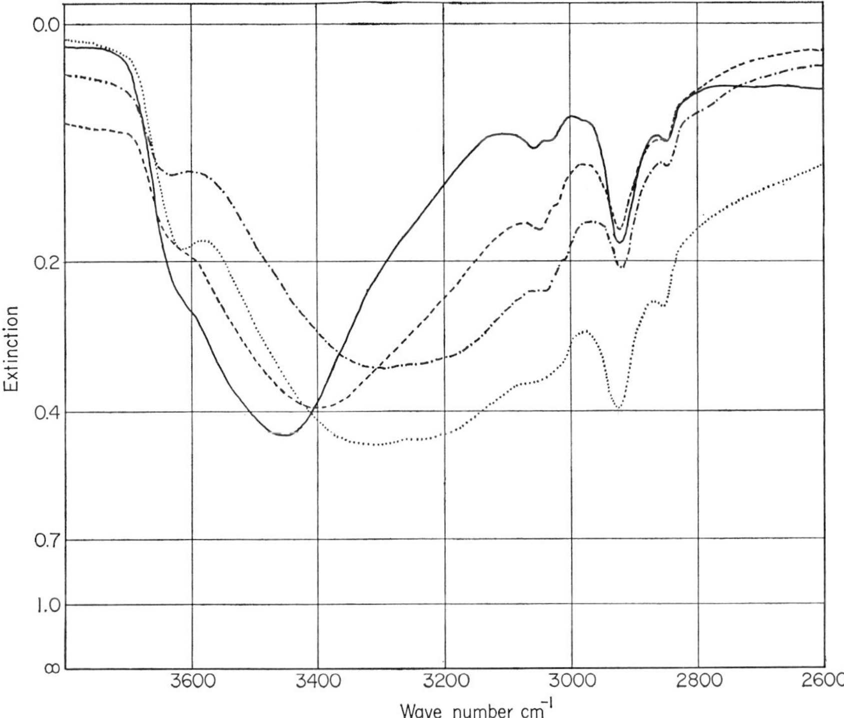

FIG. 34. Na^+ salt of: ——, polystyrenesulfonic acid; – – – –, polystyreneselenonic acid; –·–·, polystyreneseleninic acid; · · · ·, polystyrenephosphinic acid. 7% cross-linked, 3 days of sulfonation of selenonation, or of phosphination; H_2O hydrated, all at 11% relative atmospheric humidity.

the nature of the anion present. The molecules of hydration water thus are in direct interaction with both cation and anion at low degrees of hydration. If, considering the position of the cation relative to its corresponding anion, we further take into account the results given in Sections IV.3 and IV.15, we obtain the picture of the attachment as shown in Fig. 35.* Hence the observed OH stretching vibration band is that of the OH groups in the hydrogen bridges by which the water molecules are linked to the anions.

General Considerations on Hydrogen Bridge Formation

The following properties of the groups involved are decisive for the nature of the hydrogen bridges formed.

* In considering such schematic representations, we should keep in mind the fact that the hydrate structures are continually being rearranged by thermal motions. Any attachment position then is in dynamic equilibrium with analogous alternative positions.

TABLE 6

DEPENDENCE OF THE POSITION OF THE MAXIMUM OF HYDRATION WATER BAND ON THE NATURE OF THE ANION

Anion	$—SO_3^-$	$—SeO_3^-$	$—P(S^{2-})(O)(O)$	$—SeO_2^-$	$—P(H^-)(O)(O)$
Summary in Fig. 34, see also figure noted at right	22	23	24	25	26
Position of the maximum of the broad, intense band of water of hydration for the respective Na^+ salt (cm^{-1}), when one or two water molecules per ion pair are present	3459	3400	3395	3335	3335
Hydrogen bridge acceptor property of the oxygen atoms of the anions	—Becomes stronger→				

(a) Steric conditions (see Sections IV.13 and IV.5.1).

(b) The hydrogen bridge acceptor property of the acceptor group (see Section IV.8).

(c) The hydrogen bridge donor property of the donor group, here the OH group of the water molecules. This hydrogen bridge donor property becomes stronger, the stronger the dipole moment μ of the OH or OD groups of the molecules of hydration water becomes (see Section IV.9).

We note that the band of the stretching vibration of the OH or OD groups is further shifted toward smaller wave numbers, the stronger the bridge is. Conversely, the scissor vibration of the water molecule is shifted toward larger wave numbers under the same conditions, since the more strongly the OH groups are bound by hydrogen bridges, the greater is the force constant of their scissor vibration.

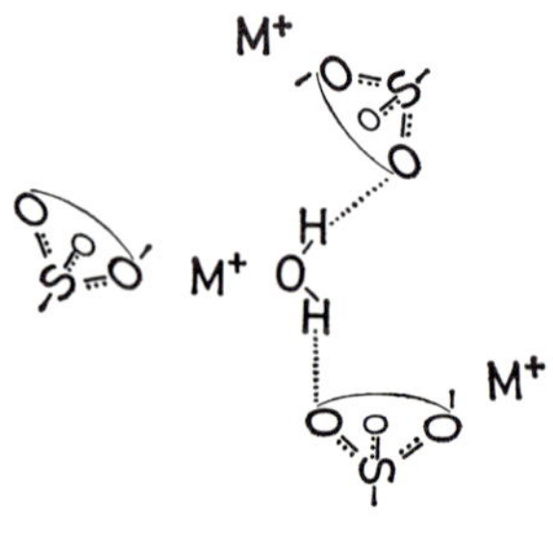

FIG. 35. The attachment of the hydration water molecules at low degree of hydration.

IV.8. Hydration Water Band Position and Hydrogen Bridge Acceptor Property of Anions

In Section IV.6.2 we have seen that the band of the OH stretching vibration of the water of hydration is progressively shifted toward smaller wave numbers in the series of anions from $\left(-\mathrm{S}\overset{\mathrm{O}^-}{\underset{\mathrm{O}}{\cdots\mathrm{O}}}\right)$ to $\left(-\mathrm{P}\overset{\mathrm{H}^-}{\underset{\mathrm{O}}{\cdots\mathrm{O}}}\right)$ (Result 23). Hence it follows that the hydrogen bridges by which the water of hydration is bound to the oxygen atoms of the anions become stronger in the same order.

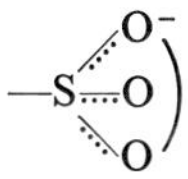

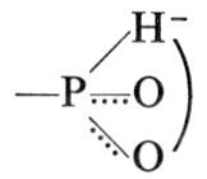

By definition, a stronger hydrogen bridge acceptor property leads to stronger bridges and hence to a stronger shift of the band of the stretching vibration of the OH or OD groups in these bridges toward smaller wave numbers.* From the course of the bands of the OH stretching vibrations, it follows[65] that the oxygen atoms become progressively stronger hydrogen bridge acceptors in the series $\left(-\mathrm{S}\overset{\mathrm{O}^-}{\underset{\mathrm{O}}{\cdots\mathrm{O}}}\right)$ to $\left(-\mathrm{P}\overset{\mathrm{H}^-}{\underset{\mathrm{O}}{\cdots\mathrm{O}}}\right)$.

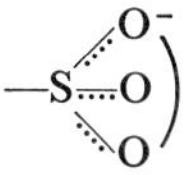

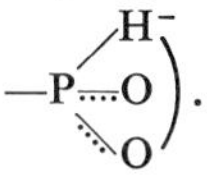

IV.9. Cation–Water Interaction, Stretching Vibration, and Hydrogen Bridge Donor Property of OH Groups in Hydrogen Bridges

It is assumed that the interaction of the water molecules with the cation changes the hydrogen bridge donor property of the OH groups of the water molecules. With this in mind, we have accurately determined the position of the band maximum† of the OH stretching vibration of the water of hydration in the presence of 1 to 2 water molecules per cation (see footnote †, p. 61) for 29 different hydrated salts of polystyrenesulfonic acid.[66]‡ A comparable degree of hydration is essential, since, as we shall see in Section IV.16, the positions of these bands are strongly dependent on the degree of hydration.

* Glemser and Hartert[63] report a similar conclusion from spectra of salts containing water of crystallization.[64]

† See Section VII.4.

‡ To avoid the presence of hydroxyl ions, the solutions used for ion exchange were slightly acidified when necessary. (For further details see Section VI.3, p. 255 ff.)

[63] O. Glemser and E. Hartert, *Naturwissenschaften* **42**, 534 (1955).

[64] F. A. Miller and C. H. Wilkins, *Anal. Chem.* **24**, 1253 (1952).

[65] G. Zundel, *Zh. Strukt. Khim.* **6**, 384 (1965).

[66] G. Zundel and A. Murr, *Z. Physik. Chem.* (*Frankfurt*), **54**, 49 (1967).

The membranes used were all 5% cross-linked and 3 days sulfonated. The positions of the band maxima are given in Table 7, column 3.*

On the basis of data given in Table 7, we must now consider how the cation–water interaction can change the structure of the water molecule so markedly as to produce a shift of the OH stretching vibration of the hydrogen bridges of the molecules of hydration water. Cations differ in their masses, their electrostatic fields (determined by valence and ionic radius), and their electron structure.

We must look to see if the cation masses exert an appreciable influence on the position of the band of the OH stretching vibration of the water of hydration. This could be assumed, since the cation could possibly take part to some extent in this vibration. We compare the cases of Zr^{4+} and Hf^{4+} in Table 7 (see footnote, p. 44). In these two cases the band of the OH stretching vibration of the water of hydration has the same position. The masses of the cations, however, are very nearly in the ratio 1:2. The mass of the cation is, in these cases, particularly strongly bound to the water molecule. If the cation mass does exert an essential influence on the position of the band of the water of hydration, this must be particularly true with these cations. Since, however, the band has the same position in both cases (see also Fig. 33), it follows that the influence of the mass of the cation on the stretching vibration of the water of hydration is insignificant.

We will now consider the influence of the electrostatic field of the cation. $(1 - H_K)E_{KH}$ is the field of the cation in the direction of the OH bond at the hydrogen nuclei of the water molecule. H_K takes into account the screening of the hydrogen nucleus by the electrons of the water molecule. The quantity E_{KH} is the field strength in the direction of the OH bond at the position of the hydrogen nuclei of the water molecule, which can be calculated from purely geometrical considerations by Coulomb's law. These E_{KH} values can serve as a relative measure for the field at the position of the hydrogen nuclei.

* Reference may be given here to some IR studies of water of hydration in crystals (see footnotes 63, 64, 67–80a).

[67] L. H. Jones, *J. Chem. Phys.* **22**, 217 (1954).
[68] P. J. Lucchesi and W. A. Glasson, *J. Am. Chem. Soc.* **78**, 1347 (1956).
[69] J. Van der Elsken and D. W. Robinson, *Spectrochim. Acta*, **17**, 1249 (1961).
[70] E. Hartert and O. Glemser, *Z. Elektrochem.* **60**, 746 (1956).
[71] J. Fujita, K. Nakamoto, and M. Kobayashi, *J. Am. Chem. Soc.* **78**, 3963 (1956).
[72] C. Duval and J. Lecomte, *Z. Elektrochem.* **64**, 582 (1960).
[73] I. Gamo, *Bull. Chem. Soc. Japan* **34**, 760 (1961).
[74] I. Gamo, *Bull. Chem. Soc. Japan* **34**, 1430 (1961).
[75] E. Schwarzmann and O. Glemser, *Z. Anorg. Allgem. Chem.* **312**, 45 (1961).
[76] E. Schwarzmann and O. Glemser, *Naturwissenschaften* **50**, 330 (1963).
[77] G. Blyholder and S. Vergez, *J. Phys. Chem.* **67**, 2149 (1963).
[78] I. Nakagawa and T. Shimanouchi, *Spectrochim. Acta* **20**, 429 (1964).
[79] Y. Kermarrec, *Compt. Rend.* **258**, 5836 (1964).
[80] J. R. Ferraro and A. Walker, *J. Chem. Phys.* **42**, 1278 (1965).
[80a] C. Postmus and J. R. Ferraro, *J. Chem. Phys.* **48**, 3605 (1968).

TABLE 7

BAND MAXIMUM POSITION OF OH GROUPS STRETCHING VIBRATIONS IN HYDROGEN BRIDGES OF HYDRATION WATER MOLECULES IN POLYSTYRENESULFONIC ACID SALTS AT LOW DEGREE OF HYDRATION

Ion	Figure number	Maximum of the band of the OH stretching vibration $\bar{\nu}^a$ in cm^{-1}	Ionic radius (Å) (after footnote 4)	E_{KH} (ESU · cm^{-2} × 10^6)
1	2	3	4	5
Li^+	29	3458 ± 2	0.68	1.49
Na^+		3459 ± 2	0.97	1.08
K^+		3460 ± 2	1.33	0.76
Rb^+		3462 ± 2	1.47	0.67
Cs^+		3459 ± 2	1.67	0.57
$(CH_3)_4N^+$	—	3449 ± 5	—	—
$(C_2H_5)_4N^+$	—	3430 ± 5	—	—
Be^{2+}	55	3201 ± 8	0.35	4.46
Mg^{2+}	29	3394 ± 5	0.66	3.07
Ca^{2+}		3406 ± 10	0.99	2.11
Sr^{2+}		3412 ± 10	1.12	1.85
Ba^{2+}		3440 ± 10	1.34	1.51
Al^{3+}	30	3195 ± 8	0.51	5.53
Ga^{3+}		3315 ± 6	0.62	4.78
In^{3+}		3338 ± 6	0.81	3.83
Tl^{3+}		(3400 ± 10)[b]	0.95	3.32
Sc^{3+}	31	3350 ± 5	0.81	3.83
Y^{3+}		3364 ± 6	0.93	3.40
La^{3+}		3373 ± 6	1.14	2.73
Ce^{3+}	32	3369 ± 6	1.07	2.93
Gd^{3+}		3372 ± 6	0.94[c]	3.34
Zr^{4+}	33	3285 ± 8	0.79	5.29
Hf^{4+}		3280 ± 8	0.78	5.33
Mn^{2+}	37	3406 ± 6	0.80	2.59
Co^{2+}		3381 ± 6	0.72	2.87
Ni^{2+}		3373 ± 6	0.69	2.94
Cu^{2+}	38	3314 ± 8	0.72	2.87
Zn^{2+}		3317 ± 8	0.74	2.80
Fe^{3+}	39	3310 ± 8	0.64	4.71

[a] In estimating the maximum error of these wave-number values, the following points play a part: number of measurements present in each case, superposition of interferences, overlapping of other bands, and the band width. The deviations given are probably too large.

[b] The value is in parentheses, since $Tl(OH)_2^+$ is always present along with Tl^{3+} in the membrane (on this point see Section IV.20).

[c] Personal communication from R. C. Weast, editor-in-chief, "Handbook of Chemistry and Physics," Chemical Rubber Co., Cleveland, Ohio.

The geometry assumed in calculating these E_{KH} values is shown in Fig. 36.* We obtain:

$$E_{KH} = \frac{Ze_0 \cos \beta}{R^2}$$

where Z is the valence of the cation and e_0 the elementary charge. To determine R values of r_{OH} and 2γ, given in Table 2, were used. Within the accuracy of this discussion, c could be taken as 0.25 Å for all cations. The figure of 0.25 Å is obtained by representing the lone electron pairs of the water molecule according to Brill [81] and the attached cation on the same scale. A different c value is then obtained for each individual cation. Since small changes in c have no marked effect, we can take 0.25Å as the average value of c for all cations. The values of E_{KH} calculated according to the given equation are given in Table 7, column 5.

The comparison of the positions of the bands with the E_{KH} values in Table 7 shows:

Result 24: The maximum of the band of the stretching vibration of the OH groups of the hydration water in the hydrogen bridges is shifted progressively toward smaller wave numbers from Ba^{2+} to Be^{2+}, from In^{3+} to Al^{3+}, and from La^{3+} to Sc^{3+}. The band of this OH stretching vibration generally† is more strongly shifted toward smaller wave numbers within all

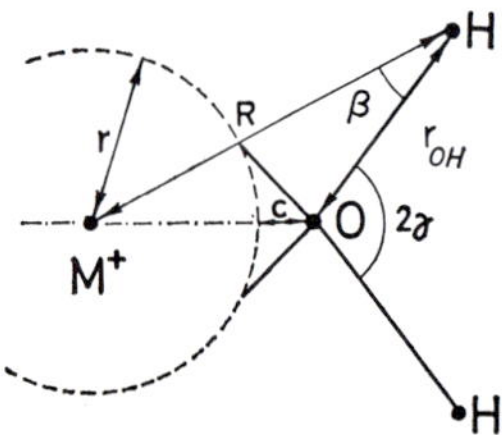

FIG. 36. The geometry of the attachment of water of hydration to the cation. (The dimensions of the lone electron pairs of the water molecule and their center of electrical charge are taken from a publication by Brill.[81] This author refers to Schneider[82] who in turn refers to Pople[83] and Dunham and Pople.[84])

* It could be conjectured that the cation is not symmetrically attached to the lone electron pairs of the water, but preferentially to one of the two (also see footnote 85). In this case, however, the calculated values would all be affected in the same fashion, so that our discussion would still be valid.

† In the following we shall encounter some special cases. The hydration of the ions of the transition elements is discussed in Section IV.10. The anomaly in the presence of the alkali metal ions will be discussed in Section IV.14.

[81] R. Brill, *Angew. Chem.* **74**, 895 (1962).

[82] W. G. Schneider, *J. Chem. Phys.* **23**, 26 (1955).

these cationic groups, the greater E_{KH} is, i.e., the stronger the electrostatic field of the cation at the hydrogen nuclei of the water molecule.

From this it follows that the hydrogen bridges by which the water molecules are bound to the oxygen atoms of the anions become progressively stronger with increasing cation field.

We shall now explain the reasons for this band shift. The OH bonds of the water molecules are polarized by the cationic field.* This was shown by Fajans and Joos[12] in their refraction studies. These bonds are stretched and loosened, on the one hand because the hydrogen nuclei of the water molecule are repelled by the cation, and, on the other hand because the bonding electrons are removed from the OH bonds by the cation field. Hence, the hydrogen atoms of the OH groups of the water become increasingly positively charged with increasing cation field, i.e., the hydrogen bridge donor property of these groups increases. In addition, the hydrogen bridge acceptor property of the oxygen atoms, serving as acceptors at the neighboring anions, also increases with increasing cation field as they are also polarized by the cations. One can see, in the case of polystyrenesulfonic acid (Section V.2.1), that the anion electrons are rearranged during polarization in such a fashion that the oxygen atoms become stronger acceptors. Both effects strengthen the hydrogen bridges through which the molecules of hydration water are bound to the oxygen atoms of the anions. This in turn leads to the observed shift of the maximum of the band of the OH stretching vibration of the hydration water molecules toward smaller wave numbers. They must of course be greater, the stronger the cation field, which is indeed observed.

The increase in the dipole moment of the OH groups of the water molecules by the cation field is closely associated with the shift of the band. Hence, this shift is closely connected with the ion-induced dipole interaction between the cation and water, or more specifically with the interaction between the cation and the induced dipole. The ion–dipole interaction cannot influence the position of the band essentially, for in this interaction the structure of the water molecule is not changed.

Table 7 and, particularly, Fig. 32 show further:

Result 25: If we compare the spectra and the E_{KH} values for Mg^{2+} with those for Ce^{3+} and La^{3+}, we see that the electrostatic cation field at the hydrogen nuclei of the water molecule is stronger for Mg^{2+} than for Ce^{3+}

* The polarization of the water molecules by the ions has also been discussed by Vdovenko and Suglobov[86] in connection with IR-spectroscopic and solubility studies of uranyl salt solutions.

[83] J. A. Pople, *Proc. Roy. Soc.* (*London*) **A202**, 323 (1950).

[84] A. B. F. Duncan and J. A. Pople, *Trans. Faraday Soc.* **49**, 217 (1953).

[85] F. Vaslow, *J. Phys. Chem.* **67**, 2773 (1963).

[86] V. M. Vdovenko and D. N. Suglobov, *Zh. Fiz. Khim.* **34**, 51 (1960).

and La^{3+}. Nevertheless, the maximum of the band is shifted further toward smaller wave numbers for Ce^{3+} and La^{3+} than for Mg^{2+}.

It follows that for Ce^{3+} and La^{3+} the hydrogen bridges are stronger than those for Mg^{2+}. This occurs even though the electrostatic field of these ions is smaller at the hydrogen nuclei of the water molecule. Hence the discussed interaction cannot be solely responsible for the enhancement of the hydrogen bridge donor property and for the shift of the band of the OH stretching vibration of the hydrogen bridges toward smaller wave numbers. This is produced by a covalent bonding component in the interaction between the cation and water, since it also changes the structure of the water molecule. This can easily be seen, since this kind of interaction comes about through the lone electron pairs of the water molecule, interacting more or less strongly with partially occupied *d* orbitals of the cation. For the covalent bonding component in Ce^{3+} and La^{3+}, the partially occupied 4*f* and 5*d* orbitals are available. The electron density in the OH bond of the water molecule is somewhat reduced. The OH groups of the water molecules are progressively stretched and the hydrogen atoms in them become progressively positive with increasing covalent interaction between the cation and water. In other words, the hydrogen bridge donor property of these groups increases. The hydrogen bridges which link the water to the anions become stronger. The maximum of the band of the OH stretching vibration must in this case also be shifted toward smaller wave numbers by the covalent cation–water interaction, which is observed (see also Section IV.10).

Result 26: If we compare the spectra and E_{KH} values for Be^{2+} and Al^{3+} with those for Zr^{4+} and Hf^{4+}, we see that in the cases of Zr^{4+} and Hf^{4+} E_{KH}, the electrostatic cation field at the hydrogen nuclei of the water molecules, is much stronger than with Be^{2+}. Even in the case of Al^{3+} it is but a little stronger than for Zr^{4+} and Hf^{4+}. Nevertheless the maximum of the band of the OH stretching vibration of the hydration water is considerably shifted further toward smaller wave numbers than it is for Zr^{4+} and Hf^{4+}.

It follows that the hydrogen bridges for Be^{2+} and Al^{3+} are stronger than for Zr^{4+} and Hf^{4+}, even though the electrostatic field of these ions at the hydrogen nuclei of the water molecules is smaller. The Be^{2+} and Al^{3+} ions have extremely small radii. They can approach very closely the lone electron pairs of the water molecule. Since the field of these ions is very strong, they can attract the lone electron pairs of the water to themselves. Presumably the lone electron pairs of the water molecule interact with the unoccupied orbitals of the Be^{2+} and Al^{3+} ions. A covalent bonding component arises with these extremely small ions. In the case of Be^{2+} the 2*s* and 2*p* orbitals—in Al^{3+}, the 3*s*, 3*p*, and 3*d*—are available for this purpose. The Be^{2+} and Al^{3+} ions accordingly attract the electrons from the OH bonds of the water molecule more strongly than do the relatively large Hf^{4+} and Zr^{4+} ions.

Therefore the Be^{2+} and Al^{3+} ions increase the hydrogen bridge donor property of the OH group of their molecules of water of hydration more strongly than do the Hf^{4+} and Zr^{4+} ions. Thus this leads to stronger hydrogen bridges and to the observed larger shift of the band maximum.

In summary, we see that the band of the OH stretching vibration shows that the hydrogen bridges by which the hydration water molecules are bound to the anions become stronger if the polarizing effect of the cation field, on the one hand, and the covalent interaction between the cation and water, on the other hand, increase. This is explained by the hydrogen bridge donor property of the OH groups of these water molecules becoming more pronounced, as a result of interaction with the cations. The shift of the band of the OH stretching vibration of the hydration water molecules due to the cations gives information on the changes which the structure of the water molecules undergoes through cation action.

IV.9.1. Coupling of the Cation–Water and Anion–Water Interactions

At low degrees of hydration the strength of the hydrogen bridges which bind the water molecules to the anions depends on the interaction between the cations and the water molecules.

The strength of the dipole moment μ of water, which is essential both for the interaction between the cation and water and for that between the anion and water, is not solely determined by the dipole moment of the free water molecule. Rather, there is added to this dipole moment a component due to the interaction of the water molecule with both cation and anions.

Since this dipole moment of the water molecule is increased in particular by the formation of the hydrogen bridges to the anions, the interaction of the cation with the water molecule must be strengthened by the formation of the hydrogen bridges to the anion.

Thus the interactions of the cation and the anion, respectively, with water molecules at low degrees of hydration are not independent, but they mutually reinforce one another.

IV.9.2. Structure of Pure Liquid Water—A Consequence

We have seen that the hydrogen bridge donor property of the OH groups of the water molecules increases when ions are attached to the lone electron pairs of these water molecules. The same is also the case when OH groups, e.g., from other water molecules, are attached to these lone electron pairs. That is to say, if the lone electron pairs of the water molecules function as hydrogen bridge acceptors, the hydrogen bridge donor property of the OH groups is enhanced. This is significant for the concept of the structure of pure liquid water.

In pure liquid water at 20°C, most of the water molecules are present in a tetrahedral arrangement, i.e., the two OH groups are effective as hydrogen bridge donors and the two lone electron pairs as hydrogen bridge acceptors. The number of free OH groups has been estimated by numerous authors [87-95a] using various methods. Even if all these quantitative estimates are rather tenuous, they still show one thing, namely that the number of free OH groups is much smaller than was formerly supposed.[97,98] This number is about 12% in liquid water at 0°C and increases slowly as the temperature increases. Frank and Wen [99] have advanced the hypothesis that the free OH groups do not occur singly in the water structure, but that fissures occur in this structure.

The previously demonstrated decrease in the hydrogen bridge donor property of the OH groups, when the lone electron pairs of the water are no longer effective as acceptors, brings about a loosening of the hydrogen bridges formed by the water molecules adjoining these fissure surfaces. Hence the water molecules at these fissures are more easily rearranged, not only because they are no longer attached by four hydrogen bridges, but also because the hydrogen bonds are weaker than those of completely coordinated water molecules.

IV.10. Hydration in the Presence of Transition Element Ions

The position of the maximum of the band of the stretching vibration of the OH groups in the hydrogen bridges of the molecules of water of hydration can be obtained from Table 7, for the salts of polystyrenesulfonic acid with the ions of the transition elements Mn^{2+}, Co^{2+}, Ni^{2+}, Cu^{2+}, Zn^{2+}, and Fe^{3+}.[100]

[87] G. H. Haggis, J. B. Hasted, and T. J. Buchanan, *J. Chem. Phys.* **20**, 1452 (1952).
[88] L. Pauling, "The Nature of Chemical Bond," 3rd Ed. Cornell University Press, Ithaca, New York, 1960.
[89] M. O. Bulanin, *Opt. i Spektroskopiya* **2**, 557 (1957).
[90] W. Luck, *Ber. Bunsenges. Physik. Chem.* **66**, 766 (1962).
[91] W. Luck, *Ber. Bunsenges. Physik. Chem.* **67**, 186 (1963).
[92] W. Luck, *Ber. Bunsenges. Physik. Chem.* **69**, 626 (1965).
[93] W. Luck, *Naturwissenschaften* **52**, 49 (1965).
[94] W. Luck, *Fortschr. Chem. Forsch.* **4**, 653 (1964).
[95] W. Luck, *Angew. Chem.* **75**, 691 (1963).
[95a] W. Luck, *Discussions Faraday Soc.* **43**, 115 (1967).
[96] T. T. Wall and D. F. Hornig, *J. Chem. Phys.* **43**, 2079 (1965).
[97] A. Eucken, *Nachr. Akad. Wiss. Goettingen Math. Physik. Kl. IIa. Math. Physik. Chem. Abt.* p. 38 (1946).
[98] A. Eucken, *Z. Elektrochem.* **52**, 6 (1948).
[99] H. S. Frank and W.-Y. Wen, *Discussions Faraday Soc.* **24**, 133 (1957).
[100] G. Zundel and A. Murr, *Z. Physik. Chem.* (*Frankfurt*) **54**, 59 (1967).

Figures 37, 38, and 39, as well as Table 7, show:

Result 27: E_{KH}, the electrostatic cation field in the direction of the bond at the hydrogen nuclei of the water molecules, is larger in the case of Mg^{2+} than for any divalent ion of a transition element. However, for all these cations, with the exception of Mn^{2+}, the band of the stretching vibration of the OH groups in the hydrogen bridges of the water hydration molecules is more strongly shifted toward smaller wave numbers than it is in the case of Mg^{2+}.

Result 28: The wave-number values for the maxima of these bands of hydration water show the following shift as a function of the position of the cation in the periodic table.

Ion	Mn^{2+}		Co^{2+}		Ni^{2+}		Cu^{2+}		Zn^{2+}
Band position	3406	>	3381	>	3373	>	3314	<	3317

The shift of the band of the OH stretching vibration of the water of hydration toward smaller wave numbers increases progressively from Mn^{2+} to Cu^{2+} and then decreases but slightly for Zn^{2+}

Result 29: If the membranes are hydrated at comparable atmospheric humidity, the intensities of the bands of the OH stretching vibrations of the hydration water for the various salts are related to one another as follows:

Intensity for Mn^{2+} < for Co^{2+} < for Ni^{2+} and intensity for Cu^{2+} and Zn^{2+} < for Co^{2+} and Ni^{2+}.

For a low degree of hydration the intensity is for Zn^{2+} < for Cu^{2+} and for a high degree of hydration the intensity is for Cu^{2+} > for Zn^{2+}.

The same intensity variations are found for the scissor vibration—though less pronounced.

Conclusions

(a) For the ions of the transition elements, the hydrogen bridges by which the hydration water molecules are bound to the anions are stronger than would be expected, taking into account the electrostatic field of these cations at the hydrogen nuclei of the water molecules; this is shown by Result 27. Hence, for the ions of the transition elements, an additional interaction between the cation and water also must exist, apart from the polarization of the water molecules by the cation fields. This additional interaction must enhance the hydrogen bridge donor property of the OH groups of the hydration water molecules and shift the band of the OH stretching vibration further toward smaller wave numbers.

The following results from the literature,[101,102] concerning these peculiarities in the interaction of the ions of the transition elements are known. The

[101] L. E. Orgel, "An Introduction to Transition-Metal Chemistry." Methuen, London and Wiley, New York, 1963.

[102] C. J. Ballhausen, "Ligand Field Theory." McGraw-Hill, New York, 1962.

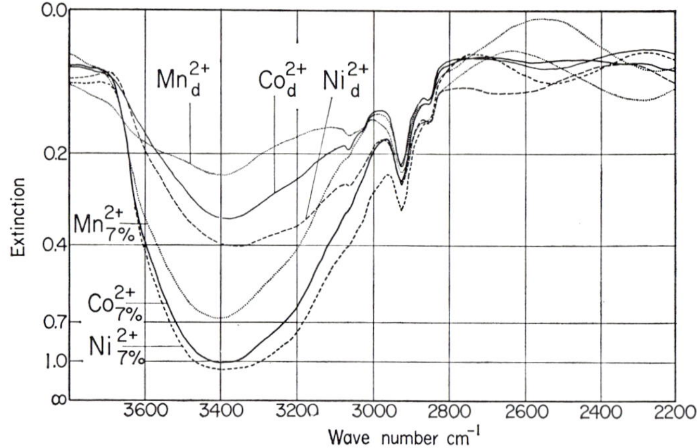

FIG. 37. Mn^{2+}, Co^{2+}, and Ni^{2+} salts of polystyrenesulfonic acid, 5% cross-linked, 3 days of sulfonation; H_2O hydrated: at 7% relative atmospheric humidity (7%), thoroughly dried membrane (d).

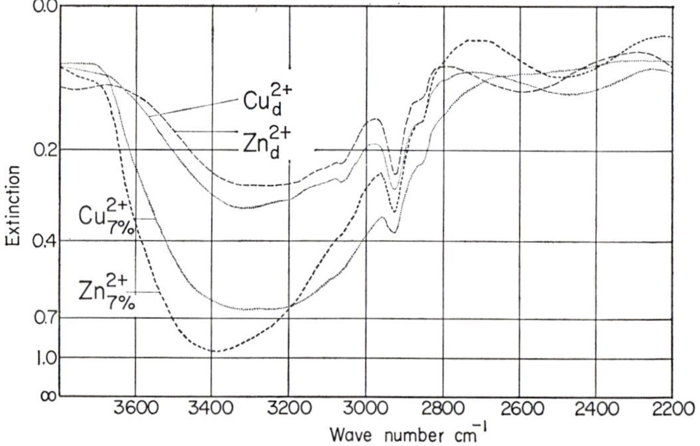

FIG. 38. Cu^{2+} and Zn^{2+} salts of polystyrenesulfonic acid, 5% cross-linked, 3 days of sulfonation. H_2O hydrated: at 7% relative atmospheric humidity (7%), thoroughly dried membrane (d).

d electrons of the cations are rearranged by the dipole fields of the water molecules. In this way—with the exception of Mn^{2+} and Fe^{3+}—so-called crystal field energy is made available. The bonding orbits of these cations, which are formed through participation of the *d* electrons, enter into interaction with the lone electron pairs of the water molecules, whereby a covalent bonding component arises. This leads, as explained above, to an

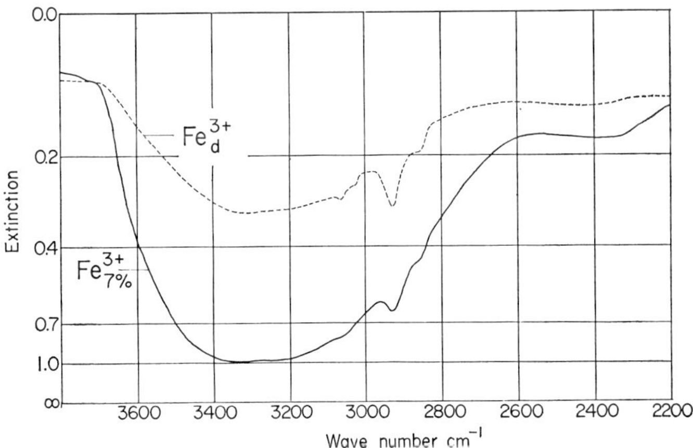

FIG. 39. Fe^{3+} salt of polystyrenesulfonic acid, 5% cross-linked, 3 days of sulfonation. H_2O hydrated: at 7% relative atmospheric humidity (7%), thoroughly dried membrane (d).

enhancement of the hydrogen bridge donor property of the OH groups of the water molecules.

(b) Irving and Williams [103] investigated the stability of the numerous complexes which these cations form and found the following cation dependence: $Mn^{2+} < Co^{2+} < Ni^{2+} < Cu^{2+} > Zn^{2+}$. If we compare this variation with our observation of the course of the maximum of the OH stretching vibration band of the hydration water (Result 28), it follows that the variation of the stability of the complexes with the nature of the cation found in the cases investigated by Irving and Williams corresponds to the variation in the position of this maximum. The observed band shift hence affords us a starting point with respect to the magnitude of the covalent bonding component of the cation–water bond.

(c) The Cu^{2+} and Zn^{2+} ions shift the band of the OH stretching vibration of the hydration water further toward smaller wave numbers than do the Co^{2+} and Ni^{2+} ions. Hence they are in a stronger covalent interaction with their molecules of water of hydration (Result 28). However, the number of water molecules which are attached under the same condition of comparable atmospheric humidity is smaller for Cu^{2+} and Zn^{2+} than for Co^{2+} and Ni^{2+} (Result 29). We shall see later that with Cu^{2+} and Zn^{2+} at low degrees of hydration, complexes are formed with the coordination number 4, with Co^{2+} and Ni^{2+} complexes with the coordination number 6. It follows that under the same condition of comparable atmospheric humidity over the membrane,

[103] H. Irving and R. J. R. Williams, *Nature* **162**, 746 (1948).

the number of water molecules attached per cation is determined not only by the strength of the cation–water bond, but to an equal extent by the coordination number.*

We will now consider the hydration in the particular cases Mn^{2+}, Co^{2+}, Ni^{2+}, and Fe^{3+}:

The hydrate complexes of these cations are octahedral high-spin complexes (footnote 102, pp. 245, 252, 255, and 261). In this case—as explained in part (c)—it is understandable that Co^{2+} and Ni^{2+} attach more water molecules per ion than do Cu^{2+} and Zn^{2+}. This occurs under the same conditions, regardless of the fact that the interaction between the cation and the water of hydration is less than with Cu^{2+} and Zn^{2+} since the number of ligands is greater in octahedra than in the hydrate complexes formed.

Cu^{2+}:

It is known (footnote 101, p. 81 ff. and footnote 102, p. 268 ff.) that, as a result of the Jahn-Teller effect, the Cu^{2+} ion attaches four groups, quadratically arranged in a plane, with strong bonds and two more, on the surfaces of the square, with much weaker bonds. The complexes are hence strongly disturbed octahedra.

According to Results 27 and 28, in Cu^{2+} the band of the stretching vibration of the hydration water is particularly strongly shifted toward smaller wave numbers, in contrast to Mg^{2+}. Further, according to Result 29, the intensity of this band increases with increasing atmospheric humidity much less strongly than does that of the corresponding band for Co^{2+} and Ni^{2+}.

From this it follows that the band observed at 3314 cm^{-1} is the band of the OH stretching vibration of the water of hydration which is strongly bound to the Cu^{2+} in a quadratic arrangement.

At low degrees of hydration the Cu^{2+} is in strong interaction not only with the water molecules, but also with the oxygen atoms of the $—SO_3^-$ ions, as is shown by Result 5 in Section IV.2. Hence if the corner sites are not occupied by molecules of water of hydration, they are oriented toward oxygen atoms from $—SO_3^-$ ions.

Zn^{2+}:

Orgel (see footnote 101, p. 83), writes on the hydration complex of Zn^{2+}: "The Zn^{2+} ion more than any other of the group is stable in tetrahedral environments. It seems that in aqueous solutions $[Zn\ (H_2O)_6]^{2+}$ is the predominant species (perhaps due to the enormous excess of water)."

The band of the OH stretching vibration of the water of hydration for Zn^{2+}

* Next we will discuss the formation of octahedral, tetrahedral, and square arrangements. If at a low degree of hydration not enough water molecules are available for the completion of such a complex, the complex probably will be somewhat deformed.

is much more strongly shifted toward smaller wave numbers than is the case for Co^{2+} and Ni^{2+} (Result 28), and hence the covalent bonding component between Zn^{2+} and water is stronger. Under the same conditions with low atmospheric humidity surrounding the membrane, fewer water molecules per cation are attached (Result 29). Hence the coordination number given by the cation must be smaller than for the octahedral arrangement.

With a higher atmospheric humidity surrounding the membrane, Zn^{2+} attaches more water molecules per ion than Cu^{2+}, according to Result 29. We know that under these conditions there are only four coordination possibilities available per ion. Accordingly, with an increasing amount of water available, the coordination number of Zn^{2+} increases. The tetrahedral complex is then increasingly transformed into an octahedral one.

This confirms Orgel's hypothesis, that the octahedral arrangement of the water molecules of Zn^{2+} in homogeneous liquid electrolyte solutions is determined by the large number of water molecules available, while tetrahedrons are the prevailing form in the presence of smaller amounts of water.

Zn^{2+}, like Cu^{2+} at very low degrees of hydration, is in strong interaction, not only with the water molecules, but also with the $—SO_3^-$ ions. This follows from Result 5 in Section IV.2. Hence the corners of the tetrahedron are turned toward the oxygen atoms of the $—SO_3^-$ ions when they are not occupied by water molecules.

We shall discuss why the covalent bonding interaction between water and Zn^{2+} which in the nonhydrated condition has 10 *d* electrons should be so extraordinarily strong. The energy required to bring one of the *d* electrons into the next higher shell is only 9.7 eV (see footnote 101, p. 67). Hence, hybrids between *d* orbitals and orbitals of the 4th shell, with tetrahedral arrangement, can be formed,[104] which interact with the lone electron pairs.

IV.11. The Band of the Scissor Vibration

In Table 8, column, 2, the positions of the band of the H_2O scissor vibration are given. Each value was measured at least four times on at least two different membranes. This accurate determination of the band positions was necessary, since, as is to be seen from Table 8, the greatest difference between salts of polystyrenesulfonic acid is only 15 cm^{-1}. This means that the shift of the scissor vibration under the influence of the cations is only 5% of the corresponding shift of the band of the OH stretching vibration, for which the largest difference observed is 267 cm^{-1}. The membranes were cross-linked and sulfonated to the same degree as those examined in Section IV.9.

[104] H. Hartmann, "Theorie der chemischen Bindung," p. 167. Springer, Berlin, 1954.

TABLE 8

POSITIONS OF THE BAND OF THE H_2O SCISSOR VIBRATION OF WATER HYDRATION MOLECULES IN POLYSTYRENESULFONIC ACID SALTS (LOW DEGREE OF HYDRATION)

Ion	Band of the scissor vibration (cm^{-1})	Ionic radius (Å) (after footnote 4)	E_{KH} ($ESU \cdot cm^{-2} \times 10^6$)
1	2	3	4
Li^+	1642	0.68	1.49
Na^+	1641	0.97	1.08
K^+	1641	1.33	0.76
Rb^+	1641	1.47	0.67
Cs^+	1644	1.67	0.57
Be^{2+}	1651	0.35	4.46
Mg^{2+}	1648	0.66	3.07
Ca^{2+}	1639	0.99	2.11
Sr^{2+}	1639	1.12	1.85
Ba^{2+}	1636	1.34	1.51
Al^{3+}	1648	0.51	5.53
Ga^{3+}	1636	0.62	4.78
In^{3+}	1637	0.81	3.83
Tl^{3+} [a]	—	0.95	3.32
Sc^{3+}	1646	0.81	3.83
Y^{3+}	1642	0.93	3.40
La^{3+}	1638	1.14	2.73
Ce^{3+}	1637	1.07	2.93
Gd^{3+}	1643	0.94 [b]	3.34
Zr^{4+}	1643	0.79	5.29
Hf^{4+}	1645	0.78	5.33
Mn^{2+}	1638	0.80	2.59
Co^{2+}	1639	0.72	2.87
Ni^{2+}	1643	0.69	2.94
Cu^{2+}	1636	0.72	2.87
Zn^{2+}	1641	0.74	2.80
Fe^{3+}	1638	0.64	4.71

[a] $Tl(OH)_2^+$ is present together with Tl^{3+} in the membrane (on this point see Section IV.20).

[b] Personal communication, R. C. Weast, editor-in-chief, "Handbook of Chemistry and Physics," Chemical Rubber Co., Cleveland, Ohio.

From Table 8 and Fig. 40 we obtain Result 30.

Result 30: Compared with free water (1595 cm^{-1}), the band of the scissor vibration of water of hydration is shifted toward larger wave numbers.

Result 31: Within the groups of the cations, the band of the scissor vibration of the water of hydration is progressively shifted toward larger wave numbers from Ba^{2+} to Be^{2+}, from In^{3+} to Al^{3+}, and from La^{3+} to Sc^{3+}. This means that this band is usually shifted within the cation groups toward larger wave numbers with increasing cation field at the hydrogen nuclei of the water molecules. This is shown by a comparison with the E_{KH} values given in column 4 of Table 8.

This result can easily be understood on the basis of the results given in Section IV.9. Thus, within these groups of cations the hydrogen bridges by which the molecules of water of hydration are bound to the oxygen atoms of the anions become stronger. The stronger the hydrogen bridges become, the greater the force constant of the scissor vibration, and the further the band of the scissor vibration is shifted toward larger wave numbers.

If, however, we compare the positions of the bands not only within the individual groups, we see the following result.

Result 32: With the scissor vibration, generally, there is a less readily apparent connection between the position of the band and the interaction with the cation.

This shows that the connection between the interaction of the ions with water and the force constant of the scissor vibration is considerably more complicated than in the case

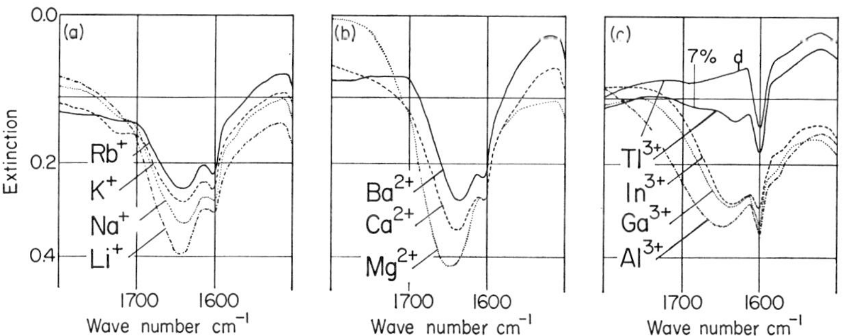

FIG. 40. H_2O scissor vibration. (a) Li^+, Na^+, K^+, and Rb^+ salts of polystyrenesulfonic acid (taken with model 21, NaCl prism, linear in wave numbers) 6% cross-linked, 3 days of sulfonation, H_2O hydrated: 33% relative atmospheric humidity. (b) Mg^{2+}, Ca^{2+}, and Ba^{2+} salts of polystyrenesulfonic acid (taken with model 21, NaCl prism, linear in wave numbers), 6% cross-linked, 3 days of sulfonation, H_2O hydrated: 33% relative atmospheric humidity. (c) Al^{3+}, Ga^{3+}, In^{3+}, and Tl^{3+} [$Tl(OH)^{2+}$ also present (see Section IV.20)] salts of polystyrenesulfonic acid, 5% cross-linked, 3 days of sulfonation, H_2O hydrated: 7% relative atmospheric humidity, the Tl^{3+} salt also thoroughly dried membrane (d).

of the stretching vibration. Presumably the change of the valence bond angle of the water molecule as a consequence of the interaction plays an important role.

Further, in connection with the scissor vibration, we obtain Result 33.

Result 33: The intensity of the band of the scissor vibration increases progressively from Rb^+ to Li^+, from Ba^{2+} to Mg^{2+}, and from Tl^{3+} to Al^{3+} if the membranes have been hydrated at the same relative atmospheric humidity, according to Fig. 40.

Result 34: The dependence of the band of the H_2O scissor vibration on the nature of the anion present cannot be determined exactly, since other bands are sometimes superimposed (see Tables A.1–A.5). In the salts of polystyrenesulfonic acid, the band of the D_2O scissor vibration is masked by the band of the antisymmetric stretching vibration of the $—SO_3^-$ ions [e.g., Fig. 1(e, f)]. It can be clearly seen in the Na^+ salt of polystyreneselenonic acid [Fig. 4(c, d)] at 1206 cm^{-1}, and in the Na^+ salt of polystyreneseleninic acid [Fig. 5(c, d)] at 1209 cm^{-1}. In the Na^+ salt of polystyrenethiophosphonic acid [Fig. 7(c, d)] it is masked. In the Na^+ salt of polystyrenephosphinic acid [Fig. 6(c, d)] it is found at high degrees of hydration as a shoulder at approximately 1210 cm^{-1}.

The Band of the Torsion Vibration of the Water Molecule

Result 35: In the region 800–650 cm^{-1} an increase in the absorption toward smaller wave numbers is observed in the spectra of the membranes hydrated with H_2O. This is caused by a vibration of the H_2O molecule. It is shown in Fig. 41 by the dependence on the degree of hydration and, in particular, by the comparison of the spectra of the membranes hydrated with H_2O and D_2O, respectively. This band can only be due to the torsion vibration of the water molecules (see Section III.2.2).

IV.12. The Band at 3615 cm^{-1} and the Free OH Groups of Water Molecules

We have seen in Section IV.5 that the band at 3615 cm^{-1} cannot be that of the stretching vibration of OH groups bound by hydrogen bridges, by virtue of its position. This band must be ascribed to the stretching vibration of free OH groups. This assignment is confirmed by the dependence of the intensity of this band on the nature of the ions present. If these free OH groups belonged to water molecules possessing two free OH groups, two bands, namely, those of the antisymmetric and the symmetric stretching vibrations, should be expected; for the interaction of these H_2O molecules with their surrounding is not so large as to cause the appearance of only a single band (cf. Section IV.5.1). The fact that just one band is found in the region of large wave numbers thus shows that these water molecules only possess one free OH group, while they are linked by the other OH group to an oxygen atom of an anion (cf. Mohr *et al.*[47]).

The intense broad band of the OH stretching vibration of the hydrogen bridges of the hydration water molecules affords information on the interaction between the ions and water and indirectly on the hydrate structure (Section IV.7; Fig. 35). On the other hand, the shoulder at 3615 cm^{-1} gives us essential information on the hydrate structure.[105]

[105] G. Zundel and A. Murr, *Z. Naturforsch.* **21a**. 1391 (1966).

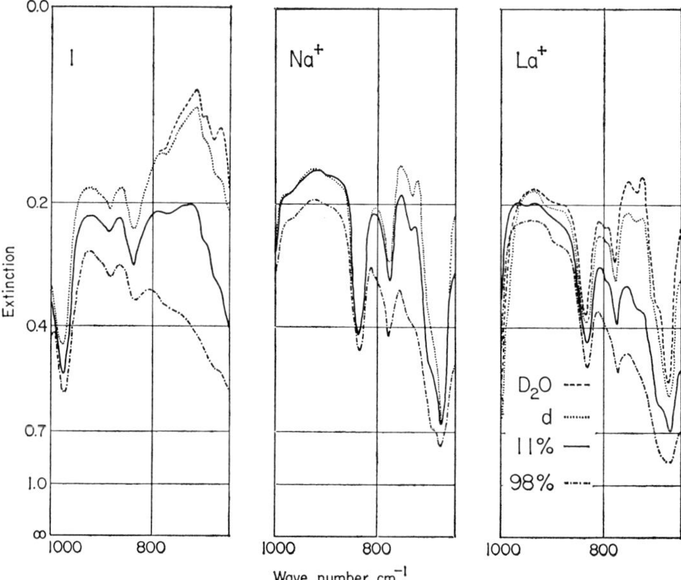

FIG. 41. Na^+ salt of polystyrenethiophosphonic acid (1), Na^+ and La^{3+} salts of polystyrenesulfonic acid, H_2O hydrated: –·–·, at 98% relative atmospheric humidity; ——, at 11% relative atmospheric humidity; –· ·–, at 1% relative atmospheric humidity; · · · ·, thoroughly dried membrane. D_2O hydrated: – – – –.

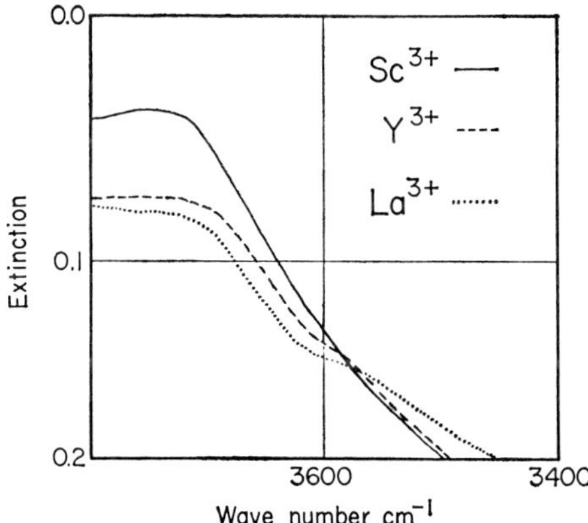

FIG. 42. Sc^{3+}, Y^{3+}, and La^{3+} salts of polystyrenesulfonic acid, 5% cross-linked, 3 days of sulfonation, thoroughly dried membrane.

IV.12.1. Dependence on the Nature of the Cations

We will now consider the dependence of the band at 3615 cm^{-1} on the nature of the cation present.

From Fig. 42 we obtain Result 36.

Result 36. The smaller the radius of the cation in the polystyrenesulfonic acid scandium-group salts, i.e., the more strongly the cation field polarizes the hydration water molecules, the weaker is the intensity of this shoulder at 3615 cm^{-1}.

Concerning this shoulder in the other salts, we examine Figs. 30(a), 31–33, and 37–39. In Table 9, column 3 gives the hydrated salts of polystyrenesulfonic acid with which this shoulder at 3615 cm^{-1} can be observed. If we compare the data in this table with the results in Section IV.9 and IV.10, we see that the following result holds more generally:

Result 37: The shoulder which arises from the band at 3615 cm^{-1} is not found with all salts of polystyrenesulfonic acid. It is only observed in salts in which the cations do not appreciably enhance the hydrogen bridge donor affinity of the OH groups of the molecules of water of hydration.

Hence it follows that the number of free OH groups of molecules of water of hydration in the network of hydrate structures is greater, the less the hydrogen bridge donor affinity of the OH groups is enhanced by the cations.

IV.12.2. Dependence on the Nature of the Anions

We will now consider the intensity of the shoulder at 3615 cm^{-1} in the series of Na^+ salts with the following anions

$$\left(-S\begin{matrix} O \\ \cdots O \\ O \end{matrix}\right)^-, \quad \left(-Se\begin{matrix} O \\ \cdots O \\ O \end{matrix}\right)^-, \quad \left(-P\begin{matrix} S \\ \cdots O \\ O \end{matrix}\right)^{2-}, \quad \left(-Se\begin{matrix} O \\ \\ O \end{matrix}\right)^-, \quad \left(-P\begin{matrix} H \\ \cdots O \\ O \end{matrix}\right)^-$$

in Figs. 22–26 and, in particular, in the summary in Fig. 34. We see:

Result 38: The intensity of the band at 3615 cm^{-1} is stronger in the Na^+ salt of polystyrenephosphinic acid than in the Na^+ salt of polystyrenesulfonic acid.

It follows that the number of free OH groups of hydration water molecules is larger in the Na^+ salt of polystyrenephosphinic acid than in the Na^+ salt of polystyrenesulfonic acid. Hence the hydrate structural network is considerably more disintegrated in the first case.

We have seen in Section IV.8 that, in the series of anions studied from $\left(-S\begin{matrix} O \\ \cdots O \\ O \end{matrix}\right)^-$ to $\left(-P\begin{matrix} H \\ \cdots O \\ O \end{matrix}\right)^-$, the hydrogen bridge acceptor property of the oxygen

TABLE 9

THE BAND AT 3615 CM^{-1} POLYSTYRENESULFONIC ACID SALTS (LOW DEGREE OF HYDRATION)

Ion	Figure number	Band at 3615 cm^{-1} observed as a shoulder	Ionic radius (Å) (after reference 4)	E_{KH} (ESU · cm^{-2} × 10^{6})
1	2	3	4	5
Li^{+}	29	yes	0.68	1.49
Na^{+}	29	yes	0.97	1.08
K^{+}	29	yes	1.33	0.76
Rb^{+}	29	yes	1.47	0.67
Cs^{+}	—	yes	1.67	0.57
Be^{2+}	55	no	0.35	4.46
Mg^{2+}	29	yes	0.66	3.07
Ca^{2+}	29	yes	0.99	2.11
Sr^{2+}	29	yes	1.12	1.85
Ba^{2+}	29	yes	1.34	1.51
Al^{3+}	30	no	0.51	5.53
Ga^{3+}	30	no	0.62	4.78
In^{3+}	30	no	0.81	3.83
Tl^{3+}	30	no[a]	0.95	3.32
Sc^{3+}	31	no	0.81	3.83
Y^{3+}	31	yes	0.93	3.40
La^{3+}	31	yes	1.14	2.73
Ce^{3+}	32	yes	1.07	2.93
Gd^{3+}	32	yes	0.94[b]	3.34
Zr^{4+}	33	no	0.79	5.29
Hf^{4+}	33	no	0.78	5.33
Mn^{2+}	37	yes	0.80	2.59
Co^{2+}	37	no	0.72	2.87
Ni^{2+}	37	no	0.69	2.94
Cu^{2+}	38	no	0.72	2.87
Zn^{2+}	38	no	0.74	2.80
Fe^{3+}	39	no	0.64	4.71

[a] $Tl(OH)_2^{+}$ is present together with Tl^{3+} in the membrane (on this point see Section IV.20).

[b] Personal communication, R. C. Weast, editor-in-chief, "Handbook of Chemistry and Physics," Chemical Rubber Co., Cleveland, Ohio.

atoms of the anions is enhanced. The hydrogen bridges by which the water molecules are bound to the anions become stronger. This contradicts our result, for the number of free OH groups is greatest in the Na^+ salt of polystyrenephosphinic acid. The reason for this may be found in the number of the oxygen atoms available per anion as hydrogen bridge acceptors. It is three for the $-S(O^-)(O)(O)$ ion* but only two for the $-P(H^-)(O)(O)$ ion. Thus the number of free OH groups of molecules of water of hydration depends on the number of oxygen atoms available as acceptors on the anions. The number is greater, the fewer the acceptors available.

In considering the effect of the anions on the network of the hydrate structures, the hydrogen bridge acceptor affinity of the anions as a whole must be taken into account. The greater the affinity, the less disintegrated is the network of the hydrate structures. The affinity is stronger, the stronger the hydrogen bridge acceptor affinity of the individual oxygen atoms and the greater the number of oxygen atoms on the anion; it is presumably also influenced by the presence of hydrophobic groups on the anion.

IV.12.3. The Band at 3615 cm^{-1} and the Rearrangement Processes in the Hydrate Structure Network

The hydrate structures are very rapidly rearranged by the thermal motions of the hydration water molecules. This rearrangement has been measured through studies of relaxation processes; and it has been discussed by numerous authors.[99,106-115]

The rate constants of the rearrangement of the hydration water molecules in the inner coordination sphere for various metal ions are given by Eigen[108] (see Fig. 5 of that work). We can see a connection between the rate constants

* The presence of the Na^+ ion in each case hinders one oxygen atom as a hydrogen bridge acceptor; in the Na^+ salt of polystyrenethiophosphonic acid the presence of the two Na^+ ions even hinders two oxygen atoms.

[106] O. J. Samoilow, "Die Struktur von wässrigen Elektrolytlösungen." Teubner, Leipzig, 1961.

[107] M. Eigen, *Ber. Bunsenges. Physik. Chem.* **67**, 753 (1963).

[108] M. Eigen, *Pure Appl. Chem.* **6**, 97 (1963).

[109] T. J. Swift and R. E. Connick, *J. Chem. Phys.* **37**, 307 (1962).

[110] H. G. Hertz and M. D. Zeidler, *Ber. Bunsenges. Physik. Chem.* **67**, 774 (1963).

[111] H. G. Hertz and M. D. Zeidler, *Ber. Bunsenges. Physik. Chem.* **68**, 821 (1964).

[112] H. G. Hertz. *Ber. Bunsenges. Physik. Chem.* **68**, 907 (1964).

[113] T. H. Wu, *J. Geophys. Res.* **69**, 1083 (1964).

[114] R. G. Pearson and M. M. Anderson, *Angew. Chem.* **77**, 361 (1965).

[115] D. W. G. Smith and J. G. Powles, *J. Mol. Phys.* **10**, 451 (1966).

and the appearance of the shoulder at 3615 cm^{-1}: In the case of those cations for which the rate constant is large, the shoulder at 3615 cm^{-1}, which indicates the presence of free OH groups, is encountered. From this evidence it appears that the free OH groups are those OH groups in the hydration water molecules which are temporarily freed in the thermal rearrangement processes.

Here we find a single exception, namely Cu^{2+}, in which we observe no free OH groups, although the rate constant given by Eigen[108] is fairly large. The value obtained for Cu^{2+} by Pearson and Anderson[114] shows the same irregularity. In regard to the following interpretation, it is interesting that these authors[114] remark: "there is a sign of a second, slower reaction." The deviation in the case of Cu^{2+} can be understood in the light of our results given in Section IV.10. We have seen there that Cu^{2+} binds four water molecules strongly in a quadratic arrangement. It is known that Cu^{2+} attaches two additional water molecules perpendicularly to the square at high degrees of hydration, but that the attachment of these additional water molecules is much weaker. The rearrangement rate constant given by Eigen[108] is obviously that of the rearrangement of these two weakly bound water molecules. However, in the present case the degree of hydration is so small that only the strongly attached water molecules are present.

IV.12.4. A Weak Band at 3530 cm^{-1}

In Fig. 43 are shown spectra of a membrane of the Na^+ salt of polystyrenesulfonic acid. In this case 3–4 days elapsed after each hydration step. From Fig. 43 and 23, we obtain:

Result 39: In a few cases a very weak band is found at about 3530 cm^{-1}. With the Na^+ salt of polystyrenesulfonic acid, this band is observed under conditions (a) and (b) given below. It is also observed with polystyreneselenonic acid. It has the following properties:

(a) The band is observed in polystyrenesulfonic acid only when the membranes have been very slowly dried.

(b) The band disappears only on thorough drying.

(c) Compared to the above-discussed intense band of the OH stretching vibration of the water of hydration, the band is relatively sharp.

As this band has only been observed in these few cases, it has not been possible so far to assign it or to reach any conclusions about it.

IV.13. Temporary Equilibrium Configurations of the Hydrate Structures at Low Degrees of Hydration

We have seen in Section IV.7 that the molecules of water of hydration lie between the cation and the neighboring anions (Fig. 35). The results given in Section IV.12 have shown that the number of free OH groups is in some cases so small that their bands cannot be seen at all in the IR spectrum.

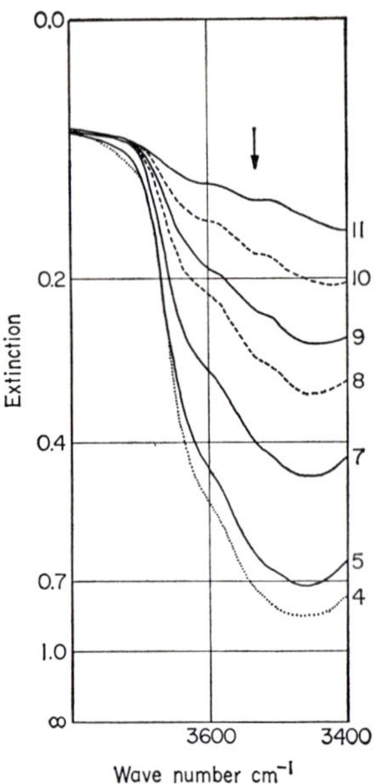

FIG. 43. Na^+ salt of polystyrenesulfonic acid, 5% cross-linked, 3 days of sulfonation, H_2O hydrated: measured after 3–4 days at each humidity; 4, 33%, 5, 22%; 7, 5%; 8, 1% relative atmospheric humidity; 9, 10, and 11, incompletely dried.

When this occurs, nearly all the hydration water molecules are bound to a cation and to oxygen atoms of neighboring anions by two hydrogen bridges.

Is this sterically possible? Luck[116–118] estimated the bonding energy of a hydrogen bridge in dependence on its degree of bending. He writes[116]: "There is a clear theoretical argument that the H bond is optimal when the partners X—H···Y lie in a straight line, whose direction should coincide with the axis of symmetry of the lone electron pairs.[82,119] However, the bonding energy decreases very slowly at first with increasing bending of the bridge." According to his estimate, the bonding energy is imperceptibly lowered when the bridge is bent by 30 degrees. Even bending by 60 degrees decreases the bonding energy less than 20%. Also, the lengths of the bridges can be slightly varied to fit the steric conditions without any rupture occurring, for this reason. The variation of the lengths has been discussed in detail in connection

[116] W. Luck, *Naturwissenschaften* **52**, 25 (1965).
[117] W. Luck, *Naturwissenschaften* **52**, 49 (1965).
[118] W. Luck, lecture at the EUCHEM conference on the hydrogen bridge, Schloss Elmau, West Germany, April 1967.
[119] H. A. Staab, "Einführung in die theoretische organische Chemie." Verlag Chemie, Weinheim, 1959.

with a Raman-spectroscopic investigation of liquid water by Wall and Hornig.[96]*

Information from our spectra regarding this point follows:

The band of the stretching vibration of the OH groups of the hydrogen bridges by which the hydration water molecules are bound to the anions is very broad. An essential cause for this is the more or less strong bending and a variation in the length of the bridges. Important results supporting this conclusion have already been given in Section IV.5.1. The thermal vibrations of the hydration water molecules also cause broadening, but it is not as important as generally assumed, because the half width of the band of water of hydration is not very strongly temperature dependent (on this point see, as an example, Fig. 118).

Figure 44† shows an instantaneous configuration of the hydrate structures.‡ The hydration water molecules link the cations with the neighboring anions, so that these water molecules each form two hydrogen bridges, more or less strongly bent and of differing lengths under the prevailing steric conditions.

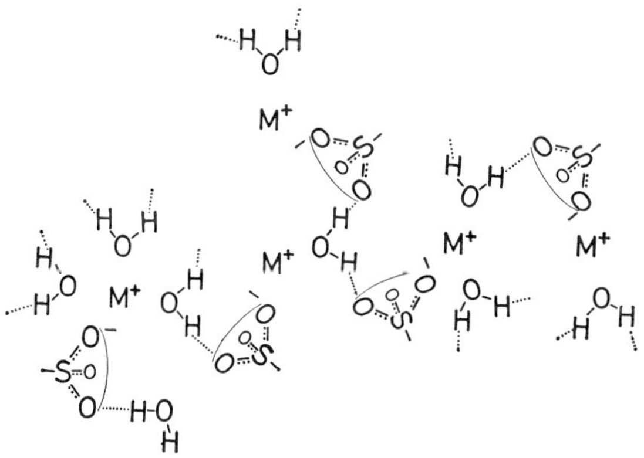

FIG. 44. Hydrate structures at low degrees of hydration.

* A connection between the O–O distance in hydrogen bridge bonds and the position of the band of the OH stretching vibration has been conjectured by numerous authors.[63,70,96,120–122] This relationship loses its validity if the hydrogen bridges are very strong. This has been pointed out by Blinc and Hadži (footnote 123, p. 862).

† See footnote, p. 67.

‡ With regard to the instantaneous configurations of the hydrate structures of salts of poly(*p*-trimethylammonium)styrene [—$N(CH_3)_3^+$ groups], see footnote 23a.

[120] O. Glemser and E. Hartert, *Z. Anorg. Allgem. Chem.* **283**, 111 (1956).

[121] E. Schwarzmann, *Z. Anorg. Allgem. Chem.* **317**, 176 (1962).

[122] E. Schwarzmann, *Naturwissenschaften* **49**, 103 (1962).

[123] R. Blinc and D. Hadži, *Spectrochim. Acta* **16**, 852 (1960).

The number of OH groups temporarily freed by the process of thermal rearrangement increases with decreasing hydrogen bridge donor affinity of the OH groups of the hydration water molecules and decreases with increasing hydrogen bridge acceptor affinity of the anions. As already discussed, they are dependent on the nature of the cations and anions.*

IV.14. Water Structure: Alkali Metal Salts $(CH_3)_4N^+$ and $(C_2H_5)_4N^+$ Salts and (Polystyrenesulfonic Acid)

We now consider an anomaly of the hydrate structure in the alkali metal salts.[124a] From Table 7, we see the following result:

Result 40: In the alkali metal ions, hardly any shift of the OH band of the H_2O molecules toward smaller wave numbers occurs with increasing electrostatic cation field at the hydrogen nuclei of the water molecules.

The position of the band maximum has been extremely carefully studied in membranes hydrated with a mixture of H_2O and D_2O.[125]† The maximal error in this investigation was $\pm 1\ cm^{-1}$. It was found that for the bands of both the OH and the OD stretching vibration the shift of these band maxima toward larger wave numbers usually observed with ions of the same valence and increasing ionic radius is reversed at Cs^+. The band for Cs^+ lies at a smaller wave number than the band for Rb^+.

A similar result is given in the literature: Bergqvist and Forslind[126] found, in NMR investigations of aqueous solutions of alkali metal chlorides, that the molar chemical shift from Li^+ to K^+ increases monotonically but decreases again from Rb^+ to Cs^+.

It seems at first possible that a kind of interaction between cation and water exists, which increases in passing from Li^+ to Cs^+ in contrast to the ion-induced dipole interaction. If, however, the magnitude and the variation of the other kinds of intermolecular interaction are estimated, it is seen that such an effect should be ruled out.

Comparison of the observations on the alkali metal salts of polystyrenesulfonic acid (Table 7) with those on pure liquid water (Section III.2.1) suggests the proper interpretation. This shows:

Result 41: In the hydrated alkali metal salts of polystyrenesulfonic acid,

* These instantaneous configurations of hydrate structures thus confirm the hypothesis on which Pople[124] based his theory of the structure of pure liquid water. He writes (see footnote 124, p. 163): "A theory is developed in which the majority of hydrogen bonds are regarded as distorted rather than as broken."

† The values given in footnote 125 are all slightly shifted toward smaller wave numbers, because a mixture of H_2O and D_2O, i.e., HDO, was also present in the membrane.

[124] J. A. Pople, *Proc. Roy. Soc.* (*London*) **A205**, 163 (1951).

[124a] G. Zundel and A. Murr, *Z. Naturforsch.* **24b**, 375 (1969).

[125] G. Zundel, A. Murr, and G.-M.Schwab, *Naturwissenschaften* **50**, 17 (1963).

[126] M. S. Bergqvist and E. Forslind, *Acta Chem. Scand.* **16**, 2069 (1962).

the band of the stretching vibration of the OH groups in the hydrogen bridges lies at about 3460 cm^{-1}, i.e., at a larger wave number than in pure liquid water, where it is found at about 3420 cm^{-1}.

If follows that the hydrogen bridges which link the molecules together in pure liquid water are stronger than those by which hydration water molecules are bound to the oxygen atoms of the $—SO_3^-$ ions in the alkali metal salts of polystyrenesulfonic acid. This favors the cross-linking of water molecules by hydrogen bridges more than the bonding of hydration water molecules by hydrogen bridges to the oxygen atoms of the anions.

The cations are located—as shown by the results in Section IV.9—at the lone electron pairs of the water molecules. Since the interaction between cation and water decreases from Li^+ to Cs^+, the bond between cation and water molecule becomes progressively weaker in this order. The OH groups of other water molecules can then displace with increasing ease the cations from the lone electron pairs and form hydrogen bridges with the latter.

Hence in passing from Li^+ to Cs^+ the association of the water molecules with one another is progressively favored as compared with the bonding of water molecules to anions, i.e., compared with the attachment of water molecules immediately between the cation and the neighboring anions.

Hence the anomalous shift of the band maximum can now be understood. In passing from Li^+ to Cs^+, the number of hydrogen bridges between the water molecules increases. The band of the stretching vibration of the OH groups in these hydrogen bridges is, however, more strongly shifted toward smaller wave numbers than is that of the hydrogen bridges by which the water molecules are linked with the $—SO_3^-$ ions. Hence on the normal shift of the band toward smaller wave numbers in passing from Cs^+ to Li^+, there is superimposed a shift of the band in the opposite direction.

If this explanation is correct, the band of the OH stretching vibration must be shifted still further toward smaller wave numbers in the presence of cations which are even more weakly bound to the water molecules than is Cs^+. With this in mind, we investigated the $(CH_3)_4N^+$ and $(C_2H_5)_4N^+$ salts of polystyrenesulfonic acid.

Result 42: In the $(CH_3)_4N^+$ salt of polystyrenesulfonic acid, the band of the OH stretching vibration is found at 3449 cm^{-1}, and in the $(C_2H_5)_4N^+$ salt it is found at 3430 cm^{-1}, under comparable conditions.

This result confirms our assumption; we see that in cations which interact only very weakly with the water molecules the band of the OH stretching vibration of the hydrogen bridges is, as expected, shifted further toward the wave number which it possessed in pure liquid water. The position of the band of the OH stretching vibrations thus shows directly the water structure formation caused by the nonpolar groups of these ions (cf. the NMR work of Herz *et al.*[110-112]).

Two reasons would lead one to expect that these alkylammonium ions have a stronger interaction with the —SO_3^- ions than the alkali ions. First, if the water molecules are not positioned between the cation and the neighboring anions, then the cation field will not be conducted so fully to the neighboring anions (Section IV.15). The cation field will then be directed preferentially to its corresponding anion. Second, since the presence of the alkylammonium ions causes the hydrogen bonds formed by the water molecules to become stronger, the rearrangement frequency of the water molecules and hence the entropy decrease. The alkylammonium ions tend, therefore, to avoid contact with the water structure; that is, a hydrophobic bonding component is present in the alkylammonium ion —SO_3^- ion bond (cf. footnote 127).

The fact that this interaction of the alkylammonium ions with the —SO_3^- ions is stronger than the interaction between alkali ions and —SO_3^- ions, can be seen in the spectra. In Section IV.2 it was seen that the stronger the interaction between the cations and the —SO_3^- ions, the greater the splitting of the band of the antisymmetric stretching vibration. At a comparable degree of hydration, this splitting is about 30 cm^{-1} for the Na^+ salt and 55 cm^{-1} for the $(C_2H_5)_4N^+$ salt. The interaction of the $(C_2H_5)_4N^+$ ions with the —SO_3^- ion is indeed much stronger than that of the Na^+ ion, just as expected.

In summary, we see, on the one hand, that the band of the OH stretching vibration of the hydrogen bridges that link the molecules in pure liquid water lies at a smaller wave number than the band of the OH stretching vibration of the hydrogen bridges by which the water molecules are bound to the oxygen atoms of the —SO_3^- ions, i.e., the hydrogen bridges are stronger in pure liquid water. On the other hand, the shift of the band of the OH stretching vibration in the series of the alkali metal cations shows an anomaly. This can be explained as follows. The interaction between the cation and water decreases in passing from Li^+ to Cs^+ and the association of water molecules is favored over their attachment among the cation and the neighboring anions. This water structure formation is decisive for the hydrophobic bonding.

Conclusions Concerning Biological Membranes

The finding that alkylammonium ions avoid contact with the water structure is important for biological membranes. Finean[127a] has postulated that the alkylammonium groups of the lecithin and sphingomyelin, which are present in large quantities in the membrane, are directed inward, i.e., away from the water phase. From the above discussion, this is immediately clear.

[127] G. Nemethy, *Angew. Chem.* **79**, 260 (1967).

[127a] J. B. Finean, "Chemical Ultrastructure in Living Tissues," p. 69, Thomas, Springfield, Ill, 1961.

IV.15. Dissociation Process, Ion Pairs, and Conductivity

The splitting of the band of the antisymmetric stretching vibration of the $—SO_3^-$ or $—SeO_3^-$ ions in consequence of the removal of degeneracy by the cation can, according to the results given in Section IV.2, serve as a relative measure for the strength of the interaction between the cation and its corresponding anions.[128]

Experimental results: In salts of polystyrenesulfonic acid, the comparison of the spectra shown on the left-hand and right-hand sides, respectively, of Figs. 12b–15, shows that the band splitting—a doublet at 1200 cm^{-1}—decreases with increasing degree of hydration (Result 3 in Section IV.2). This is shown somewhat more clearly in the Mg^{2+}, Al^{3+}, Sc^{3+}, and La^{3+} salts in Fig. 45. In Fig. 46, this splitting is plotted as a function of the degree of hydration in the Mg^{2+} and Sc^{3+} salts of polystyrenesulfonic acid. The integrated extinction of the scissor vibration of H_2O was used as a measure of the degree of hydration. The experimental points in the curve at the smallest degree of hydration are the results obtained with the thoroughly dried membranes; the experimental points at the largest degree of hydration are the results obtained when the membranes are hydrated at 98% relative atmospheric humidity. In the Sc^{3+} salt of polystyreneselenonic acid, Fig. 16(b) shows the splitting of the band of the antisymmetric stretching vibration of the $—SeO_3^-$ ion—a doublet at about 909 cm^{-1}—dependent upon the atmospheric humidity surrounding the membrane. All this leads to:

Result 43: The splitting of the band of the antisymmetric stretching vibration of the $—SO_3^-$ or $—SeO_3^-$ ion decreases continuously with increasing degree of hydration.

Result 44: The splitting becomes progressively smaller with increasing degree of hydration even at a relative atmospheric humidity of more than 90%.

If we take into account the adsorption isotherms plotted by other authors on similar systems[24–29] (see footnote †, p. 61), we obtain a rough indication as to how many water molecules are present per cation,* if the membranes are hydrated at 90% relative atmospheric humidity. We see that under these conditions at least five water molecules per cation* are present in the membranes.

Result 43 indicates that the bond between cation and anion is progressively loosened when water molecules are attached. Glueckauf and Kitt (footnote 24, p. 330), have already inferred a loosening of the cation from the anion on the attachment of the first water molecule. They reached this conclusion

* This, however, does not mean that they are all always directly attached to the cation, for with a high atmospheric humidity over the membrane, a "second hydration shell" is more or less completely built up (on this point see Section IV.18).

[128] G. Zundel and A. Murr, *Electrochim. Acta* **12**, 1147 (1967).

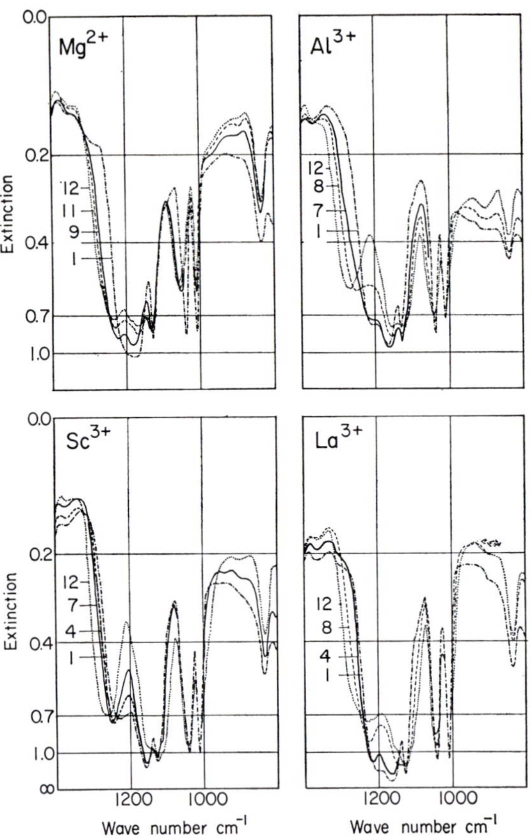

FIG. 45. Mg^{2+}, Al^{3+}, Sc^{3+}, and La^{3+} salts of polystyrenesulfonic acid, 5% cross-linked, 3 days of sulfonation; 1, 98%, 4, 33%; 7, 5%; 8, 1% relative atmospheric humidity; 9, 11, incompletely dried; 12, thoroughly dried membrane.

because they found the adsorption of this first water molecule to be proportional to $\sqrt{a_w}$, a_w being the water activity. With an increasing degree of hydration, the loosening of the cation from its corresponding anions is progressive. Result 44 indicates that, even when the membranes are hydrated at more than 90% relative atmospheric humidity (i.e., when more than five water molecules per cation* are already present in the membrane), the band of the antisymmetric stretching vibration is still split. The cation is hence preferentially located at one of the oxygen atoms of its corresponding anions,

* See footnote p. 95.

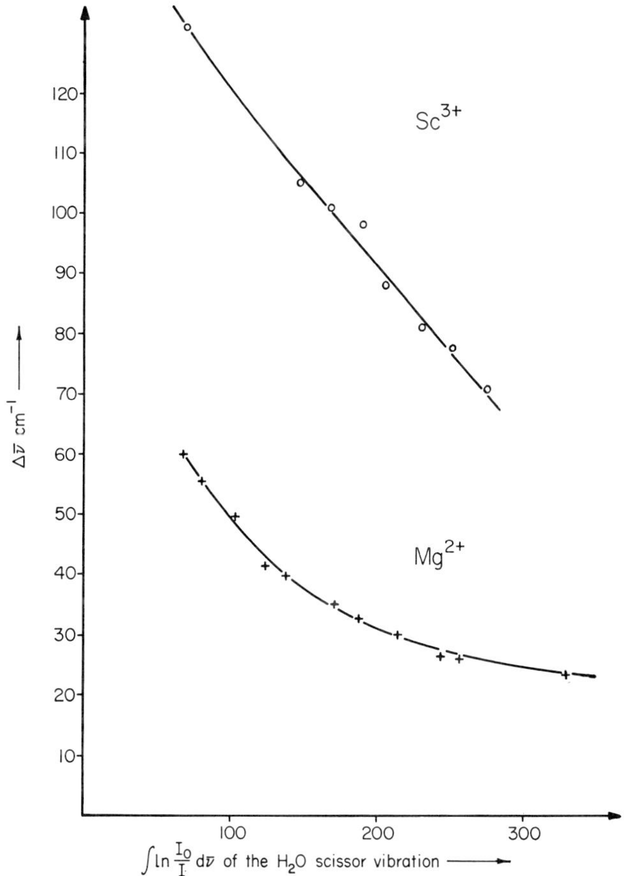

FIG. 46. The splitting-up of the antisymmetric stretching vibration of the —SO_3^- ion as a function of the degree of hydration.

the bonds between cation and anions being nevertheless considerably loosened (cf. the results in footnotes 129–132).

We shall discuss how the water molecules loosen the cation from its corresponding anions:

If no water of hydration is present, the electrostatic field extends essentially from the cation to its corresponding anions, i.e., to the anions on which the

[129] D. C. Whitney and R. M. Diamond, *Inorg. Chem.* **2**, 1284 (1963).
[130] D. C. Whitney and R. M. Diamond, *J. Phys. Chem.* **68**, 1886 (1964).
[131] C. H. Jensen and R. M. Diamond, *J. Phys. Chem.* **69**, 3440 (1965).
[132] D. C. Whitney and R. M. Diamond, *J. Inorg. Nucl. Chem.* **27**, 219 (1965).

cation is located. If now, with an increasing degree of hydration, water molecules are located among the cations and the neighboring anions, these water molecules behave as a "dielectric" among the cations and the neighboring anions. The dipole fields of the water molecules weaken the field among the cation and its corresponding anions. The weakening of this field and the associated weakening of the bonds of the cation to its corresponding anions is naturally stronger, the more water molecules are attached, and the more their dipole moments are increased by the ion fields. This dipole moment is increased, as we have seen in Section IV.9, not only by the electrostatic but also by the covalent ion–water interaction. The dissociation process is discussed from a rather different standpoint in Section V.16.1.

An additional loosening of the cations from their corresponding anions in connection with the covalent bonding of the cation and water will presumably also be caused by some discharge of the cations in the interaction. It is probable that the dipole induced in the cation by the dipole fields of water molecules when it is very easily polarizable, for example, with Cs^+, also plays a part in loosening these ions from the anions.

The foregoing discussion of the dissociation processes in aqueous media shows, apart from the results here obtained, that the dissociation process must depend very strongly on the specific structure of the solvating molecules.

Consequences for the Idea of the Ion Pair

The cation–anion groups are nothing other than ion pairs. Results 43 and 44 hence indicate that the cation–anion bond in an ion pair becomes progressively weaker with increasing degree of hydration of the pair. More generally, this bond is dependent on the degree of solvation and the specific structure of the solvating medium. Thus the behavior of a particular solvated ion pair, and hence the properties of solutions containing it, must also depend strongly on the same factors.

*On Conductivity**

Even at a relatively high degree of hydration the cation is strongly attached, preferentially, to its corresponding anions. However, the more completely the cation–anion bonds are loosened, the more easily they are temporarily ruptured. The cation "jumps" to the neighboring anion, whereby the corresponding cation of this anion also is forced to "jump." The cation passes from one anion to another so quickly that in the IR spectra no anions without attached cations can be found. Hence, even at relatively high degrees

* An investigation of the conductivity of aromatic sulfonate solutions has been carried out.[133] Here the authors discussed their investigations from the viewpoint of the interaction between ion and solvent molecules.

[133] Hatsuho Uedaira and Hisashi Uedaira, *Bull. Chem. Soc. Japan* **37**, 1885 (1964).

of hydration, the cation, in the IR spectra, is found preferentially bound to its corresponding anions.

In conclusion, it may be pointed out that homogeneous liquid electrolyte solutions and fused salts have been investigated for the removal of degeneracy of the vibration of ions by several workers. Thus Lee and Wilmshurst[134] studied nitrate, chlorate, carbonate, and sulfate solutions. Hester and Plane,[135,136] Irish and Walrafen,[137] and Steger and Pauly[138] investigated nitrate solutions, and Janz and James[139] investigated fused nitrates. In addition to the removal of the degeneracy of the vibrations of the ions, these authors in some cases also considered the intermolecular vibrations themselves. In particular, alkylammonium ion pairs were investigated through their intermolecular vibrations by Evans and Lo.[140]

In summary, we can state that if the degree of hydration increases the bond of the cation to its corresponding anions is progressively loosened. This can be understood if one considers the interaction between the ions and the water. This consideration is essential to the concepts of the dissociation process and of the ion pair. The cation–anion bond in ion pairs is thus directly dependent on the degree of solvation and on the nature of the solvate molecules. Even at relatively high degrees of hydration, the cations are still preferentially attached to the oxygen atoms of their corresponding anions, but the bond is considerably loosened. This last feature is essential to the understanding of the processes of conductivity in concentrated salt solutions.

IV.16. Dependence of the Ion–Water Interaction on the Degree of Hydration

Let us consider how the position of the maximum of the intense band of the OH stretching vibration of the water of hydration changes with increasing degree of hydration; that is, when the 1st, 2nd, 3rd, etc., water molecules are attached to the ions.[141] Spectra of membranes of the Na^+ salts at different degrees of hydration are to be found in Figs. 22–26. The spectra of the Mg^{2+}, Sc^{3+}, and La^{3+} salts of polystyrenesulfonic acid at different degrees of hydration are shown in Figs. 47–49. So we can follow the shift of the maximum of the band of the OH stretching vibration of the water of hydration, we have shown the position of this band in Fig. 50(a) as a function of the

[134] H. Lee and J. K. Wilmshurst, *Australian J. Chem.* **17**, 943 (1964).
[135] R. E. Hester and R. A. Plane, *J. Chem. Phys.* **40**, 411 (1964).
[136] R. E. Hester and R. A. Plane, *Inorg. Chem.* **3**, 769 (1964).
[137] D. E. Irish and G. E. Walrafen, *J. Chem. Phys.* **46**, 378 (1966).
[138] E. Steger and S. Pauly, *Naturwissenschaften* **53**, 154 (1966).
[139] G. J. Janz and D. W. James, *J. Chem. Phys.* **35**, 739 (1961).
[140] J. C. Evans and G. Y.-S. Lo, *J. Phys. Chem.* **69**, 3223 (1965).
[141] G. Zundel and A. Murr, *J. Chim. Phys.* **66**, 246 (1969).

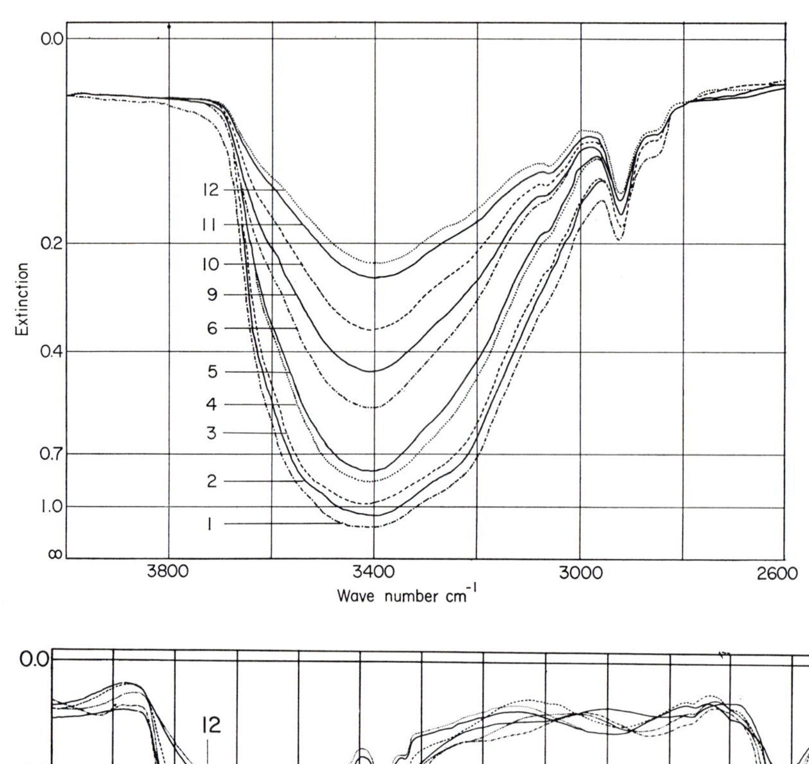
0.0
0.2
0.4
0.7
1.0
∞
Extinction
12
11
10
9
6
5
4
3
2
1
3800
3400
3000
2600
Wave number cm⁻¹

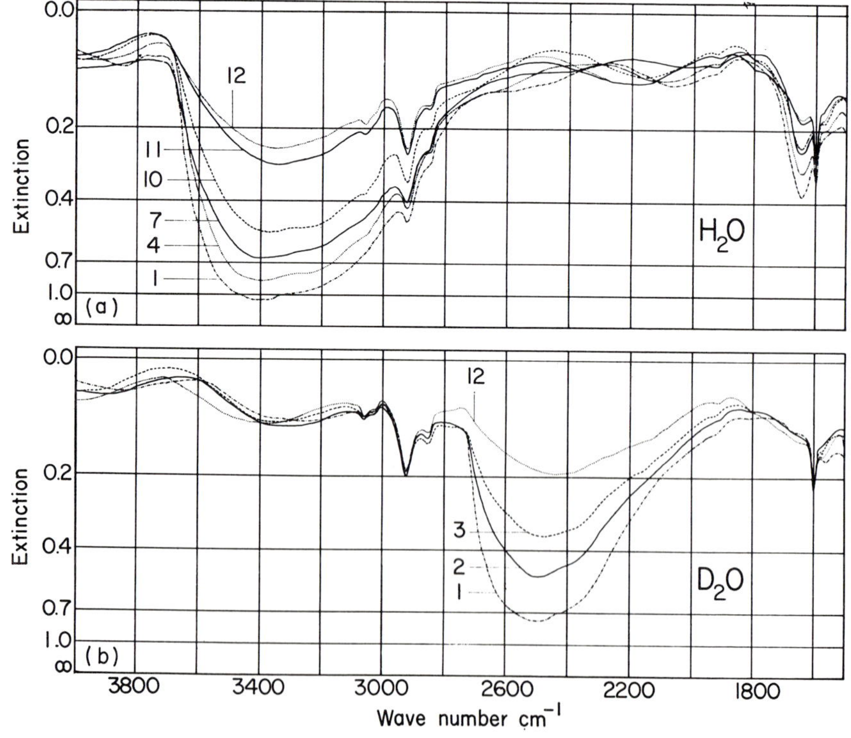
0.0
0.2
0.4
0.7
1.0
∞
Extinction
12
11
10
7
4
1
(a)
H₂O
(b)
3
2
1
D₂O
3800
3400
3000
2600
2200
1800
Wave number cm⁻¹

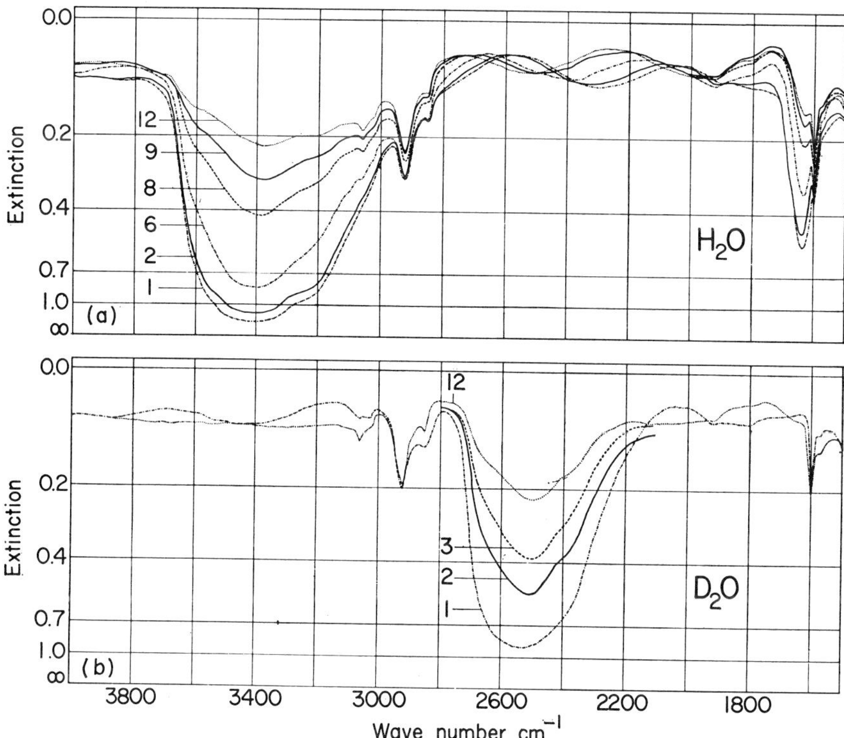

FIG. 49. La^{3+} salt of polystyrenesulfonic acid, 5% cross-linked, 3 days of sulfonation. (a) H_2O hydrated: 1, 98% relative atmospheric humidity; 2 → 9, decreasing atmospheric humidity; 12, thoroughly dried membrane. (b) D_2O hydrated: 1, atmospheric humidity over a saturated solution of $BaCl_2$ in D_2O; 2 → 3, decreasing atmospheric humidity; 12, thoroughly dried membrane.

relative humidity of the atmosphere with which the membrane was in equilibrium during the recording of the particular spectrum. In Fig. 50(b) the position of this band is shown as a function of the integrated extinction of the scissor vibration.

From these figures we obtain:

Result 45: The band of the stretching vibration of the OH groups in the

---

FIG. 47. Mg^{2+} salt of polystyrenesulfonic acid, 5% cross-linked, 3 days of sulfonation, H_2O hydrated: 1, 98%; 2, 71%; 3, 53%; 4, 33%; 5, 22%; 6, 11% relative atmospheric humidity; 9, 10, and 11, incompletely dried; 12, thoroughly dried membrane.

FIG. 48. Sc^{3+} salt of polystyrenesulfonic acid, 5% cross-linked, 3 days of sulfonation. (a) H_2O hydrated: 1, 98%; 4, 33%; 7, at 5% relative atmospheric humidity; 10, 11, incompletely dried; 12, thoroughly dried membrane. (b) D_2O hydrated: 1, 100% relative atmospheric humidity; 2 → 3, decreasing atmospheric humidity; 12, thoroughly dried membrane.

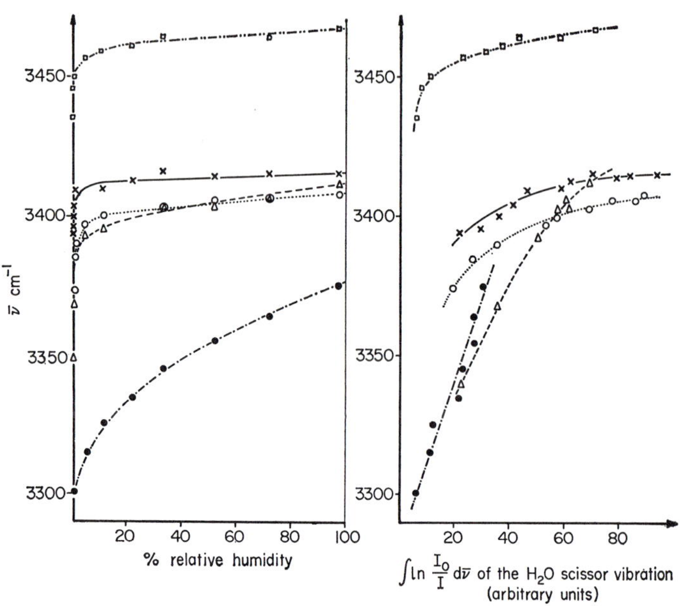

FIG. 50. Dependence of the position of the band of the OH stretching vibration of the water of hydration, observed at low degrees of hydration, on the degree of hydration. –··–, Na^+ salt of polystyrenesulfonic acid; ——, Mg^{2+} salt of polystyrenesulfonic acid; ····, La^{3+} salt of polystyrenesulfonic acid; – – – –, Sc^{3+} salt of polystyrenesulfonic acid; –·–·, Na^+ salt of polystyreneseleninic acid.

hydrogen bridges of the hydration water is shifted toward larger wave numbers with increasing degrees of hydration. In other words, the shift toward smaller wave numbers which this band undergoes as a consequence of the interaction of the water molecules with the ions decreases with increasing degree of hydration. This change is more marked at lower degrees of hydration. It is also stronger the more strongly the ions interact with the water molecules. This is shown by the Na^+ salt of polystyreneseleninic acid (strong hydrogen bridge acceptor property of the oxygen atoms of the —Se(O)(O)⁻ ions) and the Sc^{3+} salt of polystyrenesulfonic acid (strong field of the Sc^{3+} ion at the hydrogen nuclei of the water molecules).

It follows that the hydrogen bridges by which the hydration water molecules are bound to the anions are weaker, the more water molecules there are attached per ion. According to our discussion and results in Sections IV.8 and IV.9, we can understand this shift as dependent on the degree of hydration. If two or more water molecules attach to the cation, the effect of the cation is spread over two, three, etc., water molecules. If, however, the effect of the cation is spread over several water molecules, then the hydrogen bridge donor property of the OH groups of the water molecules is less strongly enhanced. The hydrogen bridges are thereby weakened. A corresponding argument holds for the anion, for when two hydrogen bridges are attached to a single oxygen atom of the anion, they are naturally weaker than when only a single one is present. The weaker the bridge, the less is the shift of the band toward smaller wave numbers. Hence the band of the OH stretching vibration of the molecule of water of hydration must shift toward larger wave numbers with increasing degrees of hydration.

The above considerations are of greater importance in passing from one to two attached water molecules per ion than in passing from two to three. It is thus understandable that the change in the position of the band with degree of hydration is greatest when the latter is small. It is also understandable that this change is more pronounced, the more strongly the ions act on the water molecules.

Comparison with Investigations of Differential Molar Enthalpy and Entropy

Glueckauf and Kitt,[24] as well as Dickel and co-workers[25–27] investigated the differential molar enthalpy $-\Delta\bar{H}$ and the differential molar entropy $-\Delta\bar{S}$ of attachment of water to corresponding substances, in dependence on the degree of hydration. Results of these investigations are shown in Fig. 51, which is taken from the paper of Dickel *et al.*[26] These authors found that $-\Delta\bar{H}$, or $-\Delta\bar{S}$, is much greater for the attachment of the first water molecule per ion pair than for that of the second, for the second greater than for the third, and so on. The explanation for this should be the same as the explanation for the shift of the band of the OH stretching vibration with the degree of hydration.

Figure 50 shows us further that the position of the band of the stretching vibration tends toward a limiting value with increasing degree of hydration, and that this limiting value in turn depends on the nature of the ions present. However, caution is advisable in drawing conclusions from this. We shall see in Section IV.18 that, at such a high degree of hydration, the "second hydration shell" is more or less completely built up. Under these conditions, the bands of the stretching vibrations of the differently attached molecules of water of hydration are already superimposed. However, it may be expected that these limiting values will also yield interesting information when more of these series of measurements are available. The comparison with the position of the band of the stretching vibration of the OH groups of the hydrogen bridges in pure liquid water should be particularly interesting.

In summary, we see that the band of the OH stretching vibration of the hydrogen bridges of the hydration water molecules shifts less and less toward smaller wave numbers as the degree of hydration increases. This can be understood if the effect of the ions on the molecules of water of hydration is considered as a function of the number of such molecules present per ion.

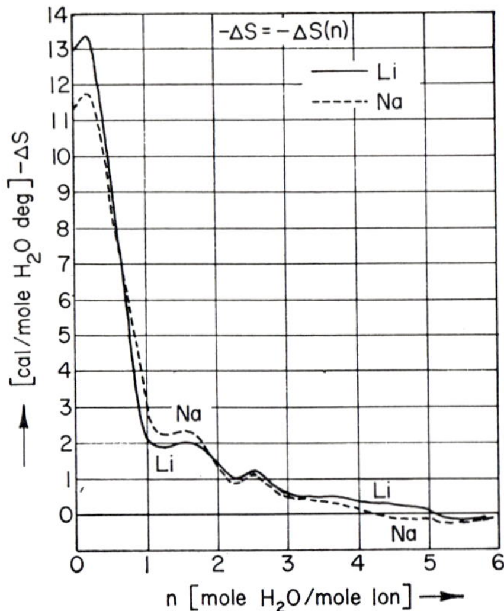

FIG. 51. Differential molar entropy $-\Delta\bar{S}$ of the adsorption of water as a function of *n*. Li and Na forms of polystyrenesulfonic acid exchanger S 100. $-\Delta\bar{S} = -\Delta\bar{S}(n)$. (Sincerest thanks are due Professor G. Dickel, of the Physical Chemistry Institute, Munich, for this figure.)

IV.17. The Overtone of the Scissor Vibration $2\bar{\nu}_2$

We have seen in Section III.2.1 that the shoulder at about 3250 cm^{-1} in the spectrum of liquid water is due to the overtone of the scissor vibration of the water molecules. We have shown in Section IV.5 (Fig. 27) that the shoulder in the spectra of water of hydration at about 3250 cm^{-1} at low degrees of hydration is to be assigned in the same way. Slightly different results are, however, obtained at higher degrees of hydration.

From Figs. 52(a, b) and 56(b) we obtain Result 46.

Result 46: In the Mg^{2+} salt of polystyrenesulfonic acid the shoulder on the side toward smaller wave numbers of the band of the OH stretching vibration in the hydrogen bridges of the water of hydration vanishes almost completely in passing from H_2O to HDO hydration [Fig. 52(a)].

Result 47: In the Sc^{3+} salt of polystyrenesulfonic acid, the shoulder on the side toward smaller wave numbers of the band of the OH stretching vibration in the hydrogen bridges of the water of hydration remains almost unchanged in passing from H_2O to HDO hydration* [Fig. 52(b)].

* If the membrane has been hydrated with HDO, no band of the H_2O scissor vibration should be found in the spectrum at about 1640 cm^{-1}. We have checked this point in all such studies.

From this result it follows that if a shoulder is observed at higher degrees of hydration on the side toward smaller wave numbers of the band of the OH stretching vibration of the hydrogen bridges of the water of hydration, this band may be the overtone of the scissor vibration.* This shoulder may also have another cause, as shown by Result 47, and this is quite important, as discussed in the following section.

IV.18. The "Second Hydration Shell"

It is supposed that at higher degrees of hydration two and subsequently more "layers" of water molecules are found among the cation and the neighboring anions, as shown in Fig. 53.†[142] This we will designate the formation of a "second hydration shell." Wc takc thc cation as reference point in what follows.

The Hydration Water Molecules of "Attachment Type II"

The band of the OH stretching vibration of the hydrogen bridges of the "attachment type II" molecules should be expected in the region 3500–3400 cm^{-1}, if $—SO_3^-$ ions are present as anions, taking into consideration the hydrogen bridge acceptor property of the oxygen atoms of these ions. We first look at the cases in which the band of the OH stretching vibration of the hydrogen bridges of the molecules, which are attached directly to the cation, is shifted particularly far toward smaller wave numbers. This can be expected with the Be^{2+} and Al^{3+} salts of polystyrenesulfonic acid.

From Figs. 54 and 55 we obtain the next result.

Result 48: In the Al^{3+} salt of polystyrenesulfonic acid, an intense band appears at 3430 cm^{-1} with increasing degree of hydration. In the spectrum of the Be^{2+} salt a clearly marked shoulder is observed at about 3430 cm^{-1}, even when the membrane has been hydrated at 7% relative atmospheric humidity.

In the spectra of the Be^{2+} and Al^{3+} salts further anomalies are observed with increasing degree of hydration, which are discussed in Section IV.19.

From Fig. 56(a) we obtain:

Result 49: The band at about 3430 cm^{-1} is still found when the Al^{3+} salt of polystyrenesulfonic acid is hydrated with HDO.

* The shoulder on the side toward smaller wave numbers of the complex of the OD stretching vibration (Fig. 28) is caused, in the cases shown in Fig. 28, by the overtone of the scissor vibration of D_2O.

† See footnote p. 67.

[142] G. Zundel and A. Murr, *Z. Physik. Chem. (Leipzig)* **233**, 415 (1966).

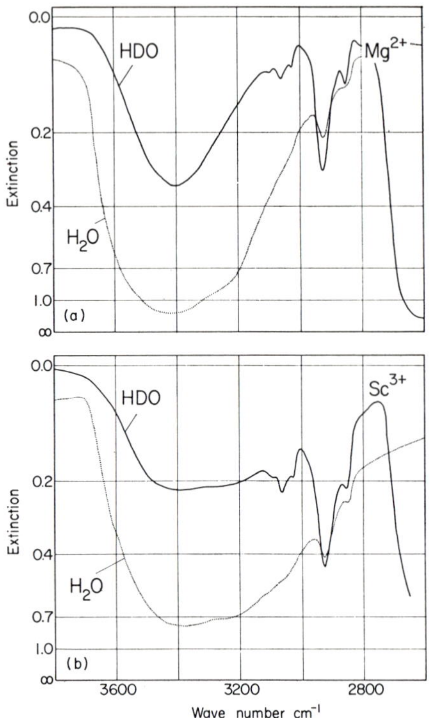

FIG. 52. Salts of polystyrenesulfonic acid, 5% cross-linked, 3 days of sulfonation. (a) Mg^{2+} salt, HDO hydrated: ——, atmospheric humidity over a saturated solution of $BaCl_2$ in H_2O–D_2O (mixture of about 7% H_2O and 93% D_2O); H_2O hydrated: ····, 98% relative atmospheric humidity. (b) Sc^{3+} salt, HDO hydrated: ——, atmospheric humidity over a saturated solution of LiCl in H_2O–D_2O (mixture of about 7% H_2O and 93% D_2O); H_2O hydrated: ····, 11% relative atmospheric humidity.

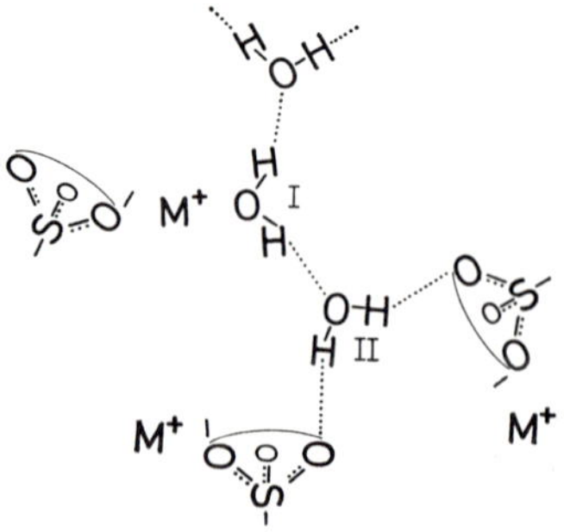

FIG. 53. The "second hydration shell."

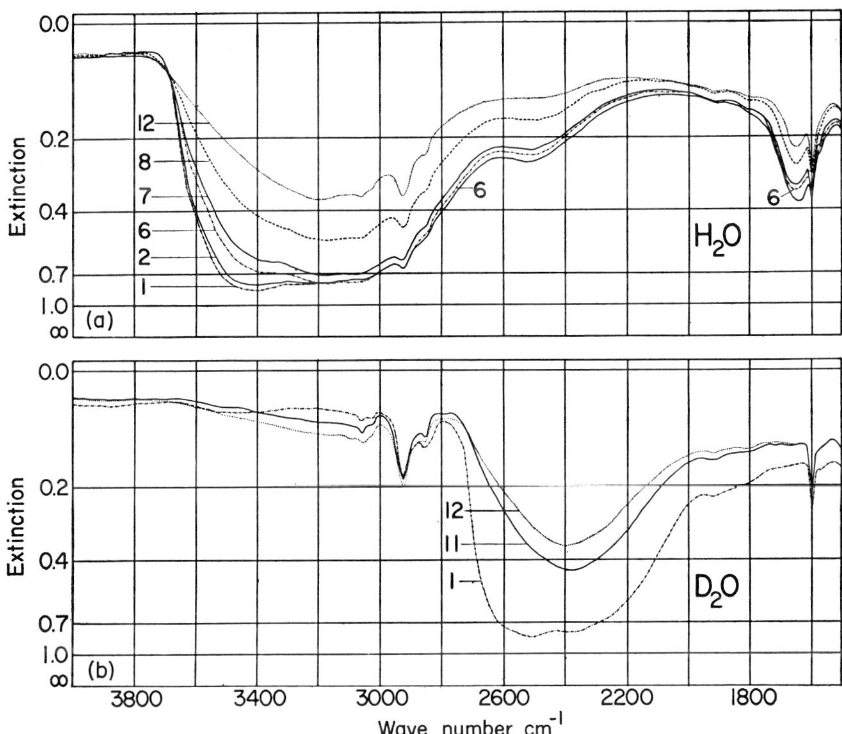

FIG. 54. Al^{3+} salt of polystyrenesulfonic acid, 5% cross-linked, 3 days of sulfonation. (a) H_2O hydrated: 1, 98% relative atmospheric humidity; 2 → 8, decreasing atmospheric humidity; 12, thoroughly dried membrane. (b) D_2O hydrated: 1, 100% relative atmospheric humidity; 11, incompletely dried, thoroughly dried membrane.

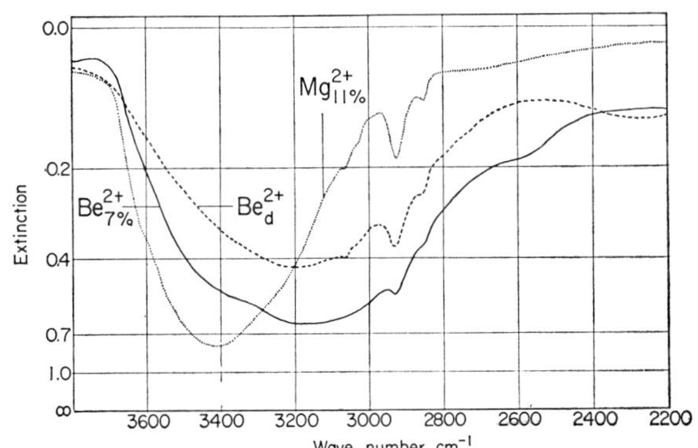

FIG. 55. Mg^{2+} and Be^{2+} salts of polystyrenesulfonic acid, 5% cross-linked, 3 days of sulfonation, H_2O hydrated: ····, Mg^{2+} salt, 11% relative atmospheric humidity; ——, Be^{2+} salt, 7% relative atmospheric humidity; – – – –, Be^{2+} salt, thoroughly dried membrane.

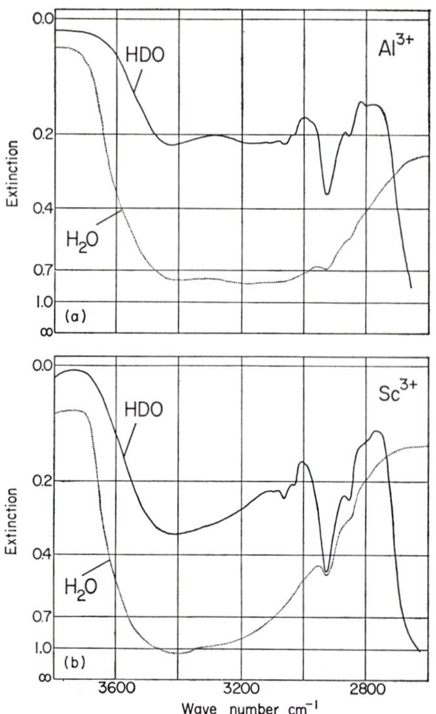

FIG. 56. Salts of polystyrenesulfonic acid, 5% cross-linked, 3 days of sulfonation. (a) Al^{3+} salt, HDO hydrated: ——, atmospheric humidity over a saturated solution of $MgCl_2$ in H_2O–D_2O (mixture of 7% H_2O and about 93% D_2O); H_2O hydrated: ···· 33% relative atmospheric humidity. (b) Sc^{3+} salt, HDO hydrated: ——, atmospheric humidity over a saturated solution of $BaCl_2$ in H_2O–D_2O (mixture of about 7% H_2O and 93% D_2O), H_2O hydrated: ····, 98% relative atmospheric humidity.

From these results it follows that the band, which in the present case arises at 3430 cm^{-1} with increasing degree of hydration, is that of the OH stretching vibration in the hydrogen bridges which bind the hydrate water molecules of "attachment type II" to the oxygen atoms of the $—SO_3^-$ ions. Hence we can follow the buildup of the "second hydration shell" with the help of this band.*

* We shall see in Section IV.19 that hydrolysis protons are also present in the Al^{3+} salt of polystyrenesulfonic acid. The band of the OH stretching vibration of the hydrogen bridges of the "external" water in the hydrate structures surrounding the hydrolysis protons also lies in the neighborhood of 3400 cm^{-1} (Section V.10). We shall, however, see in Section IV.19 that the number of hydrated hydrolysis protons is very small, so that these protons are of no consequence to this band.

The Hydration Water Molecules of "Attachment Type I"

We have seen, in the Sc^{3+} salt (Result 47), that the shoulder at about 3250 cm^{-1} cannot be ascribed solely to the overtone of the scissor vibration of H_2O. We can now understand this result. It follows that if the shoulder found on the side toward smaller wave numbers of the band of hydration water does not vanish on passing from H_2O to HDO hydration, it can be ascribed to the OH stretching vibration of the hydrogen bridge of the hydrate water molecule of "attachment type I."

This shoulder should now be more pronounced, the more strongly the cation acts on the hydrate water molecule of "attachment type I." This is because the hydrogen bridge donor property of the OH groups of the molecules of water of hydration is enhanced, making the hydrogen bridges stronger.

The prominence of this shoulder is, in fact, clearly dependent on the nature of the cation present. This is observed, e.g., in the series Sc^{3+}, Y^{3+}, La^{3+} (Fig. 31). The shoulder is more or less strongly pronounced in the salts of polystyrenesulfonic acid hydrated at 7% relative atmospheric humidity, depending on the cation [see, for example, Fig. 30(b), 31–33, and 37–39]. We have estimated the prominence of this shoulder in arbitrary units from 0 to 6 in Table 10, column 3, where 0 indicates that no shoulder at all can be seen, 1 indicates that the shoulder is very slight, and 6 that it is very pronounced. For comparison, in column 5 the already known E_{KH} values are given as a relative measure of the cation field at the hydrogen nuclei of the water molecules. Comparison of columns 3 and 5 yields:

Result 50: The shoulder is more pronounced, the stronger the cation field.

How much this shoulder is the result of the OH stretching vibration of the hydrate water molecule of "attachment type I" and how much the overtone of the scissor vibration remains to be shown by further investigation of membranes hydrated with HDO. From our results in the case of the Sc^{3+} salt (Result 47), however, it seems very probable that the OH stretching vibration of the hydrogen bridges of the hydrate water molecules of "attachment type I" is a principal cause of the progression in the prominence of the shoulder.

Do free OH groups occur?

From Figs. 30(b), 31–33, 37–39, and 47–49, we obtain:

Result 51: If any shoulder is to be found at about 3600 cm^{-1} at high degrees of hydration, it is very faint.

It follows that, even at high degrees of hydration, only a few free OH groups, i.e., groups not bound with hydrogen bridges, are present. The network of ions and molecules of water of hydration is also scarcely disintegrated at high degrees of hydration.

We have already mentioned in Section IV.6.1 that other authors in spectroscopic studies have frequently reached the conclusion that the shift of the

TABLE 10

PROMINENCE OF THE SHOULDER AT ABOUT 3250 CM^{-1} IN POLYSTYRENESULFONIC ACID SALTS (HIGHER DEGREE OF HYDRATION)

Ion	Figure number	Prominence of the shoulder at about 3250 cm^{-1} (arbitrary units, 0–6)	Ionic radius (Å) (after reference 4)	E_{KH} (ESU · cm^{-2} × 10^6)
1	2	3	4	5
Li^+	29	1	0.68	1.49
Na^+	29	1	0.97	1.08
K^+	29	1	1.33	0.76
Rb^+	29	1	1.47	0.67
Cs^+	—	1	1.67	0.57
Be^{2+}	55	6	0.35	4.46
Mg^{2+}	29	3	0.66	3.07
Ca^{2+}	29	2	0.99	2.11
Sr^{2+}	29	2	1.12	1.85
Ba^{2+}	29	1	1.34	1.51
Al^{3+}	30(b)	6	0.51	5.53
Ga^{3+}	30(b)	5	0.62	4.78
In^{3+}	30(b)	5	0.81	3.83
Tl^{3+}	30(b)	5[a]	0.95	3.32
Sc^{3+}	31	5	0.81	3.83
Y^{3+}	31	4	0.93	3.40
La^{3+}	31	3	1.14	2.73
Ce^{3+}	32	3	1.07	2.93
Gd^{3+}	32	4	0.94[b]	3.34
Zr^{4+}	33	5	0.79	5.29
Hf^{4+}	33	5	0.78	5.33
Mn^{2+}	37	3	0.80	2.59
Co^{2+}	37	4	0.72	2.87
Ni^{2+}	37	4	0.69	2.94
Cu^{2+}	38	6	0.72	2.87
Zn^{2+}	38	4	0.74	2.80
Fe^{3+}	39	6	0.64	4.71

[a] $Tl(OH)_2^+$ is present together with Tl^{3+} in the membrane (on this point see Section IV.20).

[b] Personal communication, R. C. Weast, editor-in-chief, "Handbook of Chemistry and Physics," Chemical Rubber Co., Cleveland, Ohio.

band of the OH stretching vibration under the influence of the cations is only insignificant. This seems at first to contradict our results, but is actually explained by the present findings. We have seen in Section IV.16 that the band of the OH stretching vibration of the hydration water molecules which lie immediately between cation and anion is less strongly shifted toward smaller wave numbers at higher degrees of hydration. In addition, as we have just seen, the bands of the OH stretching vibrations of hydrate water molecules of "attachment types I and II" are superimposed on this band at higher degrees of hydration. Other workers investigated, for the most part, relatively dilute solutions, which explains the apparent contradiction to our results.

In summary, we see, with increasing degrees of hydration, conditions which are shown in Fig. 53. The OH stretching vibration of the hydrogen bridges of the hydrate water molecule of "attachment type II" is found in the hydrated salts of polystyrenesulfonic acid at about 3430 cm^{-1}. If the hydrogen bridge donor property of the OH groups of the hydrate water molecules of "attachment type I" is sufficiently enhanced by interaction with the cation, the OH stretching vibration of the hydrogen bridges of these hydration water molecules is found as a shoulder on the side toward smaller wave numbers of the band of the OH stretching vibration of the hydrogen bridges of the water molecules of "attachment type II." Even at high degrees of hydration the network of the hydrate structures is but little disintegrated.

IV.19. Hydrolysis

With some salts, anomalies are found in the spectra. The salts in question have cations with particularly strong fields, namely Be^{2+} (Fig. 55); Al^{3+}, Ga^{3+}, and In^{3+} (Fig. 30) (for Al^{3+} see, in particular, Fig. 54); Zr^{4+} and Hf^{4+} (Fig. 33); and Fe^{3+} (Fig. 39). From these figures we obtain:

Result 52: A broad band is found at about 2400 cm^{-1}. This band is observed with all the above cations except In^{3+}, even when the membranes are hydrated at 7% relative atmospheric humidity.* In the case of In^{3+}, the band is only found at high degrees of hydration (spectrum not reproduced). This band vanishes on thorough drying, as is shown by comparing the spectra of the membranes hydrated at 7% relative atmospheric humidity with those of the thoroughly dried membranes.

If we observe the change of background in the spectra as a function of the degree of hydration, for example, in the Al^{3+} salt in Fig. 54, we see:

* Caution is necessary in deducing this result from the spectra since the interferences often have an appearance very similar to that of this band. However, if interferences are involved, not a single peak but a regularly spaced series of peaks is found, (see as an example Fig. 37). Further, the intensities of such interferences increase with progressive drying.

Result 53: Toward larger wave numbers from the bands of the OH stretching vibration the background is independent of the degree of hydration. However, in the region toward smaller wave numbers a slight decrease of background with increasing drying is observed. That is to say, in this region a weak continuous absorption is found, which vanishes with drying.

Interpretation of these results:

In Sections V.2 and V.3 we shall find the following results: bands at about 2950 and at 2405 cm^{-1} are to be ascribed to vibrations of the OH groups in the hydrogen bridges of the group

$$\begin{array}{c} \diagup OH\cdots O \diagdown \\ -S{=}O \qquad O{=}S- \\ \diagdown O\cdots HO \diagup \end{array}$$

. Sections V.6.2.1 and V.11 will also show us that a continuous absorption of the kind observed is closely connected with excess protons which tunnel in the network of the hydrogen bridges of their hydrate structures.

From this follows that, in the above cases there are also excess protons in the membrane. These are partly unhydrated, i.e., fixed in hydrogen bridges of the groups

$$\begin{array}{c} \diagup OH\cdots O \diagdown \\ -S{=}O \qquad O{=}S- \\ \diagdown O\cdots HO \diagup \end{array}$$

(Result 52), but are partly mobile in their hydrate structures through tunneling (Result 53). The fact that these excess protons vanish with progressive drying of the membranes affords us certainty that the excess protons do not arise through incomplete exchange of the required cations in the preparation of the membranes. Hence these excess protons can only arise by protolytic splitting of the water by the ions. The protolytic splitting of water molecules is the limiting case of the stretching, discussed in Section IV.9, of the OH bonds of the molecules of water of hydration as a result of the interaction of the water molecules with the ions.*

Result 52 indicates that some of these hydrolysis protons are unhydrated and fixed in hydrogen bridges. This is easily understandable, since in the very cases considered here, cations with a particularly strong tendency to attract water of hydration to themselves are involved. As a consequence, no further water molecules are left for the protons, and they are therefore fixed in the strong hydrogen bridges of

$$\begin{array}{c} \diagup OH\cdots O \diagdown \\ -S{=}O \qquad O{=}S- \\ \diagdown O\cdots HO \diagup \end{array}$$

groups.

* Roev and Terenin[144,145] also found a continuous absorption at the IR spectroscopic investigation of water, ethanol, and methanol adsorbed on chromium oxide. It appears probable that this is caused by tunneling protons and defect protons, which have arisen under these conditions through protolytic splitting of water molecules.

[143] L. G. Sillen and A. E. Martell, *Chem. Soc. (London), Spec. Publ.* **17** (1964).

[144] L. M. Roev and A. N. Terenin, *Proc. Acad. Sci. USSR Phys. Chem. Sect. (English Transl.)* **124,** 77 (1959).

[145] L. M. Roev, *Dokl. Akad. Nauk SSSR* **133,** 561 (1960).

We must now, however, put the following questions: In the first place, where is the second band of the vibrations of the OH groups of the

$$\begin{array}{l} \quad\;\; \diagup OH\cdots O \diagdown \\ -S{=}O \qquad\quad O{=}S- \\ \quad\;\; \diagdown O\cdots HO \diagup \end{array}$$

, i.e., the band at about 2950 cm^{-1}? Second, where is the band of the OH stretching vibration of the hydrogen bridges of the "external" water in the hydrate structures which are formed around the tunneling hydrolysis proton (see Section V.10)? Third, where is the band of the OH^- produced by hydrolysis?* All these bands are weak. They are therefore lost in the broad complex of the OH stretching vibrations of water of hydration. This band complex actually has a structure, which can be clearly seen in the spectrum of the Al^{3+} salt hydrated with D_2O at 100% atmospheric humidity in Fig. 54. In Section V.1 we shall see that the proton, when it approaches an $—SO_3^-$ ion, rearranges the latter in such a way that two SO double bonds and one SO single bond are produced (Section V.1). We must hence ask ourselves: where are the bands of these groups? They should be found at 1350 and 907 cm^{-1}. These bands are actually observed in the spectra of these salts also, but they are very weak.

In summary, we see that the protolytic splitting of the water molecules found in the Be^{2+}, Al^{3+}, Ga^{3+}, In^{3+}, Zr^{4+}, Hf^{4+}, and Fe^{3+} salts of polystyrenesulfonic acid arises as a limiting case of the stretching of the OH bonds of the water molecules by the ions, discussed in Section IV.9. The hydrolysis protons are either fixed in the hydrogen bridges of

$$\begin{array}{l} \quad\;\; \diagup OH\cdots O \diagdown \\ -S{=}O \qquad\quad O{=}S- \\ \quad\;\; \diagdown O\cdots HO \diagup \end{array}$$

groups or are tunneling in the hydrogen bridges of their hydrate structures.

IV.20. Anomaly of the Tl Form of the Exchanger

From Figs. 12, 30, and 40(c) we obtain:

Result 54: If the membrane of the Tl form is hydrated at 7% relative atmospheric humidity, bands of stretching vibrations of OH groups are found at 3400 and 3200 cm^{-1}.

* The position at which such a band is to be expected when the OH^- ion is not hydrated is obtained from studies of hydroxides.[63,71,120–122,146–152d] We shall see further in Section V.11 that a continuous absorption is also observed in the presence of hydrated OH^- ions.

[146] D. Krishnamurti, *Proc. Indian Acad. Sci. Sect. A* **50**, 223 (1959).
[147] H. A. Benesi, *J. Chem. Phys.* **30**, 852 (1959).
[148] R. M. Hexter, *J. Chem. Phys.* **34**, 941 (1961).
[149] J. K. Wilmshurst, *J. Chem. Phys.* **35**, 1800 (1961).
[150] W. Vedder and R. S. McDonald, *J. Chem. Phys.* **38**, 1583 (1963).
[151] A. Cornelis-Benoit, *Spectrochim. Acta* **21**, 623 (1965).
[152] E. Schwarzmann, O. Glemser, and H. Marsmann, *Naturwissenschaften* **52**, 344 (1965).
[152a] V. M. Bhatnagar, *Experientia* **23**, 10 (1967).
[152b] V. M. Bhatnagar, *Experientia* **23**, 697 (1967).
[152c] E. Schwarzmann and L. Lange, *Z. Naturforsch.* **23b**, 874 (1968).
[152d] E. Schwarzmann and H. Sparr, *Z. Naturforsch.* **23b**, 767 (1968).

On thorough drying, the band at 3200 cm^{-1} remains [Fig. 30(a)]. In the other salts of polystyrenesulfonic acid, if a band of an OH stretching vibration is observed at all, the band of an H_2O scissor vibration is also observed. This, however, is not the case with the Tl form. The band at 3200 cm^{-1} is still found after thorough drying, but no band of an H_2O scissor vibration is encountered [Fig. 40(c)]. From this it follows, that in the Tl form OH groups are still present in the membrane when all H_2O molecules have been removed.

Result 55: As in all cases in which hydrolysis occurs, a continuous absorption, though very weak, indicating the presence of tunneling protons (Section V.11) is still found with the hydrated Tl form. However, even after thorough drying strongly marked bands still sometimes occur at 1350 and 907 cm^{-1} (Fig. 12); these indicate the presence of $—S(=O)(=O)—OH$ groups (Section V.1)

Result 56: These anomalies depend on the concentration of the aqueous $TlCl_3$ solution from which the cations have been exchanged into the membrane. If the exchange is carried out with a 2 *N* instead of a 0.1 *N* aqueous solution of $TlCl_3$, the bands at 3200, 1350, and 907 cm^{-1} are not as intense. If we carry out the exchange starting from the Na^+ salt instead of from polystyrenesulfonic acid itself, no essential differences appear.

Result 57: These observations are not always reproducible; presumably they depend on the speed with which the membrane is dried. However, this has not been studied in detail.

In addition, these membranes, unlike all others investigated by us, become cloudy with time.

Published results supplement our findings as follows:

The logarithm of the acidic constant[143] $^*K_1$ is -1.14 in the case of Tl^{3+}. This means that Tl(III) salts are strongly hydrolyzed in aqueous solution (see below). Hydrated H^+ ions are then present as well as Tl^{3+}.

In interpreting the kinetics of the redox reaction

$$Tl^{3+} + Fe^{2+} \rightleftharpoons Tl^{2+} + Fe^{3+}$$

(footnote 153, p. 323), the assumption was made that $TlOH^{2+}$ is present in the solution as a further positively charged ion, as well as a product of hydrolysis.

We shall see in the following that the d^{10} configuration of Tl^{3+} easily passes into the d^9s configuration, and Orgel has already expressed the conjecture (footnote 101 and footnote 154) that Tl^{3+} readily forms linear complexes for this reason. Hence it appears very probable that $Tl(OH)_2^+$ ions are also present in the solution. This is confirmed by the logarithm of the acidic constant, for (according to footnote 143) the logarithm of the second acidic constant in the case of Tl^{3+} is $^*K_2 = -1.49$, i.e., not much less than $^*K_1$.

We now take as the starting point for the explanation of our results the presence of the positive cations Tl^{3+}, $TlOH^{2+}$, $Tl(OH)_2^+$, and H^+ in the $TlCl_3$ solution. If one membrane is placed in such a solution, we have a very complicated system. How many individual cations are in the membrane depends on the following factors; first, on the $TlCl_3$ concentration of the external solution, as this determines the relative number of these kinds of cations in the external solution; second, on the selectivities and specifities of these ions; third, on how the Tl^{3+} ions behave with regard to hydrolysis in the membrane.

[153] F. Basolo and R. G. Pearson, "Mechanisms of Inorganic Reactions." Wiley, New York, 1958.

[154] L. E. Orgel, *J. Chem. Soc.* (*London*) p. 4186 (1958).

Our former results can now be explained: Result 54, because the band at 3200 cm^{-1} arises in this case from the OH stretching vibration of the $TlOH^{2+}$ or $Tl(OH)_2^+$ ions, or possibly also from that of $Tl(OH)_3$; Result 55 (the occurrence of the bands at 1350 and 907 cm^{-1}), because the protons exchanged attach themselves to $—SO_3^-$ ions on drying, causing rearrangement of the latter (Section V.1); Result 56, because the dependence of the observed phenomena on the concentration of the $TlCl_3$ solution is now directly understandable; Result 57, because our assumption that the spectra also depend on the speed with which the membranes are dried now becomes directly understandable, the protolytic splitting of water molecules in the membrane being naturally dependent on the humidity.

$$Tl^{3+} + H_2O \rightleftharpoons TlOH^{2+} + H^+$$

The speed of attainment of this equilibrium and the speed of drying thus compete with one another.

In summary, we see that the anomalies which we have observed in investigating the Tl form are explained by the presence of H^+ and of $TlOH^{2+}$ or $Tl(OH)_2^+$ ions in the aqueous medium and hence are evidence for the presence of these cations.

Why is the hydrolysis constant of Tl^{3+} salt solutions so large?

The logarithm of the acidic constant $*K_1$ of Tl(III) salts in an aqueous medium is -1.14 and is thus much larger than in the presence of the other cations which we have considered. In the aqueous $AlCl_3$ solution, for example, $*K_1$ is -4.29, after Sillen and Martell.[143]

Tl^{3+} has a radius of 0.95 Å. According to our discussion in Section IV. 9 and IV.19, the Tl^{3+} cannot split water molecules protolytically under any circumstances, if its electron shells have the inert gas structure. Either our thesis, in Section IV.19, that the protolytic splitting of the water molecules occurs as a limiting case of the stretching of the bonds was incorrect, or else some special mechanism must be effective here.

The energy which transforms the d^{10} configuration into the d^9s configuration is only 9.3 eV for Tl^{3+} (see footnote 101, p. 67). Hence, in the electron configuration of the Tl^{3+} ion, there is certainly an admixture of the d^9s configuration; that is, no inert gas ion is present but a much more reactive ion. Thus this anomalous behavior in the interaction of Tl^{3+} with water is explained.

CHAPTER V

THE ACIDS

We have investigated polystyrenesulfonic acid [Fig. 1(c, d)], polystyreneselenonic acid [Fig. 4(a, b)], polystyrenethiophosphonic acid [Fig. 7(a, b)], polystyreneseleninic acid [Fig. 5(a, b)], and polystyrenephosphinic acid [Fig. 6(a, b)].

V.1. The Bands of the Anions

V.1.1 Bonding Electron Rearrangement by the Proton

The bands of the acid groups and the ions are collected in Table 11. These results[1] have been taken from the figures specified in columns 2 and 5 of this table and from Tables A.1–A.5.

If we compare the bands of the thoroughly dried acids with those of the corresponding salts, i.e., column 4 with 6 (Table 11) we obtain:

Result 58: If the acids are thoroughly dried, bonds with the character of double or single bonds arise in place of the XO bonds which—in the salts—are almost identical.

Salts:

$$-S(\cdots O)_3^{-},\quad -Se(\cdots O)_3^{-},\quad -P(\cdots S)(\cdots O)_2^{2-},\quad -Se(\cdots O)_2^{-},\quad -P(H)(\cdots O)_2^{-}$$

Acids (thoroughly dried):

$$-S(=O)_2(OH),\quad -Se(=O)_2(OH),\quad -P(SH)(=O)(OH),\quad -Se(=O)(OH),\quad -P(H)(=O)(OH)$$

If the proton approaches its anion, it rearranges the bonding electrons of the

[1] G. Zundel, *Z. Naturforsch.* **22a**, 199 (1967).

TABLE 11

BANDS OF ACID GROUPS AND ANIONS

Compound	Acid			Na$^+$-Salt	
	See figure	Bands at high degree of hydration	Bands after thorough drying	See figure	Bands
1	2	3	4	5	6
Polystyrene-sulfonic acid and its salt, respectively	56A(a), 57, and 1(c, d)	at about 1200 cm^{-1} antisymmetric stretching vibration (doublet caused by removal of degeneracy) 1034 cm^{-1} symmetric stretching vibration of the $\left(-S(O)(O)O^-\right)$ ion	1350[a] and 1172 cm^{-1} antisymmetric and symmetric stretching vibrations of the double bonds 907 cm^{-1}[b] stretching vibration of the single bond of the $-S(=O)(=O)OH$ group	56A(a), 12(a), and 1(e, f)	at about 1200 cm^{-1} antisymmetric stretching vibration (doublet caused by removal of degeneracy) 1040 cm^{-1} symmetric stretching vibration of the $\left(-S(O)(O)O^-\right)$ ion
Polystyrene-selenonic acid and its salt, respectively	56A(b), 58, and 4(a, b)	879 cm^{-1} antisymmetric stretching vibration 847 cm^{-1} symmetric stretching vibration of the $\left(-Se(O)(O)O^-\right)$ ion also nearby—but only weakly—the bands observed on thorough drying	965 and 911 cm^{-1} antisymmetric and symmetric stretching vibrations of the double bonds 725 cm^{-1} stretching vibration of the single bond of the $-Se(=O)(=O)OH$ group	56A(b) and 4(c, d)	909 cm^{-1} antisymmetric stretching vibration 860 cm^{-1} symmetric stretching vibration of the $\left(-Se(O)(O)O^-\right)$ ion
Polystyrene-thiophosphonic acid and its salt, respectively	56A(c), and 7(a,b)[c]	1170 cm^{-1} stretching vibration of the PO double bond (for details on the band of the stretching vibration of the PO single bond, see p. 26) of the $-P(=O)(SH)OH$ group independent of the degree of hydration		56A(c), 21, and 7(c, d)	1218 cm^{-1} antisymmetric 1037 cm^{-1} symmetric PO stretching vibration of the $\left(-P(S)(O)O\right)^{2-}$ ion
Polystyrene-seleninic acid and its salt, respectively	56A(d), 59, and 5(a, b)	839 cm^{-1} stretching vibration of the SeO double bond is slightly shifted toward larger wave numbers on thorough drying 658 cm^{-1} stretching vibration of the SeO single bond of the $-Se(=O)OH$ group independent of the degree of hydration		56A(d), 19, and 5(c, d)	803 cm^{-1}[d] symmetric stretching vibration 776 cm^{-1}[d] antisymmetric stretching vibration of the $\left(-Se(O)O^-\right)$ ion
Polystyrene-phosphinic acid and its salt, respectively	56A(e), and 6(a, b)[c]	1170 cm^{-1} stretching vibration of the PO double bond 965 cm^{-1} stretching vibration of the PO single bond of the $-P(=O)(H)OH$ group independent of the degree of hydration		56A(e), and 6(c, d)	1183 cm^{-1} antisymmetric 1056 cm^{-1} symmetric PO stretching vibration of the $\left(-P(H)(O)O\right)^-$ ion

[a] In the case of the $-S(=O)(=O)OD$ group, this band lies at 1340 cm^{-1}.

[b] The band position given here is that at 25°C. With reduction of temperature, this band is slightly shifted toward smaller wave numbers, with a temperature increase toward larger wave numbers.

[c] In Figs. 6 and 7, (a) and (b), in each case, only the spectrum of the membrane at a high degree of hydration is given, since the spectra in the region of the bands of the stretching vibrations of the anions are not dependent on the degree of hydration.

[d] Measured at a high degree of hydration with D_2O (see Section IV.4).

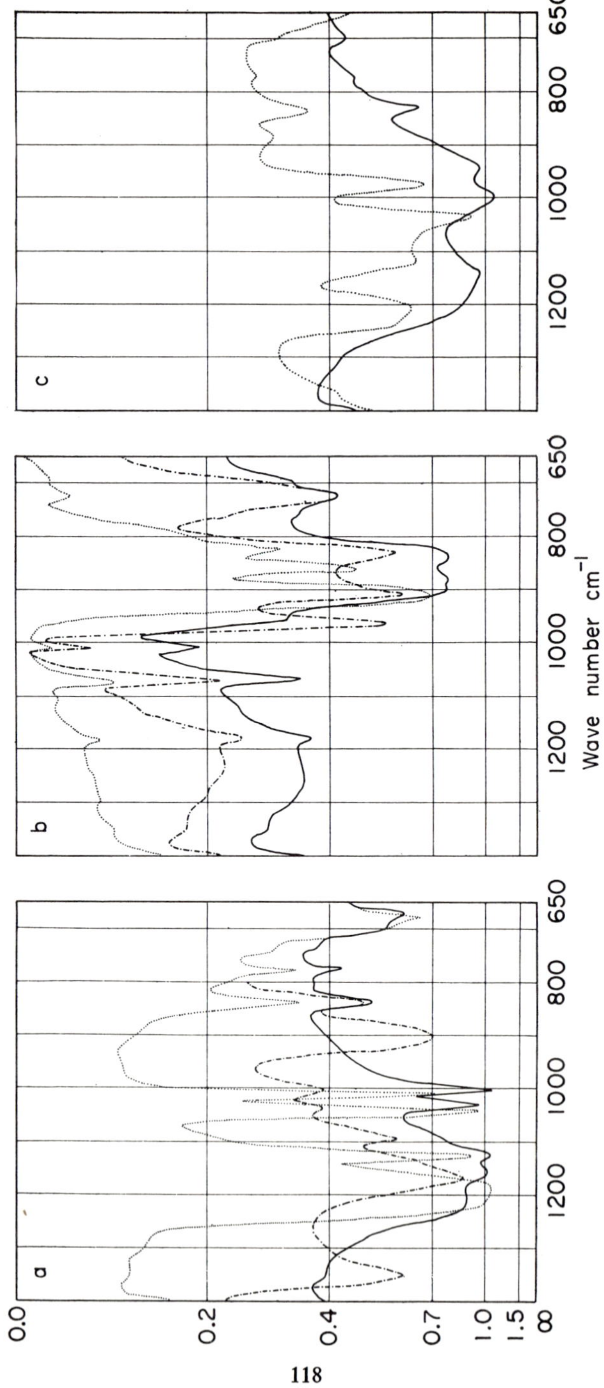
a
b
c
0.0
0.2
0.4
0.7
1.0
1.5
∞
1200
1000
800
650
Wave number cm⁻¹

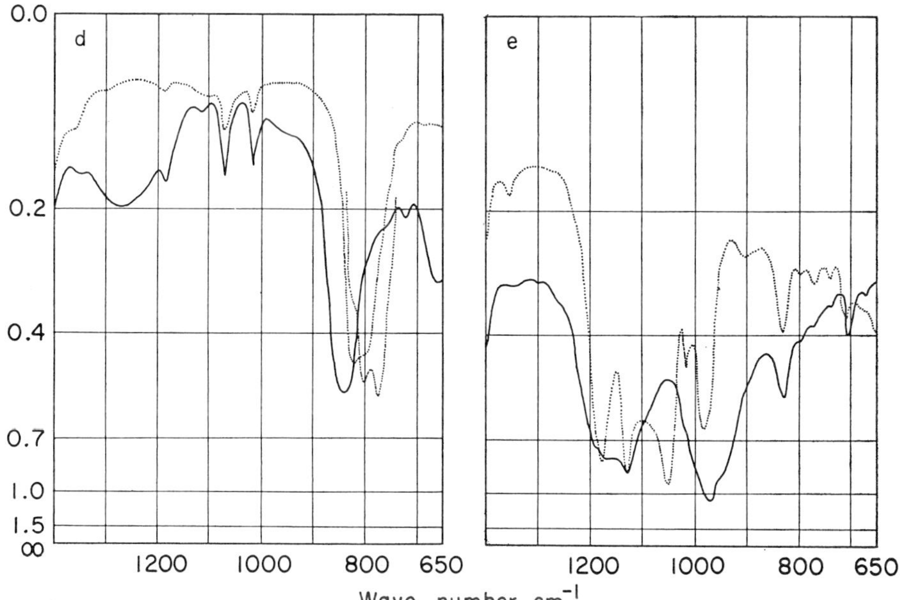

FIG. 56A. Spectra: (a) Polystyrenesulfonic acid and Na^+ salt; (b) Polystyreneselenonic acid and Na^+ salt; (c) Polystyrenethiophosphonic acid and Na^+ salt; (d) Polystyreneseleninic acid and Na^+ salt; (e) Polystyrenephosphinic acid and Na^+ salt; · · · · ·, the spectrum of the Na^+ salt [In (d) the curve with a doublet structure is the spectrum of the D_2O-hydrated membrane. The other is the spectrum of the thoroughly dried membrane]; –·–·–, spectrum of the thoroughly dried acid; ——, spectrum of the acid hydrated at 98% relative atmospheric humidity. [In (c), (d), and (e), only one spectrum of the acid is given, since the bands in this region are independent of hydration.]

bonds in these anions. Hence it considerably reduces the mesomerism of its anions (Section IV.1).

In polystyrenephosphinic acid, the PH bond is also somewhat affected by this rearrangement because the PH stretching vibration is shifted from 2304 cm^{-1} (salt) to 2365 cm^{-1} (acid), i.e., the PH bond becomes a little stronger. A more or less pronounced mesomeric bond resonance between PO and PS bonds was also found by Kabaznik *et al.*[2] (also see footnote 3) in IR investigations of salts with the anion $R_2P\langle{}^{S}_{O}\rangle^{-}$.

[2] M. I. Kabaznik, T. A. Mastrjukowa, E. I. Matrosow, and B. Fischer, *Zh. Strukt. Khim.* **6**, 691 (1965).

[3] T. Gramstad, *Spectrochim. Acta* **19**, 829 (1963).

V.1.2. Removal of the Proton from the Anion

The integral extinction of the IR bands of the anions which appear or disappear as a result of this rearrangement is thus a measure of the number of the protons removed from the anion. The degree of dissociation determined in this way we term the true degree of dissociation α_t (on this point see Kortuem, footnote 4, pp. 137 and 212). The dependence on the degree of hydration is obtained from Figs. 57–59, or from comparison of columns 3 and 4 in Table 11:

Result 59: Polystyrenesulfonic acid—the $-S(=O)(=O)OH$ groups present on thorough drying are transformed with increasing degree of hydration into

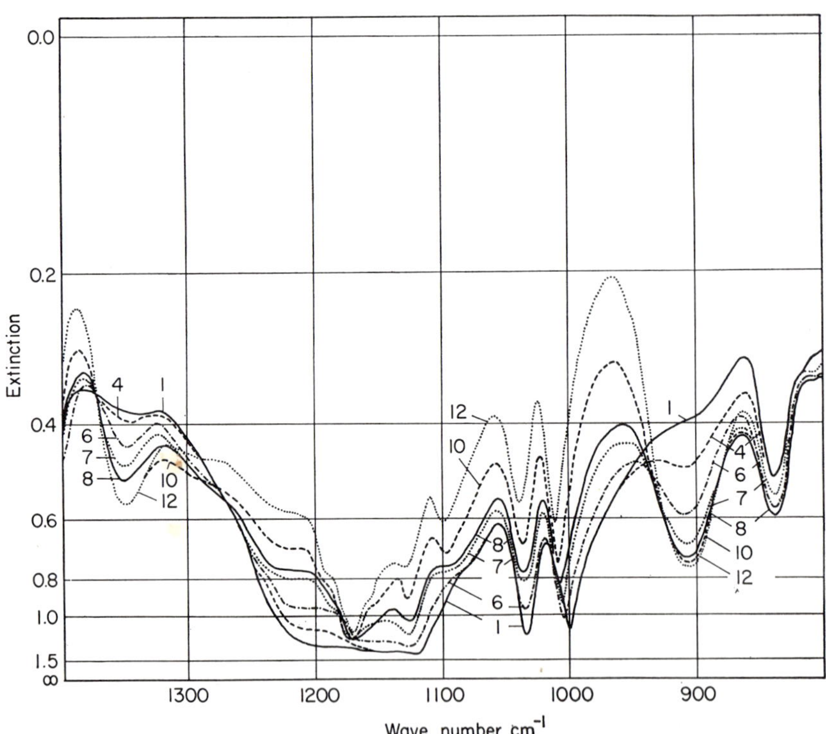

Fig. 57. Polystyrenesulfonic acid, 1% cross-linked, 4 days of sulfonation. H_2O hydrated: measured after 3 days at each humidity, 1, 98%; 4, 33%; 6, 11%; 7, 5%; 8, 3% relative atmospheric humidity; 10, incompletely dried; 12, thoroughly dried membrane.

[4] G. Kortuem, "Lehrbuch der Elektrochemie," 3rd Ed. Verlag Chemie, Weinheim, 1962.

$\left(—S\overset{O^-}{\underset{O}{\cdots O}}\right)$ ions with almost* identical SO bonds. At high degrees of hydration, almost all protons have been removed from the anions, and only approach them on increasingly thorough drying. The true degree of dissociation is nearly one, i.e., very large.

As appears from the literature,[5,6] the apparent degree of dissociation, as determined by classical electrochemical methods, is considerably smaller and, in addition, is dependent on the method of determination. This is understandable, for, on closer consideration of these methods, it is seen that the apparent degree of dissociation is a very complex quantity in view of molecular processes. Kortuem (see footnote 4, p. 137) emphasized this particularly. The same differences were found by Clarke and Woodward[7] in a Raman-spectroscopic investigation of methanesulfonic acid. The differences between the true and apparent degrees of dissociation are discussed in detail (in footnote 8).

In polystyreneselenonic acid, the $—Se\overset{O}{\underset{OH}{=O}}$ groups present on thorough drying are transformed with increasing degree of hydration into $\left(—Se\overset{O^-}{\underset{O}{\cdots O}}\right)$ ions with almost identical SeO bonds. The spectrum of the membrane hydrated at 98% relative atmospheric humidity—in particular the band of the SeO single bond at 725 cm^{-1}—shows, however, that a considerable number of $—Se\overset{O}{\underset{OH}{=O}}$ groups is still present under these conditions. Some protons are still present at the anion, even at relatively high degrees of hydration. The true degree of dissociation is hence smaller than in polystyrenesulfonic acid, but is still quite large.

In polystyrenethiophosphonic,† -seleninic, and -phosphinic acids, the bands of the PO or SeO stretching vibrations do not depend on the degree of hydration. Thus, $—P\overset{SH}{\underset{OH}{=O}}$, $—Se\overset{O}{\underset{OH}{}}$, and $—P\overset{H}{\underset{OH}{=O}}$ groups are still present at a high

* See Result 99, p. 172.

† Some $—P\overset{S}{\underset{OH}{—OH}}$ groups are also present in tautomeric equilibrium (see Section V.14).

[5] R. A. Mock, C. A. Marshall, and T. E. Slykhouse, *J. Phys. Chem.* **58**, 498 (1954).

[6] R. A. Mock and C. A. Marshall, *J. Polymer Sci.* **13**, 263 (1954).

[7] J. H. R. Clarke and L. A. Woodward, *Trans. Faraday Soc.* **62**, 2226 (1966).

[8] G. Zundel, "Die Hydratation der Ionen." Habilitationsschrift, University of Munich, 1966.

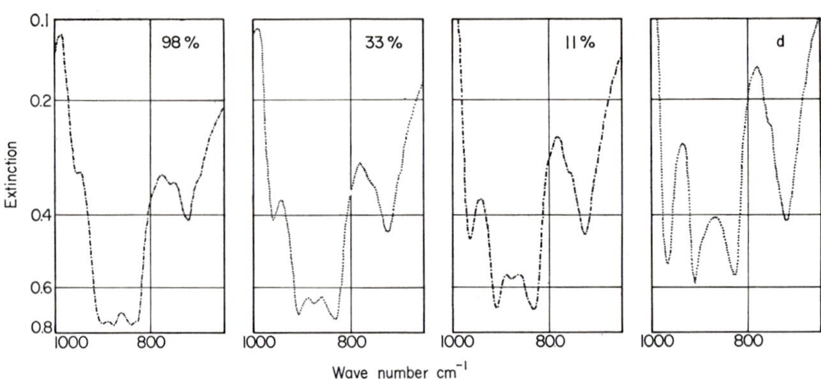

FIG. 58. Polystyreneselenonic acid, 7% cross-linked, 3 days of selenonation. H_2O hydrated: 98, 33, and 11% relative atmospheric humidity, thoroughly dried membrane (d).

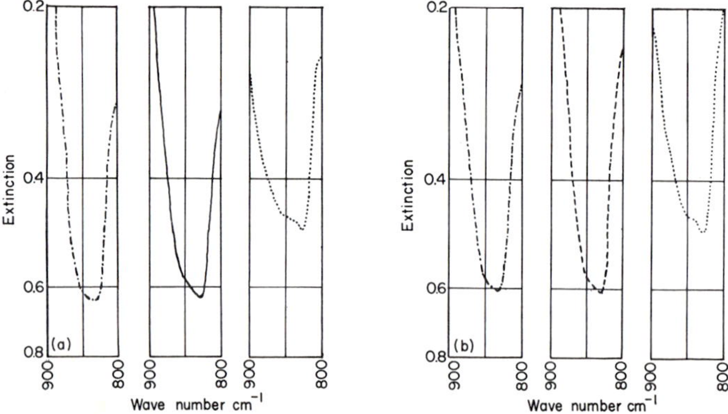

FIG. 59. Polystyreneseleninic acid, 7% cross-linked, 3 days of selenonation, region 900–800 cm^{-1}. (a) H_2O hydrated: on the left, 98%; in the middle, 22% relative atmospheric humidity; and on the right, thoroughly dried membrane. (b) D_2O hydrated: on the left, atmospheric humidity over a saturated solution of $BaCl_2$ in D_2O; in the middle, intermediate atmospheric humidity; on the right, a thoroughly dried membrane.

degree of hydration. In polystyreneseleninic and -phosphinic acids, nearly all the protons are bound to their anions, not only after thorough drying, but also at a high degree of hydration. The degree of true dissociation is very small. Polystyrenethiophosphonic acid is diprotic. The behavior is in this case somewhat more complicated. A full discussion is therefore postponed to Section V.14.

Result 60: The true degree of dissociation accordingly shows a characteristic course, dependent on the nature of the anion. It decreases in passing from polystyrenesulfonic acid to polystyrenephosphinic acid.

This course is closely related to the mesomeric bond resonance in the anions, which arises on removal of the proton. This is discussed in detail in Section V.16.

V.1.3. Other Changes Dependent on the Degree of Hydration

In polystyrenesulfonic acid, bands are found at 1128, 1097, and 1011 cm^{-1}. The bands at 1128 and 1097 cm^{-1} are in-plane skeleton vibrations of the benzene ring, with strong participation of the substituents. That at 1011 cm^{-1} is an in-plane bending vibration of the $\gtrdot$CH groups of the benzene ring.

Concerning the bands at 1128 and 1097 cm^{-1}, if the continuous and dotted curves in Fig. 1(c) are compared, we see:

Result 61: The band at 1128 cm^{-1} is observed at a high degree of hydration and vanishes with increasing drying. Conversely, the band at 1097 cm^{-1} only arises with increasing drying.* Accordingly, the band at 1128 cm^{-1} is that of the in-plane skeleton vibration of the benzene ring when the substituent has the structure $-S(\cdots O)_3^-$ (resonant $-SO_3^-$), and that at 1097 cm^{-1}† when it has the structure $-S(=O)_2OH$.

Since the substituent takes part in these vibrations,[9] this result is understandable. This change is probably caused by displacement of the electrons in the benzene ring also.

Concerning the band at 1011 cm^{-1}, we obtain from Fig. 2:

Result 62: This band is found at 1011 cm^{-1} after thorough drying of the membrane, i.e., when $-S(=O)_2OH$ groups are present. With increasing degree of hydration, i.e., when the $-S(=O)_2OH$ is transformed into $-S(\cdots O)_3^-$, it is shifted to 1001 cm^{-1}. This shift goes hand in hand with the appearance of the band

* This is particularly clearly apparent when dealing with membranes which have been rapidly dried (see Section V.8).

† When $-S(=O)_2Cl$ groups are present, a corresponding band is observed at 1081 cm^{-1} [Fig. 1(b)].

[9] G. Kresze, E. Ropte, and B. Schrader, *Spectrochim. Acta* **21**, 1633 (1965).

of the symmetric stretching vibration $\bar{\nu}_s$ of the $\left.-\mathrm{S}\begin{matrix}\cdot\cdot\mathrm{O}^{-}\\ \cdots\mathrm{O}\\ \cdot\cdot\mathrm{O}\end{matrix}\right)$ ion which lies at 1034 cm^{-1}.

The dependence of this band shift on the degree of hydration can be understood, if it is assumed that this vibration of the benzene ring and the symmetric SO stretching vibration $\bar{\nu}_s$ of the $\left.-\mathrm{S}\begin{matrix}\cdot\cdot\mathrm{O}^{-}\\ \cdots\mathrm{O}\\ \cdot\cdot\mathrm{O}\end{matrix}\right)$ ion are coupled. Such coupling shifts the band in the observed manner. This coupling appears plausible, considering the symmetry of the two vibrations. The following observations favor this coupling as the cause of the observed band shift: First, the fact that the band of the symmetric stretching vibration of the $\left.-\mathrm{S}\begin{matrix}\cdot\cdot\mathrm{O}^{-}\\ \cdots\mathrm{O}\\ \cdot\cdot\mathrm{O}\end{matrix}\right)$ ion is found at 1016 cm^{-1} in perdeuterated polystyrenesulfonic acid [Fig. 1(h, i)], as in this case it cannot be coupled with the in-plane bending vibration of the $\rangle$CH groups. Second, the fact that for polystyreneselenonic acid this band is found at 1011 cm^{-1}, independent of the degree of hydration. Third, comparison of the bands of *p*- and *o*-toluenesulfonic acid shows the same effect [Fig. 3(a, b)]. Neither in the polystyreneseleninic acid nor in the *o*-toluenesulfonic acid is a coupling of the in-plane bending with the SO stretching vibration possible. In both these cases, the in-plane bending vibration is therefore found at a larger wave number.

The result, that a vibration of a substituent is coupled with a vibration of the benzene ring, is not unusual. Mecke and Rossmy [10] have reported similar coupling phenomena with phenols (OH bending vibrations with an in-plane bending vibration of the $\rangle$CH groups of the benzene ring).

In summary, we conclude that the structural transformation of the ions under the influence of the proton can lead to considerable changes of intensity, as well as to shifts of the bands of the benzene ring. Consequently, changes do occur in the bands of the benzene ring which are dependent on the degree of hydration.

V.2. Association of Acid Groups

In Table 12 the bands which give information on the association of the acid groups are collected together.[11]

[10] R. Mecke and G. Rossmy, *Z. Elektrochem.* **59**, 866 (1955).
[11] G. Zundel, H. Metzger, and I. Scheuing, *Z. Naturforsch.* **22b**, 127 (1967).

BANDS OF THE OH OR OD GROUPS IN ACID GROUP HYDROGEN BRIDGES

Acid	Group	See figure	Bands of the OH acid groups in the hydrogen bridges		Bands of the OD acid groups in the hydrogen bridges		Change with degree of hydration	See figure	Band of the OH bending vibration (cm^{-1})	Band of the OD bending vibration (cm^{-1})
1	2	3	4	5	6	7	8	9	10	11
Polystyrene-sulfonic acid	O···HO —S=O O=S— OH···O	60	2950 intense broad	2405	2240 intense broad	1805 weak	not present at high degree of hydration increases on progressive drying	1(c, d)	masked	masked
Polystyrene-selenonic acid	O····HO —Se=O O=Se— OH····O	113	2880 intense broad	2385	2160 intense broad	masked	weak at high degree of hydration increases on progressive drying	4(a, b)	1260 broad	masked
Polystyrene-thiophosphonic acid	SH —P=O····HO OH····O=P— HS	120	2855 intense broad	2290 weak broad	2155 intense broad	masked	independent of degree of hydration	(7a, b)	masked	masked
Polystyrene-seleninic acid	O····HO —Se Se— OH····O	115	2960 intense broad	2420	2195 intense broad	masked	independent of degree of hydration	5(a, b)	1260 broad	shoulder at 940 cm^{-1}
Polystyrene-phosphinic acid	H —P=O····HO OH····O=P— H	116	2750 intense broad	2180 weak broad	2060 intense broad	masked	independent of degree of hydration	6(a, b)	masked	masked

V.2.1. Polystyrenesulfonic and -selenonic Acids

From Figs. 60* and 113 (Section V.11.5.4), spectra 12, and Table 12, columns 4–7, we see:

Result 63: In the spectra of thoroughly dried membranes of polystyrenesulfonic acid we find a broad and intense band at about 2950 cm^{-1} and a weaker band at 2405 cm^{-1}; corresponding bands are found for polystyreneselenonic acid at about 2880 and at 2385 cm^{-1}.

Result 64: If, before drying, the membranes were hydrated with D_2O, these bands are found for polystyrenesulfonic acid at about 2240 and at 1805

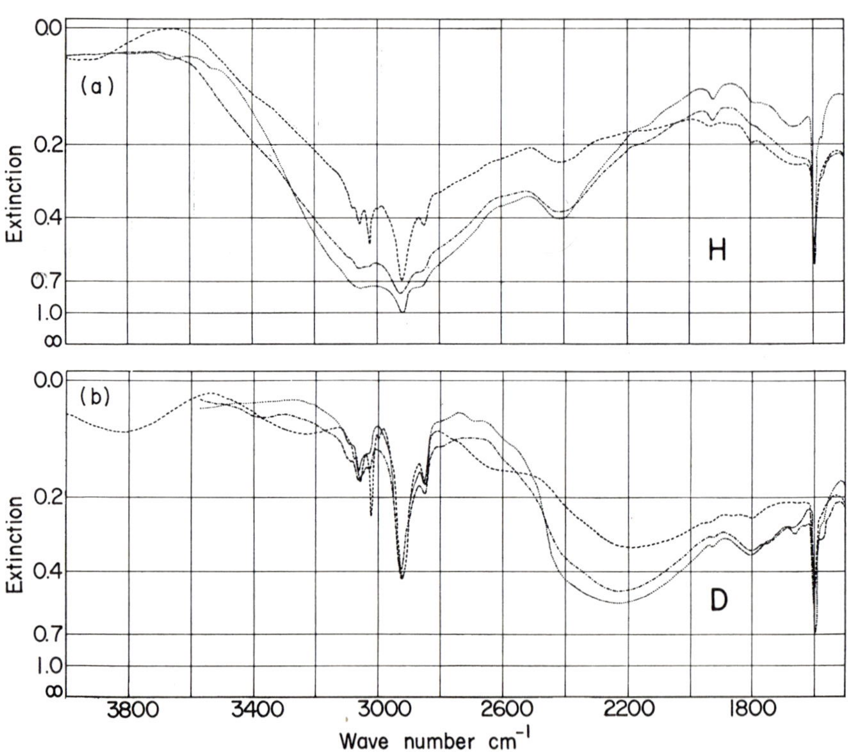

Fig. 60. Polystyrenesulfonic acid. All membranes immediately thoroughly dried from a high degree of hydration (two membranes, one over the other). ----, 10% cross-linked, 0.48 acid group per benzene ring; -·-·, 1% cross-linked, 1.06 acid groups per benzene ring; ····, 1% cross-linked, 1.50 acid groups per benzene ring. (a) Membrane was hydrated with H_2O before drying. (b) Membrane was hydrated with D_2O before drying.

* To obtain more intense bands, we have in this case examined two superimposed membranes of the usual thickness.

cm^{-1}. For polystyreneselenonic acid the first band is found at about 2160 cm^{-1}, but the other is masked.

In the case of polystyrenesulfonic acid, from Figs. 70 and 71 (Section V.4), and in the case of polystyreneselenonic acid, from Fig. 113 (Section V.12), we obtain the following:

Result 65: The band with the smaller wave number in both cases becomes stronger with progressive drying.

This process unfortunately cannot be observed in the case of the band with the larger wave number. On this band is superimposed a band of water of hydration. As we shall see in Section V.8–V.10, the intensity of this band increases with increasing degree of hydration, i.e., it changes in the opposite direction, thus concealing the intensity change of the band at about 2950 cm^{-1}.

Corresponding bands have been found by numerous authors in investigating many different acids.[12-44] The authors mostly assign the bands to vibra-

[12] F. A. Miller and C. H. Wilkins, *Anal. Chem.* **24**, 1253 (1952).
[13] L. J. Bellamy and L. Beecher, *J. Chem. Soc. (London)* p. 1701 (1952).
[14] L. J. Bellamy and L. Beecher, *J. Chem. Soc. (London)* p. 728 (1953).
[15] L. W. Daasch and D. C. Smith, *Anal. Chem.* **23**, 853 (1951).
[16] S. Detoni and D. Hadži, *J. Chem. Soc. (London)* p. 3163 (1955).
[17] S. Detoni and D. Hadži, in *Colloq. Spectroscopicum Intern., 6th, Amsterdam, 1956*, Macmillan (Pergamon), New York, 1957, p. 601.
[18] S. Detoni and D. Hadži, *J. Chim. Phys.* **53**, 760 (1956).
[19] M. Falk and P. A. Giguere, *Can. J. Chem.* **36**, 1680 (1958).
[20] A. Simon and R. Paetzold, *Z. Anorg. Allgem. Chem.* **303**, 46 (1960).
[21] A. Simon and R. Paetzold, *Z. Elektrochem.* **64**, 209 (1960).
[22] J. T. Braunholtz, G. E. Hall, F. G. Mann, and N. Sheppard, *J. Chem. Soc. (London)* p. 868 (1959).
[23] R. Blinc and D. Hadži, *Spectrochim. Acta* **16**, 852 (1960).
[24] L. C. Thomas, R. A. Chittenden, and H. E. R. Hartley, *Nature* **192**, 1283 (1961).
[25] D. F. Peppard, J. R. Ferraro, and G. W. Mason, *J. Inorg. Nucl. Chem.* **7**, 231 (1958).
[26] D. F. Peppard, J. R. Ferraro, and G. W. Mason, *J. Inorg. Nucl. Chem.* **12**, 60 (1959).
[27] D. F. Peppard, J. R. Ferraro and G. W. Mason, *J. Inorg. Nucl. Chem.* **16**, 246 (1961).
[28] R. W. Lovejoy and E. L. Wagner, *J. Phys. Chem.* **68**, 544 (1964).
[29] L. C. Thomas and R. A. Chittenden, *Spectrochim. Acta* **20**, 467 (1964).
[30] L. C. Thomas and R. A. Chittenden, *Spectrochim. Acta* **20**, 489 (1964).
[31] S. Detoni and D. Hadži, *Spectrochim. Acta* **20**, 949 (1964).
[32] D. Chapman, *J. Chem. Soc.* p. 225 (1956).
[33] A. N. Lasarev, *Izvest. Akad. Nauk. SSSR Ser. Fiz.* **21**, 322 (1957).
[34] A. N. Lasarev, *Zh. Tekhn. Fiz.* **27**, 426 (1957).
[35] R. Blinc and D. Hadži, *Mol. Phys.* **1**, 391 (1958).
[36] Y. I. Ryskin and G. P. Stavitskaya, *Opt. Spectry.* **7**, 488 (1959).
[37] E. Steger and K. Lunkwitz, *Naturwissenschaften* **48**, 522 (1961).
[38] E. Steger and K. Stopperka, *Naturwissenschaften* **48**, 523 (1961).
[39] E. Steger, *Z. Anorg. Allgem. Chem.* **309**, 304 (1961).
[40] E. Steger, *Z. Anorg. Allgem. Chem.* **325**, 89 (1963).
[41] D. Hadži, *Pure Appl. Chem.* **11**, 435 (1965).
[42] D. Hadži and N. Kobilarow, *J. Chem. Soc. (London)* p. 439 (1966).
[43] K. Stopperka, *Z. Anorg. Allgem. Chem.* **344**, 263 (1966).
[44] K. Stopperka, *Z. Anorg. Allgem. Chem.* **345**, 264 (1966).

tions of OH or OD groups in the hydrogen bridges cross-linking two acid molecules.

Are the bands found by us also to be assigned to the OH and OD groups in the hydrogen bridges arising on association? Since no water molecules are present in the membranes under the given conditions, comparison of Result 63 with Result 64 shows that these bands are to be assigned to the OH or OD acid groups.

If free OH groups of the —S(=O)(=O)—OH or —Se(=O)(=O)—OH groups were present in the thoroughly dried acids, their stretching vibration should be found in the region 3800–3400 cm^{-1}.

Result 66: If any bands at all are found in this region of the spectra of thoroughly dried polystyrenesulfonic or -selenonic acid, they are very weak.

It follows that in these acids, after thorough drying, there are almost no free OH groups present. Nearly all the acid groups are associated under these conditions. Accordingly, from the band positions, we assume that the bands are caused by OH or OD groups in very strong bridges. It follows that groups of the type —X(=O)(OH····O)(O····HO)X(=O)— are present. Details of the assignment of these bands are discussed in Section V.3.* In addition, the assignment is confirmed by the fact that the intensities of these bands increase (Result 65) on progressive drying, i.e., under the same conditions as those under which increasing association is to be expected.

At first it appears surprising that these acid groups should be associated by such powerful bridges, for we have seen in Section V.1.2 that the acid proton is easily removed from the anion, particularly in polystyrenesulfonic acid (Result 59). It could be assumed, on this basis, that the oxygen atoms of these acid groups are only weak acceptors for hydrogen bridges. In this case, however, the formation of such strong bridges should be impossible. Klages[45] on the basis of the same conclusion, conjectures that sulfonic acids do not dimerize.

Investigations of p-*Toluenesulfonic Acid*

In view of the above objection, we investigated *p*-toluenesulfonic acid. It is known that the true degree of dissociation of *p*-toluenesulfonic acid in aqueous solution is very large, as is also that of polystyrenesulfonic acid (Result 109,

* The position of these bands depends somewhat on the degrees of sulfonation and cross-linking which slightly influence the bending and the length of the bridges.

[45] F. Klages, "Lehrbuch der organischen Chemie," Vol. 2, p. 429. de Gruyter, Berlin, 1954.

Section V.11.1). It is possible, however, not only to take IR spectra of *p*-toluenesulfonic acid but also to determine its degree of association.

In Fig. 61, the spectra of saturated solutions of anhydrous* *p*-toluenesulfonic acid are shown. This acid is highly soluble in CH_2Cl_2; it is very slightly soluble in CCl_4. In the first case the film is only a few microns thick, in the latter 0.5 mm. From Fig. 61 we obtain:

Result 67: The two bands found in polystyrenesulfonic acid at about 2950 and at 2405 cm^{-1} are also found at nearly the same wave number values in the spectra of anhydrous *p*-toluenesulfonic acid in solution in chlorinated hydrocarbons.

The ebullioscopic investigation was carried out in CCl_4 (b.p. 76.7°C). The determination gave an apparent "molecular weight" of 336. This is only 2% under the molecular weight of the dimer.

Result 68: The ebullioscopic investigation shows that the anhydrous *p*-toluenesulfonic acid is present in CCl_4 solution almost entirely as the dimer. This is in spite of conditions which in no way promote association, namely a very dilute solution and temperatures over 76°C.

Even if the number of OH groups which are not dimerized is small, it should still be possible to observe their OH stretching vibration.

Result 69: A band, even though it is extremely weak, is actually found at 3525 cm^{-1} in the spectrum of anhydrous *p*-toluenesulfonic acid solution [Fig. 61(b)].

It can be supposed that this band can be ascribed to the OH stretching vibration of the few free OH groups of unassociated acid molecules. The number of such groups will increase with an increase in temperature. For this reason we carried out spectroscopic investigations of anhydrous *p*-toluenesulfonic acid at higher temperatures. These spectra are shown in Fig. 62.

Result 70: The intensity of the band at 3525 cm^{-1} increases slightly with rising temperature. It is nevertheless still very weak at 70°C.

The increase in intensity indicates that this band is the OH stretching vibration of free OH groups. Taking into account the fact that this band is still extremely weak at 70°C, this result does not contradict Result 68 that almost all acid molecules are present as dimers at 76°C. Further, it can be assumed that some dimers associate by only one bridge, so that some free OH groups can still be present despite dimerization.

From Fig. 62 we obtain the next result.

Result 71: The intensity of the two bands which we have ascribed to vibrations of the OH groups in the hydrogen bridges decreases considerably when the temperature is raised from 25° to 70°C. The intensity of the band, which for *p*-toluenesulfonic acid in CCl_4 solution lies at about 2955 cm^{-1}, decreases by nearly 31%, and that of the band at 2445 cm^{-1} by as much as 40% (corrected for density change with temperature).

* *p*-Toluenesulfonic acid monohydrate was dried in a high vacuum at 40°C. Under these conditions, drying lasted 130 hours. The acid obtained in this way is very pure and, as already found by Meyer,[46] only slightly hygroscopic; this is understandable considering its dimerization by powerful hydrogen bridges.

[46] H. Meyer, *Ann. Chem. Liebigs* **433**, 327 (1923).

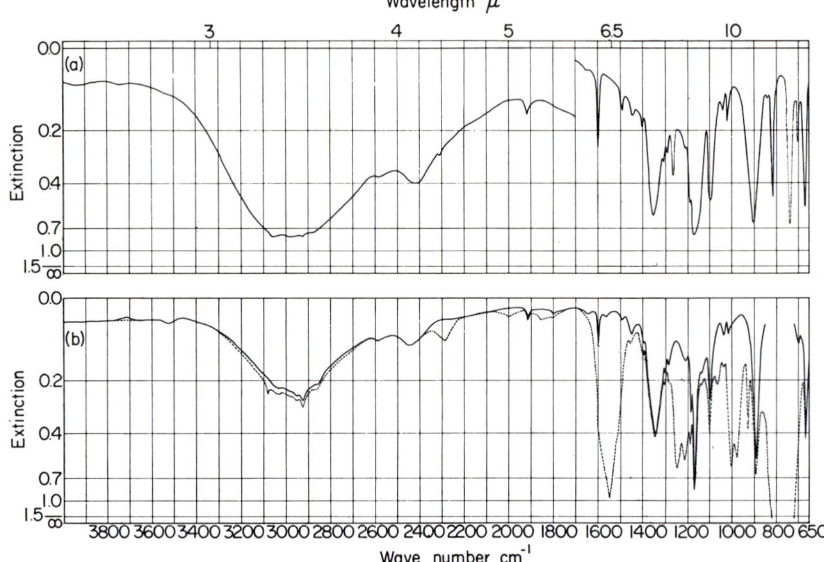

FIG. 61. *p*-Toluenesulfonic acid (anhydrous); saturated solution in each case. [Taken in an acid-resistant heated cell with variable film thickness and acid-resistant windows (germanium).] (a) in CH_2Cl_2: the bands of the solvent are shown by broken lines. (b) in CCl_4: the bands of the solvent have been compensated by a cell in the reference beam. The broken line is the spectrum in which the bands of the solvent are not compensated.

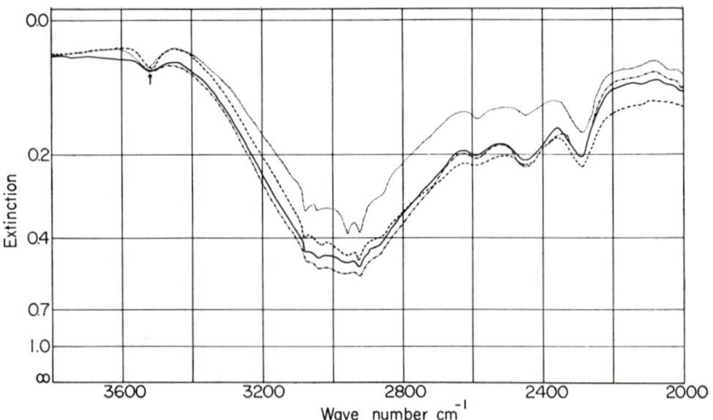

FIG. 62. Temperature dependence of the spectrum of anhydrous *p*-toluenesulfonic acid in CCl_4: —·—·, 25°C; – – – –, 55°C; ——, 40°C; · · · ·, 70°C.

The acid is also almost completely dimerized at temperatures over 76°C (Results 68, 70). Hence the basic cause of this decrease of intensity cannot be the disassociation* which occurs with a rise in temperature. Hence it follows that the extinction coefficient of the bands of the vibrations of the OH groups in the hydrogen bridge decreases with rising temperature. We shall encounter such a change of extinction with temperature again in the case of polystyrenesulfonic acid in Section V.3, Result 79.

From Results 67–70 we find that:

Almost all *p*-toluenesulfonic acid molecules in solution in chlorinated hydrocarbons are dimerized by hydrogen bridges.

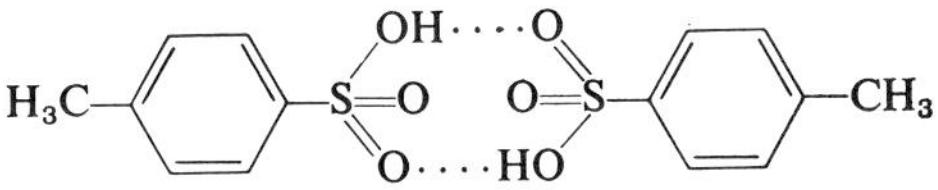

OH····O
H_3C—⟨benzene ring⟩—S═O O═S—⟨benzene ring⟩—CH_3
O····HO

The bands found for the acid dissolved in CCl_4 at about 2955 and at 2445 cm^{-1} must be ascribed to vibrations of the OH groups in these hydrogen bridges.

From a comparison of these results with those on polystyrenesulfonic and polystyreneselenonic acids it follows that, on thorough drying, the acid groups of polystyrenesulfonic and polystyreneselenonic acid associate with one another by hydrogen bridges. This is clearly apparent despite the very high true degree of dissociation (nearly one) of these acids.

OH····O OH····O
—S═O O═S— —Se═O O═Se—
O····HO O····HO

Naturally, the acid groups need not always be associated in pairs. A network of associated groups will be present.

How is the formation of the strong bridges possible?

Results 58 and 59 in Sections V.1 and V.1.2 have shown that the mesomerism in the anions is strongly reduced when the proton is attached. As we know, the electrons are regrouped so that two oxygen atoms are attached to the sulfur atom with double bonds. These oxygen atoms are thus considerably more effective acceptors for hydrogen bridges than the oxygen atoms in the

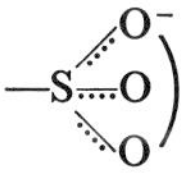

O^-
—S┈O) ions. The neighboring OH groups can then attach themselves to
O

these oxygen atoms by strong hydrogen bridges.

* We apply the word "dissociation" exclusively for the dissociation of the acid groups,

O OH····O
e.g., —S═O. The dissociation of the —S═O O═S— we term "disassociation."
OH O····HO

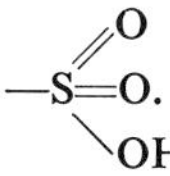

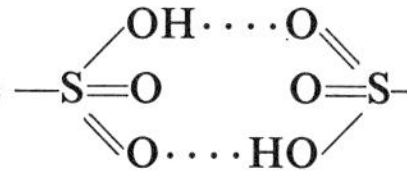

The energy balance is as follows. The energy necessary for the removal of water is supplied "externally" as heat. Energy is required for the rearrangement of the electrons in the anions. Energy is released by the formation of the bond of the excess proton to the anion, as well as by the formation of the hydrogen bridges. A decrease of entropy is connected with association. This opposes association ($\Delta G = \Delta H - T\Delta S$). The enthalpy of reaction on association is obviously large enough to outweigh the decrease in entropy.

The formation of the strong hydrogen bridges also explains the following observation, obtained from Fig. 87(d) (Section V.6.2.1):

Result 72: The band of the stretching vibration of the SO bond with single bond character at 907 cm^{-1} is slightly shifted toward larger wave numbers with decreasing temperature.

Explanation: it is known[47] that the band of the stretching vibration of the SO single bond in $-\mathrm{S}(=\mathrm{O})_2\mathrm{OH}$ groups is slightly shifted toward larger wave numbers by formation of a hydrogen bridge. This hydrogen bridge becomes somewhat stronger with decreasing temperature. In this way the SO single bond takes on a little of the character of a double bond, whereby the band of the stretching vibration of the SO single bond is slightly shifted toward larger wave numbers.

In summary, we see that, after thorough drying, the acid groups in polystyrenesulfonic and polystyreneselenonic acids are associated by very strong hydrogen bridges. These strong hydrogen bridges can form, since the proton attached to the anion reduces the mesomerism in these anions very strongly, causing two oxygen atoms to become powerful acceptors for hydrogen bridges.

V.2.2. Polystyrenethiophosphonic, Polystyreneseleninic, and Polystyrenephosphinic Acids

Figure 120 (Section V.14) shows the spectra of a membrane of polystyrenethiophosphonic acid; Fig. 115 (Section V.13), those of a membrane of polystyreneseleninic acid; and Fig. 116 (Section V.13), those of a membrane of polystyrenephosphinic acid. Spectrum 12 in each figure is the spectrum of the membrane thoroughly dried. From these figures, or from Table 12 (columns 4–7) we obtain:

Result 73: If the membranes are hydrated with H_2O before drying, a broad intense band at about 2855 cm^{-1} and a broad weak band at about 2290 cm^{-1} are observed in polystyrenethiophosphonic acid. In polystyreneseleninic acid, a broad intense band is observed at about 2960 cm^{-1} and a less intense one at 2420 cm^{-1}. In polystyrenephosphinic acid, an extremely broad intense

[47] A. Simon and H. Kriegsmann, *Z. Physik. Chem. (Leipzig)* **204**, 369 (1955).

band is observed at about 2750 cm^{-1} and a very broad, but much less intense band at about 2180 cm^{-1}.*

Result 74: If the membranes are hydrated with D_2O before drying, corresponding bands are observed. In polystyrenethiophosphonic acid, a broad intense band is observed at about 2155 cm^{-1}. In polystyreneseleninic acid, a broad intense band is found at 2195 cm^{-1}. In polystyrenephosphinic acid, a broad band is observed at about 2060 cm^{-1}.*

Comparison of Results 73 and 74 shows that these bands must be assigned to the OH or OD acid groups. These must be the groups involved, since under these experimental conditions no water molecules are present. Taking into account the positions of these bands, it must be assumed that they relate to vibrations of the OH or OD groups in very strong hydrogen bridges of groups of the type

```
        H
       /
   —P—O····HO
      \        \
       OH····O=P—
               /
              H
```

For further details on the assignment of these bands see Section V.3. This will not come as a surprise, if we consider the results given in Section V.1.2, for we have seen there that these acids are very weak. It is therefore not astonishing that these acids associate by very strong hydrogen bridges, for the doubly bonded oxygen atom, which is always present, is a strong hydrogen bridge acceptor.

From these figures we further obtain:

Result 75: In the spectra of thoroughly dried polystyreneseleninic and -phosphinic acids, no band is found in the region 3800–3400 cm^{-1}. In the spectrum of thoroughly dried polystyrenethiophosphonic acid, however, a shoulder—even if not very pronounced—is found at about 3500 cm^{-1}. In the region of the OD stretching vibrations, a corresponding shoulder, somewhat more pronounced, is found at about 2580 cm^{-1}.

It follows that no free OH groups of the acid groups are present in thoroughly dried polystyreneseleninic and -phosphinic acids. Polystyrenethiophosphonic acid is discussed more fully in Section V.14.

If we compare the spectra of the thoroughly dried† membranes in Figs. 4(a, c) and 5(a, c) (Section II.3), we obtain (Table 12, columns 10 and 11):

Result 76: In the polystyreneselenonic and the polystyreneseleninic acids a broad band is found at about 1260 cm^{-1}, if the membrane was hydrated with H_2O before drying. If these acids were hydrated with D_2O before drying this band is not observed, but instead a corresponding band is found as a shoulder at about 940 cm^{-1}. In the spectra of the other acids, these bands are masked by other bands.

* It is known (see Section II.4) that the band observed at 2365 cm^{-1} or at 1722 cm^{-1} is that of the PH or PD stretching vibration.

† It is necessary to compare the thoroughly dried membranes, as the band of the D_2O scissor vibration lies at 1210 cm^{-1}.

It follows that this band at about 1260 or at about 940 cm^{-1} must be ascribed to the bending vibration of OH or OD, respectively, in the hydrogen bridge. This must be so, since, under the given conditions, there is no other possible group with a band which could show a corresponding isotope effect.

In summary, we see that the acid groups are linked by very strong hydrogen bridges.

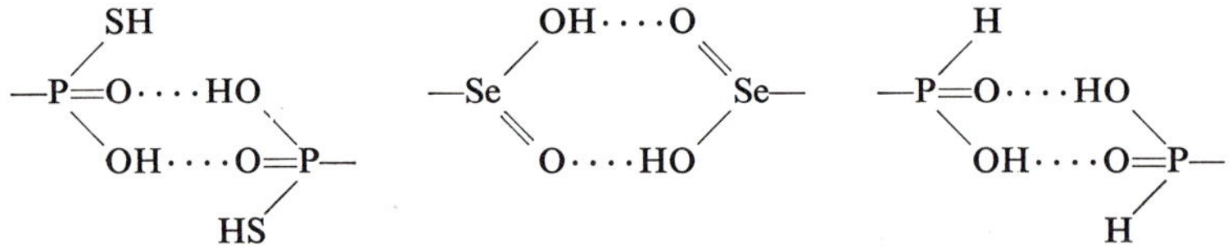

```
     SH                      OH····O                 H
    /                       /       \\              /
 —P=O····HO            —Se            Se—      —P=O····HO
    \        \              \\       /             \        \
     OH····O=P—               O····HO               OH····O=P—
            /                                              /
          HS                                              H
```

Here, too, most of the acid groups will not be associated simply as pairs; rather, there will be a network of hydrogen bridges.

V.2.3. Dependence of Acid Groups Association on Degree of Hydration

From Table 12, and from the figure given in column 3, is obtained:

Result 77: The dependence of the intensity of these bands on the degree of hydration shows a characteristic course in the order polystyrenesulfonic, -selenonic, -thiophosphonic, -seleninic, and -phosphinic acids.

From this it follows that the dependence of the association on the degree of hydration shows a characteristic course in the series of acids investigated by us. It increases in order from polystyrenesulfonic to polystyrenephosphinic acid.

V.3. Observed Bands and Potential Well in Hydrogen Bridges of Associated Acid Groups

Considering the bands collected in Table 12, we know that they must be ascribed to the OH or OD groups in the hydrogen bridges connecting the acid groups. It is apparent that in each case two bands are found.

If we conjecture that one band represents the OH stretching vibration and the other a combination of the OH stretching vibration with yet another band, we can distinguish the following cases. First, if the second vibration of the combination is a band of the resin network or of the anion, the band distance would be the same for the H acid and the D acid (i.e., according to whether the membrane was hydrated with H_2O or D_2O before drying). This is, however, not the case. Second, it could be conjectured that the OH bending vibration of the hydrogen bridge at 1260 cm^{-1} combines with another vibration and that one of the bands arises in this way.* If this were so, the isotope

* Thomas and Chittenden[30] supposed that in the phosphonic and thiophosphonic acids the band with the smaller wave number is a combination of the stretching vibration of the PO single bond and the OH bending vibration of the hydrogen bridge.

effect—i.e., the ratio $\tilde{\nu}_{OH}/\tilde{\nu}_{OD}$—should be considerably smaller than the usual value (1.37) for one of the bands. This is not the case as the data given in Table 13 show.

Finally, the following assignment for the two bands appears possible: Two hydrogen bridges are present in the

```
     /OH····O\\
—S=O          O=S—
     \\O····HO/
```

groups. Insofar as the OH stretching vibrations of these two bridges are coupled, they would split up so that two bands should be found. The splitting would, however, be greater, the stronger the coupling of the vibrations. The coupling should increase with decreasing temperature, and hence the bands should split. In fact, however, the opposite is the case (Result 78). Hence this interpretation for the appearance of two bands is not valid.

There remain two possible causes for the occurrence of two bands: First, one band could result from the stretching and the other band from the overtone of the bending vibration of the OH or OD group in the hydrogen bridge. We have already found the bending vibration at 1260 cm^{-1} in polystyreneseleninic acid. Second, several bands of stretching vibrations could occur in the region investigated by us, if a potential curve with a double minimum were present in the hydrogen bridges. Such curves have been discussed by numerous authors.[18,21,23,35,37,38,48-69]. This can be either a symmetrical potential well with a barrier [Fig. 63(b)] or an unsymmetrical potential well (Fig. 66).

[48] G. Herzberg, "Molecular Spectra and Molecular Structure," Vol. 2, p. 221 ff. Van Nostrand, Princeton, New Jersey, 1962.
[49] A. N. Baker, Jr., *J. Chem. Phys.* **22**, 1625 (1954).
[50] J. Oshida, Y. Ooskika, and R. Miyasaka, *J. Phys. Soc. Japan* **10**, 849 (1955).
[51] E. R. Lippincott and R. Schroeder, *J. Chem. Phys.* **23**, 1099 (1955).
[52] E. Heilbronner, H. H. Guenthard, and R. Gerdil, *Helv. Chim. Acta* **39**, 1171 (1956).
[53] S. Bratož and D. Hadži, *J. Chem. Phys.* **27**, 991 (1957).
[54] A. Simon and R. Paetzold, *Z. Anorg. Allgem. Chem.* **301**, 246 (1959).
[55] E. Heilbronner, H. Rutishauser, and F. Gerson, *Helv. Chim. Acta* **42**, 2286, 2304 (1959).
[56] C. Reid, *J. Chem. Phys.* **30**, 182 (1959).
[57] C. Haas and D. F. Hornig, *J. Chem. Phys.* **32**, 1763 (1960).
[58] L. J. Bellamy and P. E. Rogasch, *Proc. Roy. Soc.* (*London*) **A257**, 98 (1960).
[59] H. Zimmermann, *Z. Elektrochem.* **65**, 821 (1961).
[60] F. Gerson, *Helv. Chim. Acta* **44**, 471 (1961).
[61] R. L. Somorjai and D. F. Hornig, *J. Chem. Phys.* **36**, 1980 (1962).
[62] R. Rein and E. F. Harris, Preprint QB 12, Quantum Chemistry Group, Uppsala, Sweden, 1964.
[63] H. Zimmermann, *Angew. Chem.* **76**, 1 (1964).
[64] M. Weissmann and N. V. Cohan, *J. Chem. Phys.* **43**, 119 (1965).
[65] M. Weissmann and N. V. Cohan, *J. Chem. Phys.* **43**, 124 (1965).
[66] J. Brickmann and H. Zimmermann, *Ber. Bunsenges. Physik. Chem.* **70**, 157 (1966).
[67] J. Brickmann and H. Zimmermann, *Ber. Bunsenges. Physik. Chem.* **70**, 521 (1966).
[68] R. Rein and F. E. Harris, *J. Chem. Phys.* **45**, 1797 (1966).
[69] G. Zundel and H. Metzger, *Spectrochim. Acta* **23A**, 759 (1967).

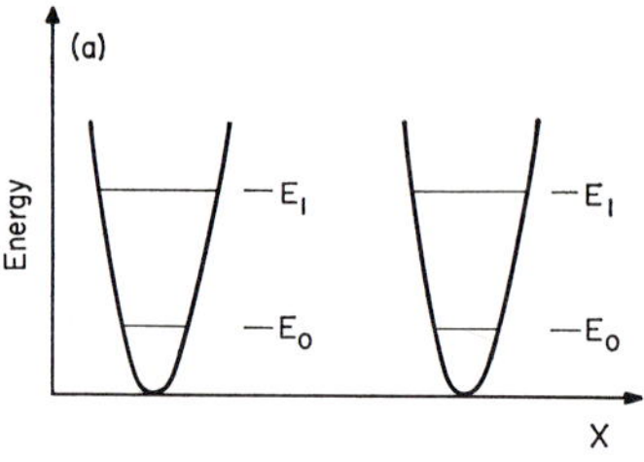

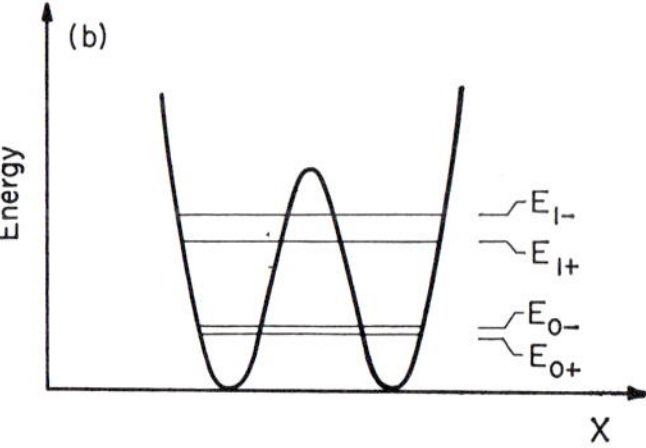

FIG. 63. Schematic representation for discussion of the symmetrical potential well with barrier in hydrogen bridges.

As for the symmetrical potential well, a potential well such as that shown in Fig. 63(b) can be regarded as made up of two parabolic potential curves, as shown in Fig. 63(a). We start with the assumption that these two potential wells are initially so far removed from one another that the wave functions of the proton do not overlap. If the potential wells are now moved together, the wave functions begin to overlap. The lower the potential barrier between the wells, the more easily the proton can tunnel through it. The energy levels of the proton in the two wells are then no longer independent of one another, and are therefore no longer equal. The energy levels E_0, E_1, etc., in the individual wells are split up; in the well as a whole the levels E_{0+}, E_{0-}, E_{1+}, E_{1-}, etc., are found. If the potential well is perfectly symmetrical, then in the IR spectrum only the transitions E_{1-}–E_{0+} and E_{1+}–E_{0-} are allowed, i.e., a pair of bands is observed. (Further details are in footnotes 61, 63, and 66; but also see Section V.11).

However, a pair of bands can also be observed if an unsymmetrical potential well with a double minimum occurs, as shown by the calculations of Somorjai and Hornig.[61] If the whole potential well is again regarded as made up of two separate wells, closely spaced energy levels, and therefore a pair of bands arises when the lowest level of one well—more precisely, of that well with the higher minimum—has approximately the same energy as a higher level of the other well. This is shown in Fig. 66.

Figure 64 shows the result of an experiment on the temperature dependence of the band pair. The exactly measured positions of the bands are collected in Table 13. In this table, the band at the smaller wave number is denoted by $\bar{\nu}_1$, that at the larger by $\bar{\nu}_2$; $\Delta\bar{\nu}_{292}$, or $\Delta\bar{\nu}_{85}$, is the distance between these two bands, and $\Delta_T\bar{\nu}$ the shift on cooling.

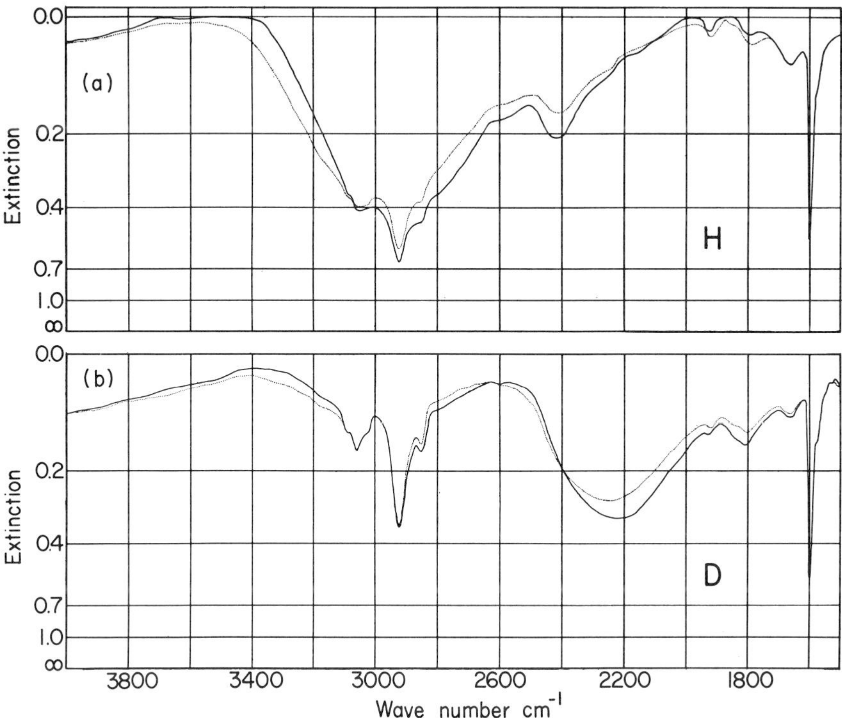

FIG. 64. Polystyrenesulfonic acid, 3% cross-linked, 3 days of sulfonation, thoroughly dried membranes: · · · ·, 292°K; ——, 85°K. (a) hydrated with H_2O before drying (H acid); (b) hydrated with D_2O before drying (D acid).

Result 78: $\Delta_T\bar{\nu}_2$ is -52 cm^{-1} for the H acid and -20 cm^{-1} for the D acid. The band ($\bar{\nu}_2$), found at about 2950 cm^{-1} is shifted on cooling from 2967 to 2915 cm^{-1}. The corresponding band of the D acid is shifted from 2254 to 2234 cm^{-1}. This band is considerably shifted toward smaller wave numbers when the temperature is lowered.

$\Delta_T\bar{\nu}_1$ is $+7$ cm^{-1} for the H acid and $+7$ cm^{-1} for the D acid. The band ($\bar{\nu}_1$), found at about 2405 cm^{-1} for the H acid, is shifted from 2407 to 2414 cm^{-1} on cooling. The corresponding band of the D acid is shifted from 1805 to 1812 cm^{-1}. This band is slightly shifted toward larger wave numbers when the temperature is lowered.

Thus with decreasing temperature the distance between the two bands ($\Delta\bar{\nu}$) decreases, both for the H and for the D acid.

Result 79: The intensity of these bands increases on cooling. The relative increase in intensity is greater for the bands at about 2405 (OH) or 1805 cm^{-1} (OD) than for those at about 2950 (OH) or 2240 cm^{-1} (OD).

TABLE 13

TEMPERATURE DEPENDENCE AND ISOTOPE FACTORS OF THE BANDS OF THE OH OR OD GROUPS OF DRY POLYSTYRENESULFONIC ACID HYDROGEN BRIDGES

Temperature	Band	Band positions and distances apart (cm^{-1})		Isotope factor	Integral extinction (abritrary units)	
		H acid	D acid	$\bar{\nu}_{OH}/\bar{\nu}_{OD}$	H acid	D acid
1	2	3	4	5	6	7
292°K	$\bar{\nu}_2$	2967	2254	1.316	1930	877
	$\bar{\nu}_1$	2407	1805	1.333	398	147
	$\Delta\bar{\nu}_{292}$	560	449	—	—	—
85°K	$\bar{\nu}_2$	2915	2234	1.304	2015	971
	$\bar{\nu}_1$	2414	1812	1.332	466	183
	$\Delta\bar{\nu}_{85}$	501	422	—	—	—
	Band shift on cooling in cm^{-1}					
	$\Delta_T\bar{\nu}_2$	−52	−20	—	—	—
	$\Delta_T\bar{\nu}_1$	+7	+7	—	—	—

This result corresponds to our Result 71 for *p*-toluenesulfonic acid. We saw there that this change of intensity must be caused by the temperature dependence of the extinction coefficient. The present result confirms this.

We now consider the question: Is a symmetrical potential well with a double minimum present? Such a well is shown in Fig. 65. The broken potential curve gives the potential curve at room temperature, the continuous one that at the lower temperature. The same is the case for the energy terms. Those at room temperature are denoted by E', those at the lower temperature by E. Permitted transitions are indicated. With decreasing temperature the oxygen atoms, linked by hydrogen bridges, move closer together, so that the potential well becomes narrower (Fig. 65). The barrier then becomes lower so that the terms are more widely split, the upper ones considerably more than the lower ones. In addition, all energy levels are elevated by the narrowing of the potential well, the upper ones slightly more than the lower ones. From Fig. 65:

$$(E'_{1-} - E'_{0+}) < (E_{1-} - E_{0+})$$

so that the band observed by us at about 2950 cm^{-1} should be shifted toward larger wave numbers on cooling. $\Delta_T\bar{\nu}_2$ should be positive. In addition,

$$[(E'_{1-} - E'_{0+}) - (E'_{1+} - E'_{0-})] < [(E_{1-} - E_{0+}) - (E_{1+} - E_{0-})]$$

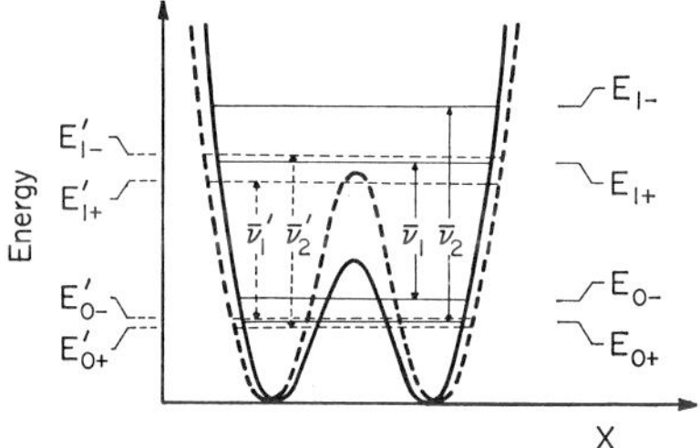

FIG. 65. The behavior of a symmetrical potential well with a barrier in a hydrogen bridge, when the oxygen atoms move together with decreasing temperature. The position coordinate X is given in arbitrary units: – – – –, higher; ——, lower temperatures. The origin of the position coordinate is, for clarity, set at the center of the potential well and thus remains fixed in passing from the dotted to continuous potential curve.

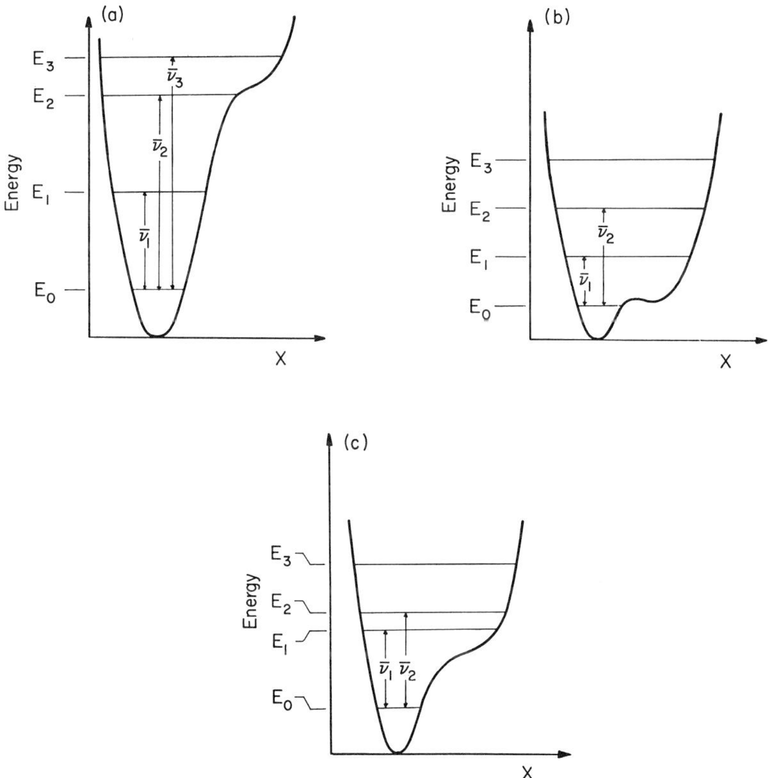

FIG. 66. Unsymmetrical potential wells in hydrogen bridges. The position coordinate X is given in arbitrary units.

the splitting should increase on cooling. Both expressions contradict our Result 78. Hence it follows that a symmetrical potential well—as shown in Fig. 65—cannot be present here.

If an unsymmetrical potential well with a double minimum is present, one well may be elevated to different extents relative to the other one. Figure 66 shows various such potential wells.

In case (a), the second well lies so high that $\bar{\nu}_2$ is approximately equal to $2\bar{\nu}_1$. Then

$$\bar{\nu}_2 - \bar{\nu}_1 \approx \bar{\nu}_1$$

However, from Table 13 we see that in our case

$$\bar{\nu}_2 - \bar{\nu}_1 \ll \bar{\nu}_1$$

Hence case (a) cannot apply.

In case (b), the second well lies so low that even the energy level E_1 lies considerably above the potential barrier of the well. Then:

$$\bar{\nu}_2 - \bar{\nu}_1 \approx \bar{\nu}_1$$

Thus case (b) cannot hold here.

In case (c) the energy level E_1 is slightly lowered, compared with conditions in case (a). $\bar{\nu}_1$ is smaller, since the well is not parabolic, but broadened. The energy levels E_1 and E_2 are only slightly split, i.e., the difference $\bar{\nu}_2 - \bar{\nu}_1$ is small. This agrees with our observations, since the band at 2405 cm^{-1} is considerably shifted, for an OH stretching vibration, toward smaller wave numbers and the distance $\Delta\bar{\nu}$ between the two bands is small compared to the wave number of an OH stretching vibration.

We have discussed in detail (see footnote 69) the explanation of the temperature dependence of the band positions (Result 78) as well as the isotopy effect (Result 81) in the presence of a potential well, as shown in Fig. 66(c). Hadži and Kobilarow[41,42] have nevertheless raised a weighty objection to the explanation of the band pair with a double potential minimum. They studied these bands in hydrogen bridges with acceptors of widely differing strengths. From the acidities of these acceptors, they estimated the relative positions of the minima and found that, in the series of compounds investigated by them, one minimum should be shifted relative to the other by more than 1000 cm^{-1}. From this, these authors concluded that the band pair cannot be explained by an asymmetrical double potential minimum. Therefore they suggest that one of the bands is that of the overtone of the bending and the other that of the stretching vibration. These are coupled together. A single potential minimum (Fig. 67) is present.

Actually, the temperature dependence of the bands (Result 78) becomes immediately understandable if it is assumed that the band with the smaller wave number is that of the overtone of the bending vibration of the OH or OD group. This results because, as the hydrogen bridge becomes stronger on cooling, the overtone of the bending vibration shifts toward larger wave numbers, while the stretching vibration shifts toward smaller wave numbers (footnote 40).

There is nevertheless a serious objection to this explanation.[69]

Result 80: Both bands are always shifted in the same direction, if the positions of the two bands are compared for the various acids (Table 12, columns 4 and 5).

In comparing the individual acids, it would also be expected that the overtone of the bending vibration should be shifted toward larger, while the stretching shifts toward smaller wave numbers. Nevertheless, if the two vibrations are strongly coupled, the stretching vibration could displace the overtone of the bending vibration toward smaller wave numbers. Such a strong coupling is plausible, if the potential well in the hydrogen bridge is anharmonic, i.e., unsymmetrical.

Result 81: The isotope factors $\bar{\nu}_{OH}/\bar{\nu}_{OD}$ are much smaller than 1.37 (Table 13). This is the value which would be expected if these groups were not bound by hydrogen bridges. This is particularly true for the isotope factor of $\bar{\nu}_1$.

From this we see that the potential well in the hydrogen bridge is actually anharmonic. This conclusion is illustrated by the curves shown in Fig. 67. The level E_1 is more strongly lowered by the anharmonicity than is E_0. Therefore $\bar{\nu}$ becomes smaller with increasing anharmonicity. The levels lie higher in the potential well for H than for D. The levels of H are therefore more strongly lowered by an increase in the anharmonicity than the levels of D. Hence $\bar{\nu}_{OH}$ is reduced more than $\bar{\nu}_{OD}$, i.e., the isotope factor becomes smaller. Thus an unsymmetrical single potential minimum is present in the hydrogen bridge, as shown in Fig. 67. The strong coupling, i.e., Fermi resonance, explains the relatively large intensity of the overtone of the bending vibration.

In summary, we see that the explanation of the band pairs with an asymmetrical double potential minimum [Fig. 66(c)] appears feasible. Nevertheless, it would be expected that these two bands would be shifted more strongly than is actually the case when the hydrogen bridge acceptor is varied.

We therefore assume that a single potential minimum is present and assign the band in the region 2960–2750 cm^{-1} to the stretching vibration, that in the

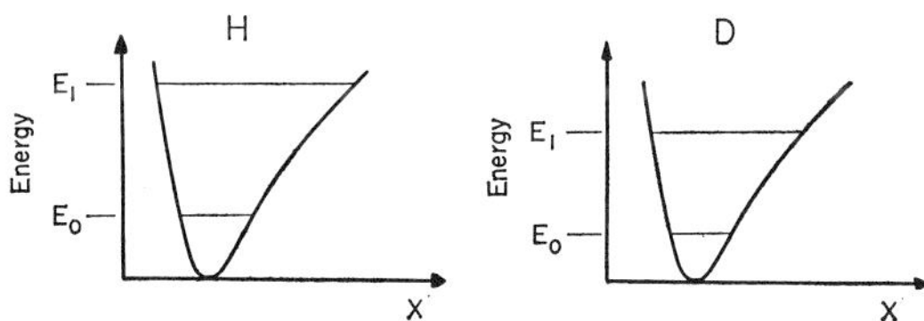

FIG. 67. The unsymmetrical potential well in the hydrogen bridges between the acid groups. The position coordinate X is given in arbitrary units.

region 2420–2130 cm^{-1} to the overtone of the bending vibration of the OH groups in the hydrogen bridges. This assignment is only possible if these vibrations are strongly coupled. The isotope factor shows that the potential is very anharmonic (Fig. 67).

V.4. Measurements on Polystyrenesulfonic Acid as a Function of Degree of Hydration

A series of measurements were conducted on membranes with various degrees of cross-linking and sulfonation.* The data on these membranes are collected in Table 14, columns 1–4. All these membranes were equally thick, to within ±3%.

The Series of Spectra

At low degrees of hydration, it appeared that the hydration depends on whether the membrane had been kept for several days or only a few hours at a definite atmospheric humidity before taking the spectrum. Hence we carried out two kinds of measurements, series I and II, these under different conditions of drying. The spectra and diagrams of these two series are illustrated by I and II in the figures appearing in the following sections. In Table 14, columns 5 and 6, the figures showing sections from these spectra are given. The symbols with which the curves of the respective membranes are plotted in these diagrams are given in Table 14, column 7.

Series I begins with the spectra of the membranes hydrated at 98% relative atmospheric humidity. Subsequently the degree of hydration of the membrane was progressively reduced in 8 to 10 steps (Figs. 68 and 69). Finally, the membrane was thoroughly dried. By "thorough drying" we understand drying carried out in a high vacuum at 20°C. At each stage the membrane was kept for about three days at a definite relative atmospheric humidity before taking the spectrum.

Series II: The differences in the conditions described below affect only low degrees of hydration. The membranes were therefore rapidly dried to a degree of hydration corresponding roughly to that at about 5% relative atmospheric humidity. In this series the first spectrum is that of the membranes at this humidity. Now the degree of hydration was again reduced stepwise, the membrane being this time kept for only about six hours at a definite relative

* The investigations described in the following sections are found, together with their results, in footnotes 70–72.

[70] G. Zundel and H. Metzger, *Z. Physik. Chem.* (*Frankfurt*) **58**, 225 (1968).

[71] G. Zundel and H. Metzger, *Z. Physik. Chem.* (*Frankfurt*) **59**, 225 (1968).

[72] G. Zundel and H. Metzger, *Z. Naturforsch.* **22a**, 1412 (1967).

TABLE 14

DATA OF POLYSTYRENESULFONIC ACID MEMBRANES OF WHICH A SERIES OF SPECTRA WERE RECORDED

Degree of cross-linking (%)	Sulfonation period at 25°C (days)	mmole H^+/gm	Acid groups per benzene ring	Measurement series I (Fig.)[a]	Measurement series II (Fig.)[b]	Symbol used in diagrams
1	2	3	4	5	6	7
1	18[c]	5.21	1.06	68(a)	70(a)	○○····
1	4	6.1	1.50	68(b) and Fig. 57	70(b) and 2	××——
10	1	3.11	0.48	—	71	●●-·-·-·
10	5	5.73	1.39	69	—	△△- - - -

[a] All spectra of thoroughly dried membranes in Fig. 79.
[b] All spectra of thoroughly dried membranes in Fig. 60.
[c] Period in hours.

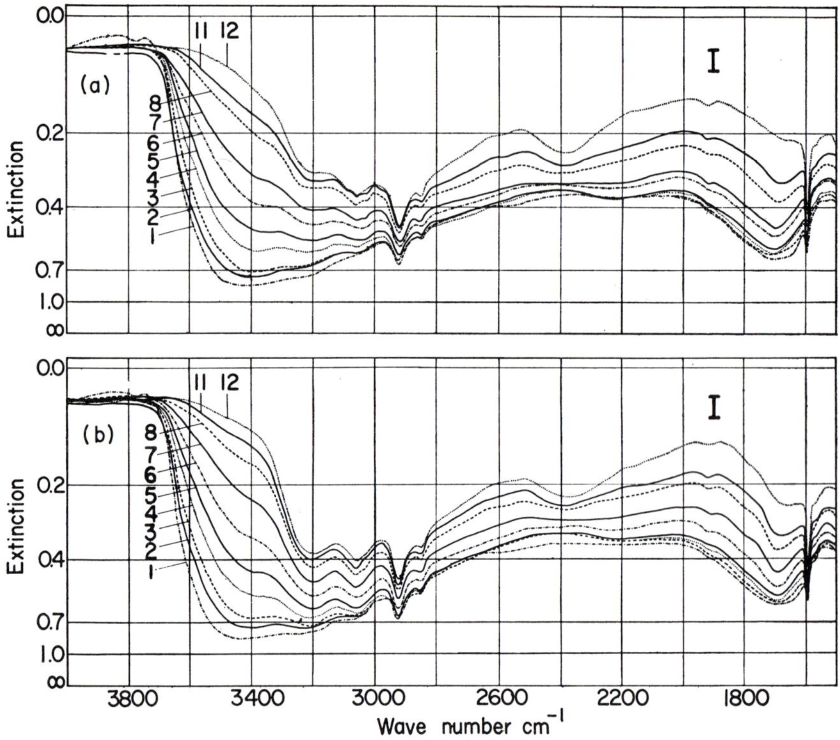

FIG. 68. Polystyrenesulfonic acid, H_2O hydrated; one drying stage every 3 days. (a) 1% cross-linked, 1.06 acid groups per benzene ring: 1, 98%; 2, 71%; 3, 53%; 4, 24%; 5, 17%; 6, 11%; 7, 5%; 8, 1% relative atmospheric humidity; 11, incompletely dried; 12, thoroughly dried membrane. (b) 1% cross-linked, 1.50 acid groups per benzene ring (the same membrane as in Fig. 57); 1, 98%; 2, 71%; 3, 53%; 4, 24%; 5, 17%; 6, 11%; 7, 5%; 8, 1% relative atmospheric humidity; 11, incompletely dried; 12, thoroughly dried membrane.

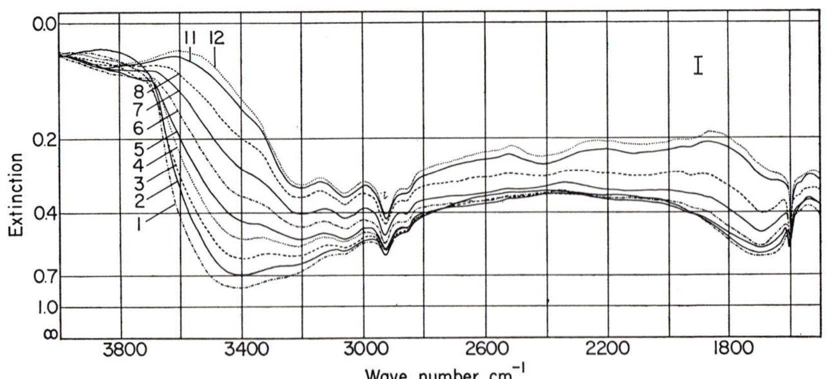

FIG. 69. Polystyrenesulfonic acid, H_2O hydrated; one drying stage every 3 days. 10% cross-linked, 1.39 acid groups per benzene ring: 1, 98%; 2, 71%; 3, 53%; 4, 33%; 5, 22%; 6, 11%; 7, 5%; 8, 1% relative atmospheric humidity; 11, incompletely dried; 12, thoroughly dried membrane.

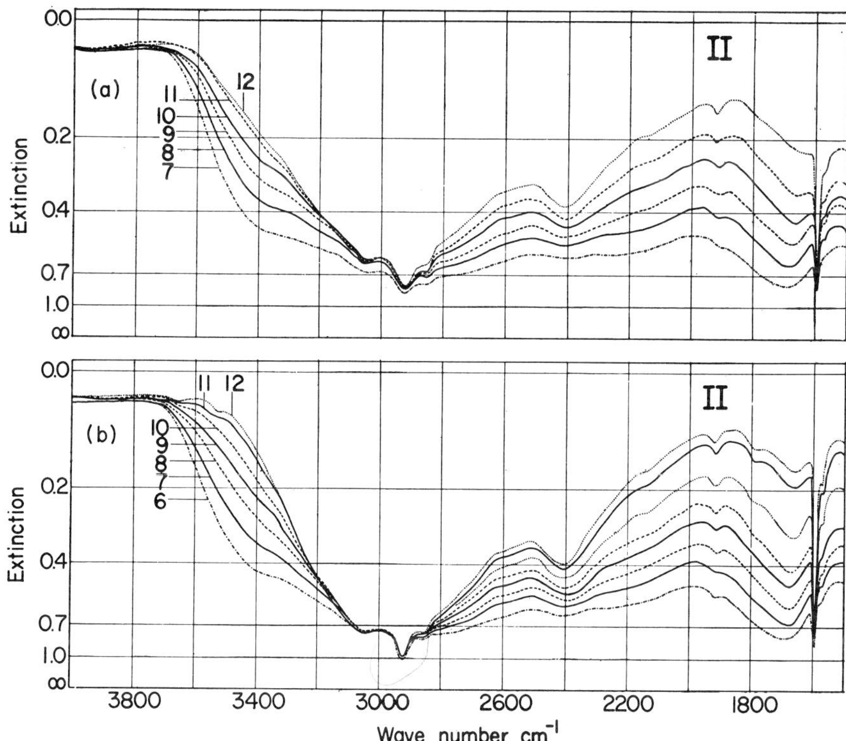

FIG. 70. Polystyrenesulfonic acid, H_2O hydrated (two membranes, one over the other). Before taking the spectra, the membrane was kept in the cell for about 6 hours at a definite atmospheric humidity. The drying stages were carried out, one after another, without interruption. (a) 1% cross-linked, 1.06 acid groups per benzene ring; 5, about 5% relative atmospheric humidity; 7 → 11, decreasing atmospheric humidity; 12, thoroughly dried membrane. (b) 1% cross-linked, 1.50 acid groups per benzene ring; 6, about 5% relative atmospheric humidity; 6 → 11, decreasing atmospheric humidity; 12, thoroughly dried membrane.

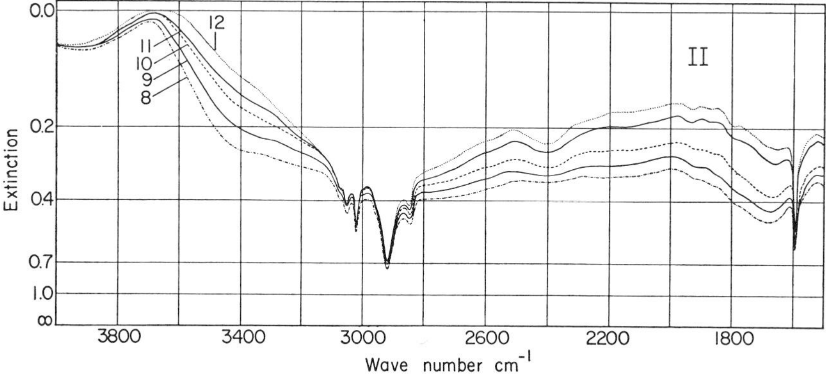

FIG. 71. Polystyrenesulfonic acid, H_2O hydrated (two membranes one over the other). Before taking the spectra, the membrane was kept in the cell for about 6 hours at a definite atmospheric humidity. The drying stages were carried out, one after another, without interruption. 10% cross-linked, 0.48 acid groups per benzene ring; 8, about 5% relative atmospheric humidity; 9→11, decreasing atmospheric humidity; 12, thoroughly dried membrane.

atmospheric humidity before the spectrum was taken. The spectra of this series were taken with a double layer sample, in order that the changes could be better observed. This was done by placing one membrane upon the other (Figs. 70 and 71).

A series of spectra with D_2O hydration is shown in Fig. 72.

The Water Adsorption Isotherms

To determine how many water molecules per acid group were present in the membrane at a definite atmospheric humidity, we weighed the membrane at different degrees of hydration; in this, the conditions of drying corresponded to those in series I. The weighing was performed with a quartz fiber balance of simple construction,[73-75], as first described by Volmer.[73] A linear quartz fiber is fixed at one end, and the membrane to be weighed is hung from the other end. The bending of the fiber under the weight is read from a mirror scale with a telescope. A quartz fiber about 30 cm long and 200 μ thick was found suitable for our membranes, with respect to the weight and its change with degree of hydration.

The number of water molecules per acid group can be calculated from the change in weight if the number of acid groups present per gram of dry polystyrenesulfonic acid is known. We determined this by Sansoni's[76,77] coulometric microtitration of ion exchangers. The values are given in Table 14, column 3, with the numbers of acid groups per benzene ring given in column 4.

Figure 73 shows the isotherms.* We can now consider the changes of the spectrum, depending on the degree of hydration, as a function of the number of water molecules present per acid group. One point must, however, be borne in mind:

Change in Membrane Length with the Degree of Hydration

The length of each membrane shows some slight dependence on the degree of hydration. The quantity of material through which the radiation is transmitted when the spectrum is taken thus changes slightly as a function of the degree of hydration. We therefore measured these lengths. For this, we

* The membranes could not be completely dried. We therefore determined the value $n = 0$ (n being the number of water molecules per acid group) with the help of our spectroscopic findings. We plotted n against the integral extinction of the band of the H_2O scissor vibration and extrapolated. When $\int \ln (I_0/I)\, d\bar{\nu} = 0$, n is 0.

[73] M. Volmer, Ph.D. Thesis, Leipzig University, 1910.

[74] A. Predwoditelew and A. Witt, *Z. Physik. Chem. (Leipzig)* **132**, 47 (1928).

[75] T. N. Rodin, *Advan. Catalysis* **5**, p. 39 (1953).

[76] B. Sansoni, *Angew. Chem.* **75**, 164 (1963).

[77] B. Sansoni, Bundesforschungsanstalt für Strahlenschutz, Neuherberg/near Munich, West Germany (private communication), 1963.

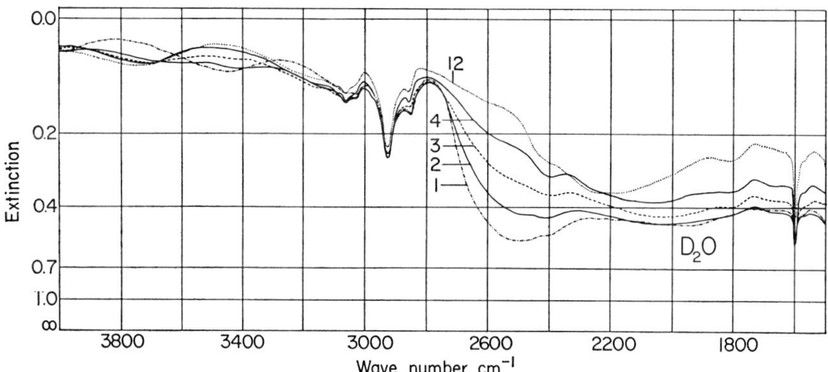

FIG. 72. Polystyrenesulfonic acid, D_2O hydrated, 5% cross-linked, 3 days of sulfonation: 1 → 4, decreasing atmospheric humidity; 12, thoroughly dried membrane.

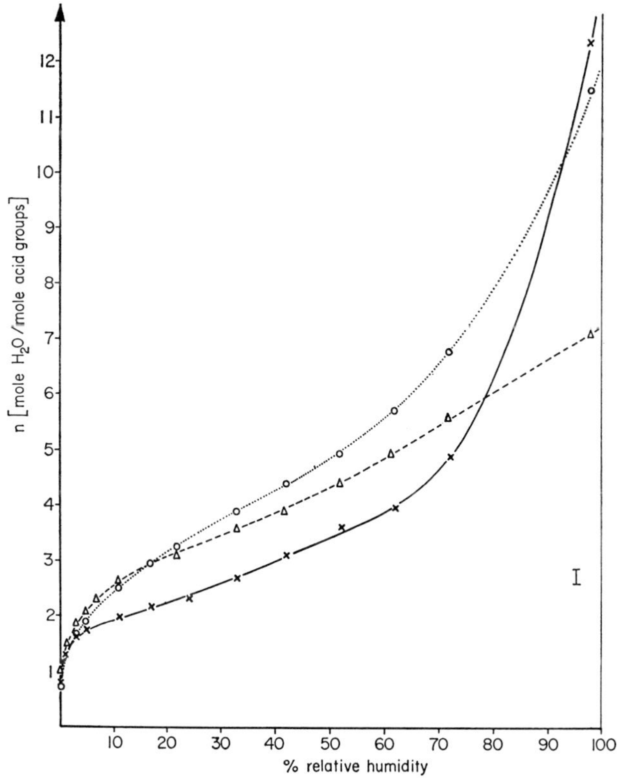

FIG. 73. Water adsorption isotherms (polystyrenesulfonic acid): ——, 1% cross-linked, 1.50 acid groups per benzene ring; · · · ·, 1% cross-linked, 1.06 acid groups per benzene ring; - - - -, 10% cross-linked, 1.39 acid groups per benzene ring.

used a chamber at a definite atmospheric humidity and placed the membranes on a scale which was covered with a transparent polyvinyl chloride membrane to prevent direct contact between the membrane and the glass. The membrane was clamped only at one end. We now followed the change of length as a function of the relative atmospheric humidity surrounding the membrane, with the aid of a microscope (for further details see footnote 78). The results are shown in Fig. 74. We found that the necessary correction amounts at most to only 5%.

The shape of the curves in Fig. 74, and, in particular, the determination of the humidity axis intersection is understandable from our results. We have seen in Section V.2 that increasingly strong hydrogen bridges between the $-\mathrm{S}(=\mathrm{O})(=\mathrm{O})\mathrm{OH}$ groups are formed with progressive drying of the membrane. These contract the exchanger network, so that the membrane contracts slightly with progressive drying. We shall see in Section V.10 that the hydrate structures become progressively larger with increasing atmospheric humidity over the membrane, since the ions tend to hydrate themselves. The pressure thus arising expands the exchanger network as far as is permitted by the mechanical stress of the cross-links in the resin network. The expansion in the case of the 1% cross-linked membrane should then be greater than in the case of the 10% cross-linked, exactly as we have observed (on this point see also footnote 79). At the inflection point of the experimental curve, the contracting effect is equal to the expansive pressure on the network. For this reason we have chosen this point as the humidity axis intersection.

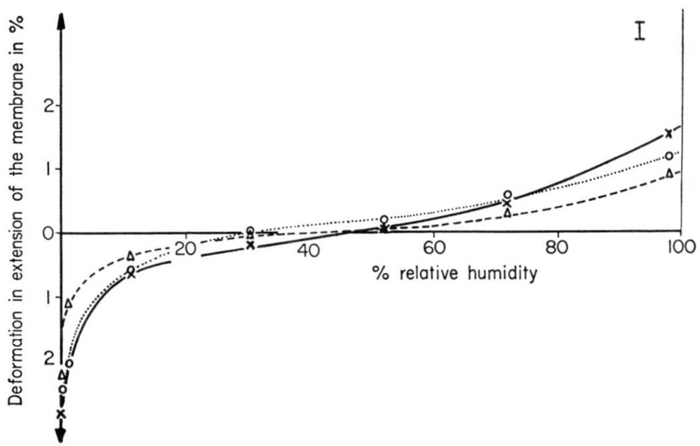

FIG.74. Changes in length of the membranes as a function of the degree of hydration (polystyrenesulfonic acid): ——, 1% cross-linked, 1.50 acid groups per benzene ring; ····, 1% cross-linked, 1.06 acid groups per benzene ring; – – – –, 10% cross-linked, 1.39 acid groups per benzene ring.

[78] H. Metzger, Ph.D. Thesis, University of Munich, 1966.
[79] G. Dickel and K. Bunzl, *Z. Physik. Chem.* (*Frankfurt*) **39**, 198 (1963).

V.5. Attachment of the Last Water Molecule to the $-\mathrm{S}(=\mathrm{O})(=\mathrm{O})\mathrm{OH}$ Group

We have seen in Section V.2.1 that, in polystyrenesulfonic acid, the band at 2405 cm^{-1} shows us the association to $-\mathrm{S}(=\mathrm{O})\begin{smallmatrix}\mathrm{OH\cdots O}\\ \mathrm{O\cdots HO}\end{smallmatrix}\mathrm{S}(=\mathrm{O})-$ groups. Further, we have seen, in Section V.1.2, that the integral extinction of the bands of the SO stretching vibrations is a measure of the true degree of dissociation of the $-\mathrm{S}(=\mathrm{O})(=\mathrm{O})\mathrm{OH}$ groups. The band of the SO single bond at 907 cm^{-1} (Fig. 57, Section V.1.2) is particularly suited for this. We must, however, bear in mind that the decrease of integral extinction of this band indicates increasing dissociation.

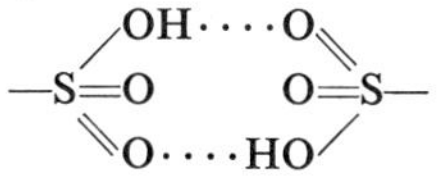

In Fig. 75, the integral extinction of the band at 2405 cm^{-1} is plotted against α_t. From this figure* we obtain:

Result 82: With decreasing degree of hydration the band at 2405 cm^{-1}, which indicates the association, appears. Even before this band appears, however, many of the $-\mathrm{S}(=\mathrm{O})(=\mathrm{O})\mathrm{OH}$ groups are no longer dissociated. The intercepts on the abscissas in Fig. 75 show directly that already 50–60% of the $-\mathrm{S}(=\mathrm{O})(=\mathrm{O})\mathrm{OH}$ groups are no longer dissociated when these groups begin to associate.

Are these "disassociated" but no longer dissociated $-\mathrm{S}(=\mathrm{O})(=\mathrm{O})\mathrm{OH}$ groups hydrated? If these $-\mathrm{S}(=\mathrm{O})(=\mathrm{O})\mathrm{OH}$ groups were not hydrated, the band of the stretching vibration of their free OH groups, i.e., OH groups not bound by hydrogen bridges, would be found in the region 3800–3500 cm^{-1} (e.g., see footnote 16). Considering the spectra of the membranes at low degree of hydration in Figs. 68–71, we see the following result.

Result 83: In the region 3800–3500 cm^{-1}, at low degrees of hydration, bands are sometimes found which are sharp but of an extremely low intensity.

* The extinction of the band at 2950 cm^{-1} cannot be considered in this discussion, because a band arises in this region with increasing degree of hydration (see Section V.8).

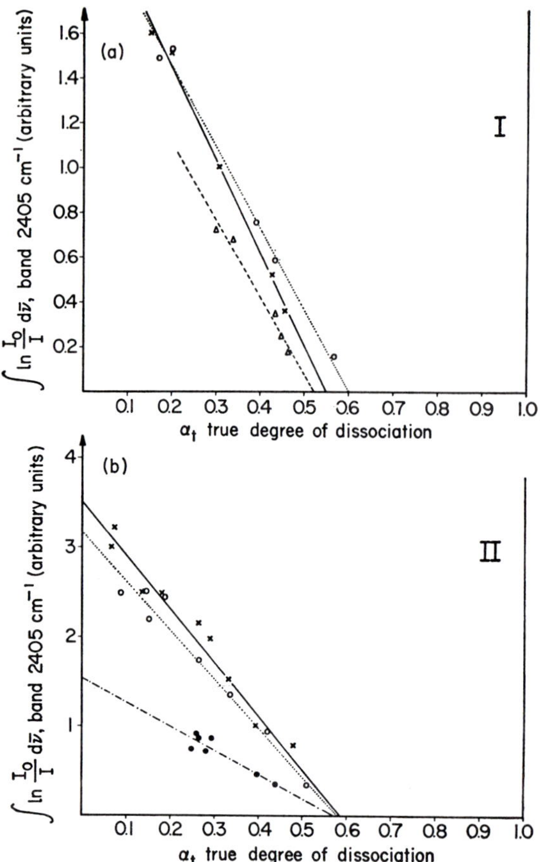

FIG. 75. The integral extinction (here and in what follows, the units for the integral extinction of the band are arbitrary but always the same) of the band at 2405 cm^{-1} (measure of the association of the —S(=O)(=O)—OH groups) as a function of the true degree of dissociation (here and in what follows, the data concerning sulfonation and cross-linking of the membrane for which each curve was obtained are to be found in Table 14). (a) slowly dried membranes, (b) rapidly dried membranes.

It follows that the number of free OH groups is very small. This is only possible if the great majority of the not yet associated —S(=O)(=O)—OH groups is hydrated. This result further shows that these molecules of water of hydration are bound to the oxygen atoms of neighboring ions through hydrogen bridges.

The last water molecule per $-\mathrm{S}\begin{smallmatrix} /\!\!/\mathrm{O} \\ =\mathrm{O} \\ \diagdown \mathrm{OH}\end{smallmatrix}$ group is then attached as shown in Fig. 78. Presumably to this water molecule there is frequently attached not only one acid OH group—as in the schematic representation in Fig. 78—but two, i.e., one to each lone electron pair of the molecule. The proton is not yet removed from the anion, a finding which is confirmed by the Results 84 and 85.

From Fig. 76(b) we obtain Result 84.

Result 84: The curves showing α_t as a function of the number of water molecules present per acid group do not pass through the origin. This confirms that the presence of one water molecule does not bring about dissociation of the $-\mathrm{S}\begin{smallmatrix} /\!\!/\mathrm{O} \\ =\mathrm{O} \\ \diagdown \mathrm{OH}\end{smallmatrix}$ groups.

How many water molecules per acid group are necessary, in polystyrenesulfonic acid, to remove the proton from the anion? In Fig. 77, $n^{\cdot}$, the number of water molecules on the proton removed from the anion, is plotted as a function of the true degree of dissociation.

The number $n^{\cdot}$ of water molecules present on the excess proton is obtained by subtracting from the quantity of water available per acid group the quantity of water of hydration on the not yet dissociated groups, and dividing the difference by α_t. The quantity of water present on the undissociated acid groups is obtained as follows: In Fig. 76(b) it is shown that, in the case of series I, 0.4 of such a water molecule is still present per acid group after thorough drying. In Fig. 75, it is shown that, when all the hydrogen bridges between the $-\mathrm{S}\begin{smallmatrix} /\!\!/\mathrm{O} \\ =\mathrm{O} \\ \diagdown \mathrm{OH}\end{smallmatrix}$ groups are broken, 0.4, 0.45, or 0.475 of such a water molecule is still present per acid group. This is based on the assumption that each undissociated group possesses only one water molecule.* From degree of dissociation zero to this degree of dissociation, we have therefore subtracted from the total quantity of water present that quantity of water molecules per acid group given by linear interpolation. In the region of higher degrees of dissociation, this water disappears linearly with increasing degree of dissociation.

Result 85: The proton is removed from the anion in the polystyrenesulfonic acid when two water molecules are available, i.e., when $H_5O_2^+$ groups are formed. This confirms that the proton is not removed from the $-\mathrm{S}\begin{smallmatrix} /\!\!/\mathrm{O} \\ =\mathrm{O} \\ \diagdown \mathrm{OH}\end{smallmatrix}$

* If more than one water molecule were present on these undissociated groups, the calculated $n^{\cdot}$ values would be too large. It could be supposed that the slight departure from the shape of the other curves shown by that of the 1% cross-linked membrane sulfonated for four days arises from this fact, assuming that it is a genuine effect; the removal of the proton could, in this case, be slightly suppressed by the high concentration of the anions.

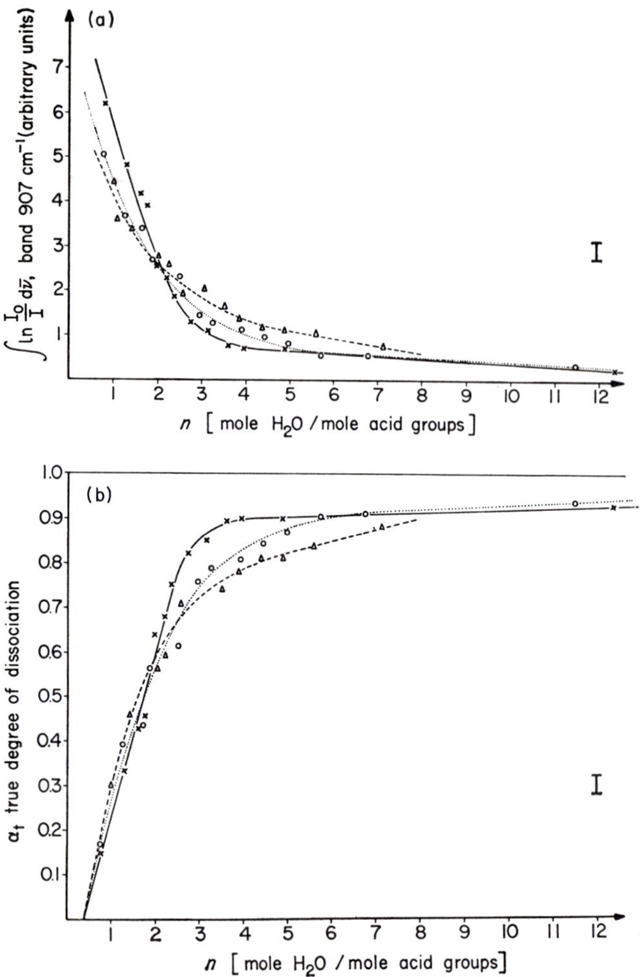

FIG. 76. (a) Integral extinction of the band of the SO single bond of the —S(=O)(=O)OH group as a function of the number of water molecules per acid group (polystyrenesulfonic acid, slowly dried). (b) The true degree of dissociation as a function of the number of water molecules per acid group (polystyrenesulfonic acid, slowly dried).

group when only one water molecule is attached to its hydrogen atom. The potential well in the hydrogen bridges, these bridges being those which the acid protons form with the last water molecule per acid group, must therefore be such that in the ground state the proton is at the water molecule. Figure 78 represents this schematically (see Section V.16).

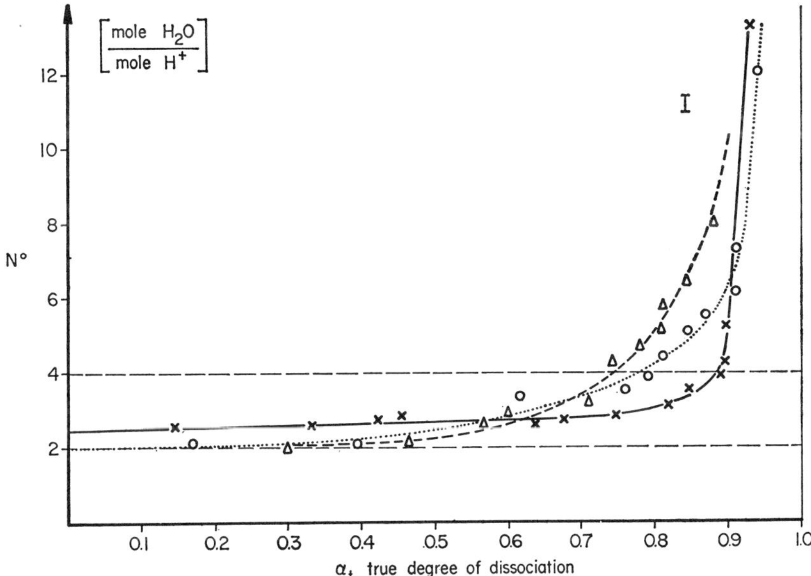

FIG. 77. The number of water molecules present at the dissociated excess proton as a function of the true degree of dissociation (polystyrenesulfonic acid, slowly dried).

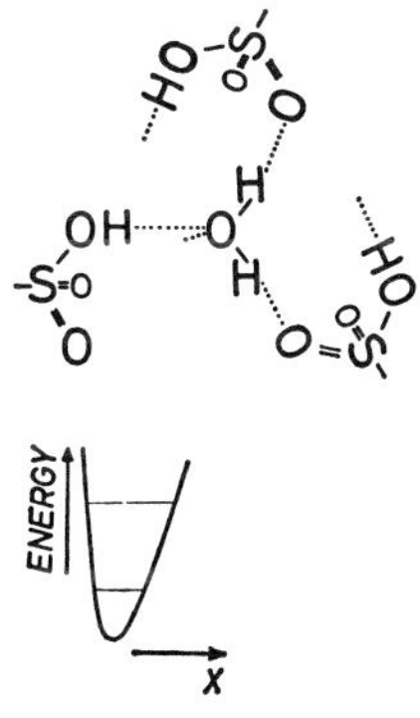

FIG. 78. The attachment of the last water molecule to the —S(=O)(=O)—OH groups and the potential curve in the hydrogen bridge formed by the acid proton (schematic representation).

OH and OD Stretching Vibration Bands, when the Last Water Molecule is Attached to the $-S(=O)(=O)OH$ *Group*

In Fig. 79 the spectra of thoroughly dried (high vacuum, 20°C) polystyrenesulfonic acid membranes are shown. All these membranes have been slowly dried. In Fig. 80 the spectrum is shown of a membrane dried, first slowly and then extensively, at 150°C. This figure also shows the spectrum of this membrane after it had become swollen again and then had been rapidly dried.* From Fig. 80, we see first of all the following result.

Result 86: Even after very extensive drying—assuming that the membrane has been dried slowly—water molecules are still present, as a weak band of a scissor vibration can still be observed at about 1640 cm^{-1}.

According to data given in Fig. 78, when the last water molecule has been attached to the $-S(=O)(=O)OH$ group, two kinds of hydrogen bridges and hence two bands of OH or OD stretching vibrations must occur. We now examine the spectra in Figs. 79 and 80, as well as those in Figs. 68 and 69, to see whether the bands of the OH or OD stretching vibrations of the two types of hydrogen bridges are to be found.

Result 87: If the polystyrenesulfonic acid membranes are dried slowly, a band is found at 3210 cm^{-1}, on hydration with H_2O, or at about 2400 cm^{-1}, on hydration with D_2O.†

We now must test whether the band at 3210 cm^{-1} is that of the overtone of the scissor vibration. From Fig. 81 we obtain:

Result 88: The band at 3210 cm^{-1} occurs also for membranes hydrated with HDO. From this follows that this band cannot be that of the overtone of the scissor vibration of a water molecule (Section III.2.1). Thus the band at 3210 cm^{-1} (OH) or at about 2400 cm^{-1} (OD) must be caused by the stretching vibration of the OH or OD group in a hydrogen bridge.

From Figs. 79 and 80 we obtain the next result.

Result 89: In the spectra of the membranes dried after hydration with D_2O, a clearly marked shoulder is found at about 2270 cm^{-1}. If the membranes are dried after hydration with H_2O, the corresponding band of the stretching

* In considering the spectra of thoroughly dried membranes, it must always be borne in mind that under these conditions of drying (high vacuum at 25°C) some $H_5O_2^+$ groups are still present, for a continuous absorption, even though weak, is still always found under these conditions (on this point see Section V.8). All $H_5O_2^+$ groups are removed only on drying at 150°C.

† In the 10% cross-linked membrane at 2405 cm^{-1}, or 2385 cm^{-1} with 1% cross-linking (see Fig. 79).

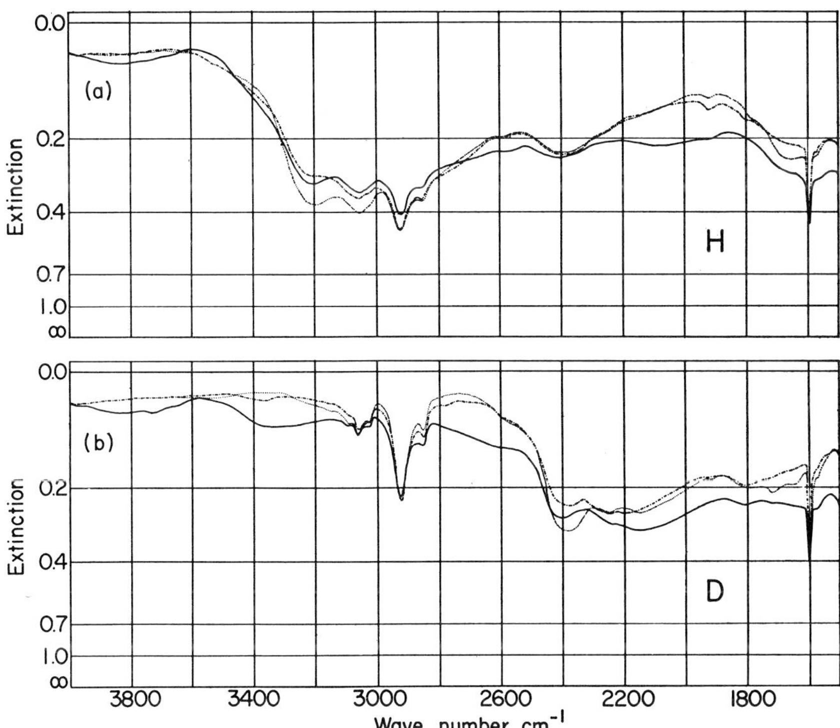

FIG. 79. Polystyrenesulfonic acid. All membranes first dried slowly, then thoroughly: ——, 10% cross-linked, 1.39 acid groups per benzene ring; · · · ·, 1% cross-linked, 1.50 acid groups per benzene ring; –·–·, 1% cross-linked, 1.06 acid groups per benzene ring; (a) membranes hydrated with H_2O before drying, (b) membranes hydrated with D_2O before drying.

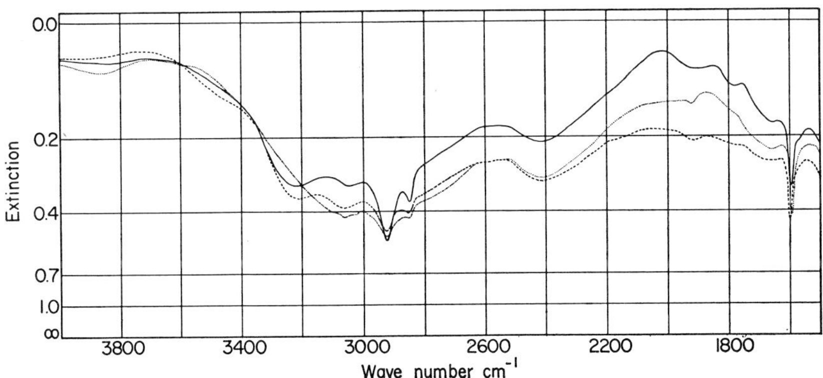

FIG. 80. Polystyrenesulfonic acid, 5% cross-linked, 3 days of sulfonation, H_2O hydrated: – – – –, first dried slowly and then thoroughly at 25°C; ——, first dried slowly and then strongly at 150°C; · · · ·, first rapidly and then thoroughly dried at 25°C.

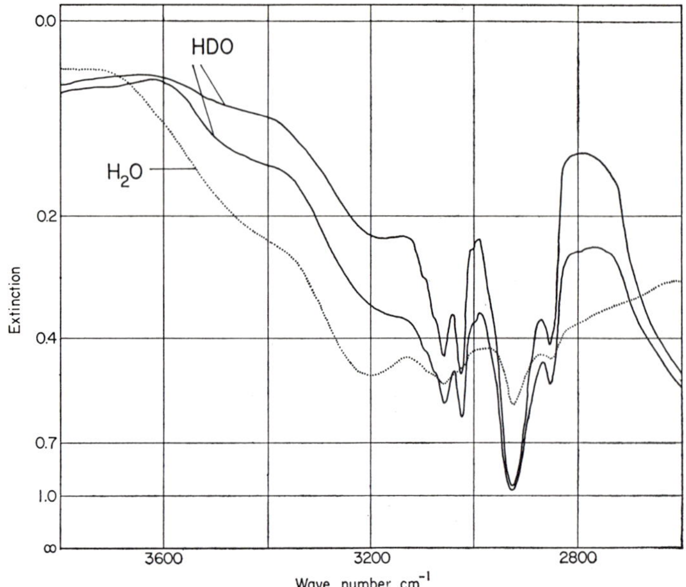

FIG. 81. Polystyrenesulfonic acid, 5% cross-linked, 3 days sulfonation: ····, H_2O hydrated; ——, HDO hydrated.

vibration of the OH groups is masked by the stretching vibrations of the ⪢CH groups at about 3050 cm^{-1}. Nevertheless, comparison of the spectra of the membranes hydrated with D_2O and H_2O respectively, in Fig. 79 shows clearly that a band arises at 3050 cm^{-1} in passing from D_2O to H_2O hydration. Figure 80 shows that, when the membrane is dried rapidly, only the weak bands of the ⪢CH stretching vibrations at about 3050 cm^{-1} are present, while, when the band at 3210 cm^{-1} is observed, a distinct band is also found at about 3050 cm^{-1}.

The conditions for observing the band at 3210 cm^{-1} (OH) or at about 2400 cm^{-1} (OD) and that at about 3050 cm^{-1} (OH) or about 2270 cm^{-1} (OD) show that these must be the bands of the stretching vibrations of the OH or OD groups in the two kinds of hydrogen bridges which arise when a water molecule is attached as shown in Fig. 78.

Figure 80 further shows Result 90.

Result 90: Even if the membrane is dried at 150°C, i.e., very strongly, the band at 3210 cm^{-1} does not vanish, i.e., this water molecule is not desorbed, in agreement with Result 86. It is thus very strongly attached. This is not strange, for, if the hydrate structures are broken down slowly, the last water molecule is attached by three strong hydrogen bridges, or perhaps even by four.

Dependence of Hydration on the Rate of Hydrate Structure Breakdown

We have considered hydration when the hydrate structures are broken down slowly. If they are broken down rapidly, essential differences appear. We see this on comparing Figs. 68 and 69 with Figs. 70 and 71, and in particular, on comparing Fig. 79 with Fig. 60.

Result 91: The band at 3210 cm^{-1} for OH, or 2400 cm^{-1} for OD, occurs only when the membrane has been dried slowly. Rapidly dried membranes do not exhibit this band.

Explanation: Only on progressive drying are $-\mathrm{S}(=\mathrm{O})(=\mathrm{O})\mathrm{OH}$ groups present. Thus, oxygen atoms with a strong hydrogen bridge acceptor property are increasingly present. Bearing this in mind, the result can be understood particularly clearly by considering Fig. 82. If only one water molecule is attached to the hydrogen atoms of some $-\mathrm{S}(=\mathrm{O})(=\mathrm{O})\mathrm{OH}$ group, the OH groups of this water molecule will generally be attached to the oxygen atoms of $\left(-\mathrm{S}{\cdots}\mathrm{O_3}\right)^-$ ions, at least at a point when not many anions have been rearranged. If now the membrane is further dried rapidly, this water molecule is desorbed. If, however, the membrane is dried more slowly, the water molecule is, to an increasing extent, offered the opportunity of finding not only an oxygen atom of an $\left(-\mathrm{S}{\cdots}\mathrm{O_3}\right)^-$ ion, but also one of an already rearranged ion, i.e., of an $-\mathrm{S}(=\mathrm{O})(=\mathrm{O})\mathrm{OH}$ group, as a hydrogen bridge acceptor. This water molecule, however, is then attached by stronger bridges, so that it can no longer be desorbed as easily. Important for the understanding of this is the conclusion to be drawn from Result 104, i.e., when the $H_5O_2^+$ groups are broken down completely, both molecules are desorbed immediately one after the other. Therefore, the last water molecule does not have time to find an oxygen atom of a $-\mathrm{S}(=\mathrm{O})(=\mathrm{O})\mathrm{OH}$ group as an acceptor.

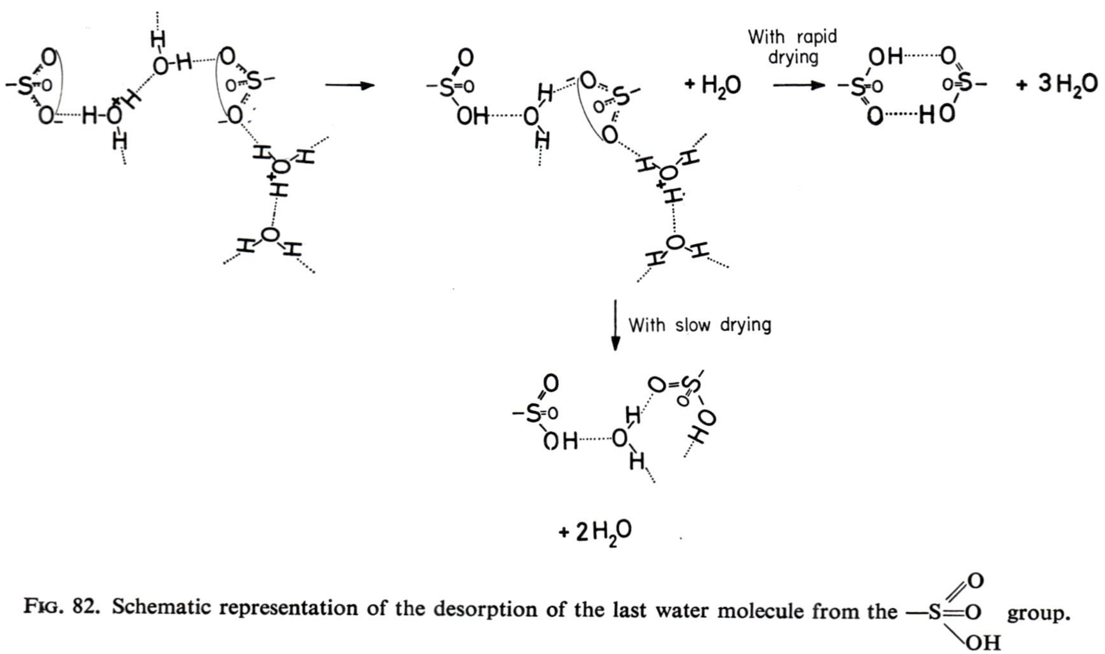

FIG. 82. Schematic representation of the desorption of the last water molecule from the —S(=O)(=O)—OH group.

In summary, we see that, if only one water molecule at the hydrogen atom of the $-S(=O)(=O)OH$ group is present, this group is not dissociated. On the one hand, this water molecule is bound to acid OH groups by hydrogen bridges. On the other hand, the OH groups of this water molecule are bound to oxygen atoms of neighboring anions. If the hydrate structure is broken down slowly, the last water molecule forms bridges, whose acceptors are oxygen atoms of $-S(=O)(=O)OH$ groups. The hydrogen bridge acceptor affinity of these oxygen atoms is much stronger than that of the oxygen atoms of the $-SO_3^-$ ions. Hence this water is so strongly bound that it cannot be removed even on very strong drying. The bands of the stretching vibration of the OH or OD groups in the two hydrogen bridges (Fig. 78) are found at 3210 cm^{-1} (OH) and about 3050 cm^{-1} (OH) or at about 2400 cm^{-1} (OD) and at about 2270 cm^{-1} (OD).

V.6. The Nature of the Group $H_5O_2^+$ and Tunnel Effect

We have seen that, in the polystyrenesulfonic acid, the proton is removed from the anion when two water molecules are available for it (Result 85), i.e., when $H_5O_2^+$ groups are formed. From Figs. 1(c, d) and 87(a), we obtain:

Result 92: In the spectra of polystyrenesulfonic acid hydrated with more than one water molecule per acid group, a continuous absorption is found. Beginning at the bands of the OH or OD stretching vibration of the water molecule, i.e., at about 3500 cm^{-1} or at about 2600 cm^{-1}, it extends toward smaller wave numbers. We now consider the nature of the groups producing this effect.

Result 93: Comparison of parts (a) and (b) in Fig. 83* shows that the extinction of the continuous absorption increases in the same way as the true degree of dissociation increases with an increase in degrees of hydration. In the Na^+ salt, no corresponding change in the spectral background with the degree of hydration is to be observed, as is shown by comparison of parts (b) and (c) of Fig. 83.

From Fig. 84(a, b) we obtain the following result:

Result 94: The extinction of the continuous absorption increases linearly

* The background in the spectra, caused by light losses—as by reflection—is always subtracted before the extinction of the continuous absorption is plotted. Further, it was supposed that the background in these spectra at 4000 cm^{-1} is independent of the degree of hydration. This had been proved by experiment.

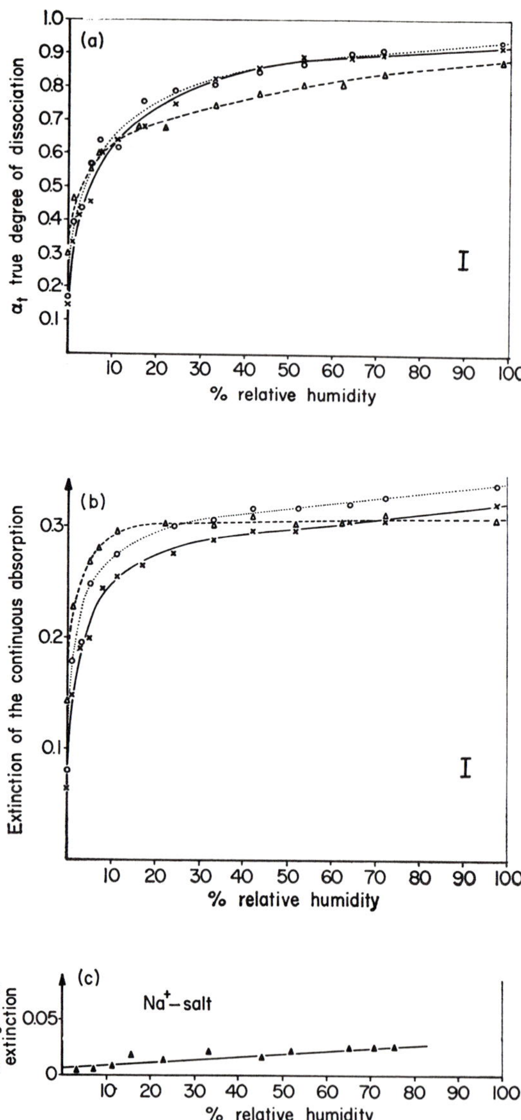

FIG. 83. (a) The true degree of dissociation as a function of the relative atmospheric humidity surrounding the membrane (polystyrenesulfonic acid, slowly dried). (b) Extinction of the continuous absorption (at about 2000 cm^{-1}) as a function of the relative atmospheric humidity surrounding the membrane (polystyrenesulfonic acid, slowly dried). (c) Extinction (background at 1950 cm^{-1}) as a function of the relative atmospheric humidity surrounding the membrane (Na$^+$ salt of polystyrenesulfonic acid).

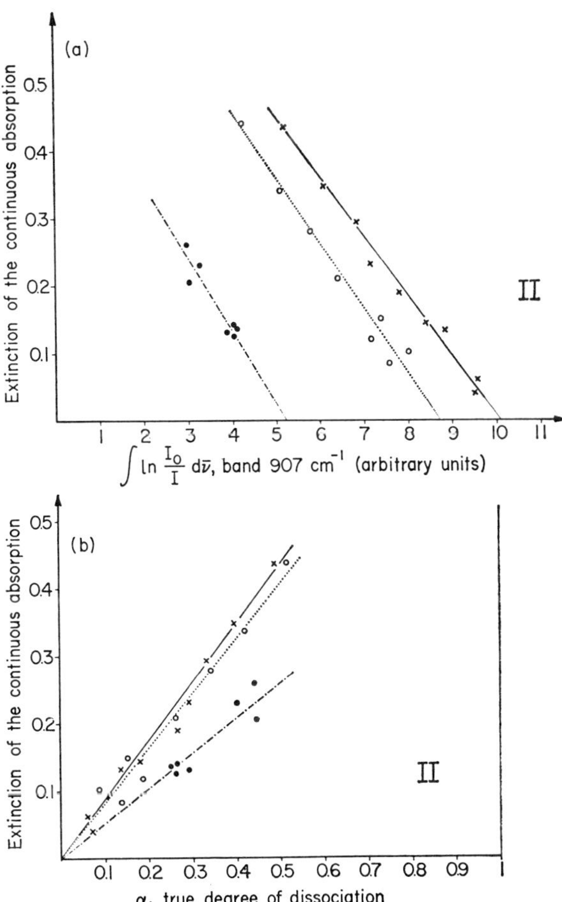

FIG. 84. The extinction of the continuous absorption at about 2000 cm^{-1} (polystyrenesulfonic acid, rapidly dried): (a) as a function of the integral extinction of the band at 907 cm^{-1}; (b) as a function of the true degree of dissociation.

with the true degree of dissociation, and hence with the concentration of the protons removed from the anion.* Accordingly, in conjunction with Result 85, two water molecules per excess proton are sufficient to produce the continuous absorption. Thus continuous absorption occurs when $H_5O_2^+$ groups are formed.

* This linear increase, however, is not found for high excess proton concentrations, as will be seen in Section V.11.2 (Result 114).

This result is confirmed by the spectra in Fig. 80 and the curves in Fig. 85.

Result 95: Even when the continuous absorption has completely disappeared with progressive drying, the band at 3210 cm^{-1} is still observed (see the spectrum drawn as a continuous line in Fig. 80). The curves in Fig. 85 do not pass through the origin of coordinates. Thus both figures show that water is still present on the $-S(=O)(=O)OH$ groups, even if the continuous absorption has completely vanished with progressive drying, as long as the membranes are dried slowly.

From this it follows that the presence of one water molecule on the hydrogen atoms of the $-S(=O)(=O)OH$ group does not give rise to continuous absorption.

The question of the processes giving rise to continuous absorption will occupy us further in the following sections. From these considerations we will obtain important information on the nature of the hydrated excess protons. First we must establish whether bands of the pyramid-shaped "ammonia-like" group H_3O^+ are found in hydrated polystyrenesulfonic acid when the $H_5O_2^+$ groups are present. To answer this question, a broader view must be taken.

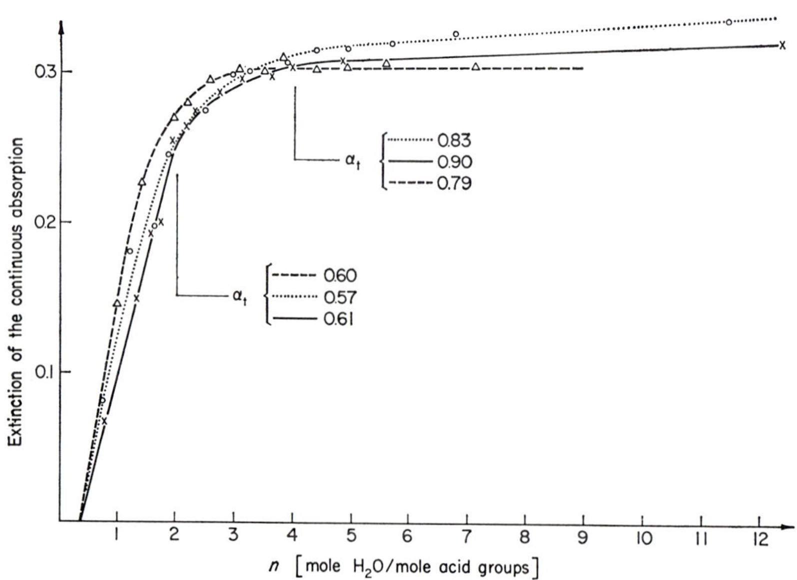

FIG. 85. The extinction of the continuous absorption at about 2000 cm^{-1} as a function of the number of water molecules per acid group (polystyrenesulfonic acid, slowly dried).

V.6.1. The Bands of H_3O^+ in Crystals

For a first indication of the ranges of wave numbers in which bands of H_3O^+ must be sought, we considered the published works on IR and Raman spectroscopy of crystals containing H_3O^+. Such crystals have been studied by numerous authors.[80-87,89] Crystals with the group $H_5O_2^+$ have been investigated by Gillard and Wilkinson,[89] and crystalline $H_9O_4^+Br^-$ by Rudolph and Zimmermann.[91]*

In these studies the following bands of H_3O^+ have been found generally: one of the antisymmetric stretching vibration in the range 3400–3100 cm^{-1}; one of the symmetric stretching vibration in the range 2650–2450 cm^{-1}; and, further, a band of the antisymmetric bending vibration at about 1700 cm^{-1}; and one of the symmetric bending vibration at about 1200 cm^{-1}. The latter is sometimes split through interactions in the crystal lattice. If these bands are compared with those of the ammonia molecules, it follows that a long-lived, pyramid-shaped H_3O^+ group is usually present in the crystals, at least in the cases investigated.

The bands of $H_5O_2^+$ in crystals† were observed by Gillard and Wilkinson[89] at about 2900, 2230, 1675, and 970 cm^{-1}.

V.6.2. The Excess Proton in $H_5O_2^+$

Are the bands of the hydrated H_3O^+ group to be found in the IR spectra of hydrated polystyrenesulfonic acid? To prove whether the H_3O^+ group vibrates as an entity, only the observation of the band of the symmetric

* Lundgren and Olovsson[92] nevertheless showed by X-ray structure investigations that in this case hydrate structures of other sizes are also present.

† X-Ray diffraction studies of the di- and trihydrates of HCl by Lundgren and Olovsson[88,88a] have elucidated the structure of the $H_5O_2^+$ group in this case, and have shown, in general, its particular stability.

[80] D. E. Bethell and N. Sheppard, *J. Chim. Phys.* **50**, C72 (1953).
[81] D. E. Bethell and N. Sheppard, *J. Chem. Phys.* **21**, 1421 (1953).
[82] C. C. Ferriso and D. F. Hornig, *J. Chem. Phys.* **23**, 1464 (1955).
[82a] C. C. Ferriso and D. F. Hornig, *J. Am. Chem. Soc.* **75**, 4113 (1953).
[83] J. T. Mullhaupt and D. F. Hornig, *J. Chem. Phys.* **24**, 169 (1956).
[84] R. C. Taylor and G. L. Vidalle, *J. Am. Chem. Soc.* **78**, 5999 (1956).
[85] N. F. Curtis, *Proc. Chem. Soc.* p. 410 (1960).
[86] P. A. Giguere and R. Savoie, *Can. J. Chem.* **38**, 2467 (1960).
[87] R. Savoie and P. A. Giguere, *J. Chem. Phys.* **41**, 2698 (1964).
[88] J.-O. Lundgren and I. Olovsson, *Acta Cryst.* **23**, 966, 971 (1967).
[88a] I. Olovsson, *J. Chem. Phys.* **49**, 1063 (1968).
[89] R. D. Gillard and G. Wilkinson, *J. Chem. Soc.* (*London*) p. 1640 (1964).
[90] M. Falk and P. A. Giguere, *Can. J. Chem.* **35**, 1195 (1957).
[91] J. Rudolph and H. Zimmermann, *Z. Physik. Chem.* (*Frankfurt*) **43**, 311 (1964).
[92] J.-O. Lundgren and I. Olovsson, *J. Chem. Phys.* **49**, 1068 (1968).

bending of the H_3O^+ at 1200 cm^{-1} can be used as a criterion, since the H_2O molecules present mask the remaining bands of the H_3O^+ group. We must also bear in mind that, in the spectra of dried polystyrenesulfonic acid,

OH····O

the band of the OH bending vibration in the —S=O O=S— lies at 1260

O····HO

cm^{-1} (Section V.2.1). Accordingly we consider spectra taken under conditions when all such bridges are broken.

Figure 86 shows spectra of polystyrenesulfonic acid: one of a membrane hydrated with H_2O and one hydrated with D_2O. If we compare these two, in passing from H_2O to D_2O hydration, a band at about 1200 cm^{-1} must vanish —insofar as there is a symmetric bending vibration of the hydrated H_3O^+

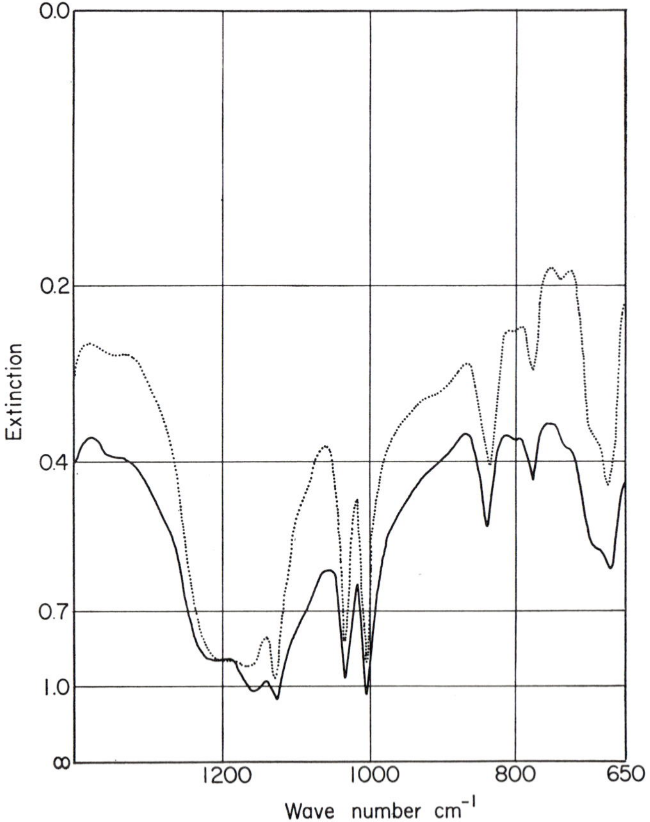

FIG. 86. Polystyrenesulfonic acid, 5% cross-linked, 3 days of sulfonation: ——, hydrated with H_2O at 98% relative atmospheric humidity; ····, hydrated with D_2O at the atmospheric humidity over a saturated solution of $BaCl_2$ in D_2O.

group—and the corresponding band of the hydrated D_3O^+ group at about 890 cm^{-1} must appear. We could not decide with complete certainty whether a band at about 1200 cm^{-1} does actually disappear on passing from H_2O to D_2O hydration,* since a band of the —SO_3^- ion lies at this position. We can be certain, however, that the bending vibration of D_3O^+ in question at about 890 cm^{-1} does not appear on hydration with D_2O. Thus Fig. 86 gives Result 96.

Result 96: In the polystyrenesulfonic acid, there is no band of the symmetric bending vibration of the hydrated hydronium ion. Also the corresponding band of $H_5O_2^+$, observed in the investigation of crystals,[89] is not found here.

The hydrated H_3O^+ does not execute vibrations as an entity in a noncrystalline medium. This—together with the other results—leads us to the hypothesis that the excess proton is extremely mobile in the $H_5O_2^+$ group. It is then very likely that the continuous absorption caused by the $H_5O_2^+$ group is closely connected with the mobility of the excess proton. We will also pose this as a hypothesis.

Suhrmann and Breyer,[93] in IR spectroscopic studies of acid solutions in the region of the overtones, found a similar continuous absorption. Wicke *et al.*,[94] discussing this work, proposed as a cause of this continuous absorption the fluctuation of the excess proton in the $H_9O_4^+$ complex (see Section V.9, also footnotes 95–102, and also the NMR investigation of inorganic ion exchangers by Ducros and Dupont[103]).

V.6.2.1. Tunnel Effect and Potential Well in $H_5O_2^+$

How does the excess proton move in $H_5O_2^+$ in the present cases? To answer this, we studied the question of whether the continuous absorption is temperature dependent.[104]

* Bethell and Sheppard (footnote 80, p. C72) investigated liquid and solid nitric acid monohydrate. In passing from the liquid to the solid state, an extremely intense band arose at 1134 cm^{-1}. This shows that the symmetric bending vibration, if it were to be observed at all, should be very intense, and should therefore appear in the spectrum as a pronounced band.

[93] R. Suhrmann and F. Breyer, *Z. Physik. Chem. (Leipzig)* **B23**, 193 (1933).
[94] E. Wicke, M. Eigen, and T. Ackermann, *Z. Physik. Chem. (Frankfurt)* **1**, 340 (1954).
[95] M. Eigen and L. De Maeyer, *Naturwissenschaften* **42**, 413 (1955).
[96] M. Eigen and L. De Maeyer, *Z. Elektrochem.* **59**, 986 (1955).
[97] M. Eigen and L. De Maeyer, *Proc. Roy. Soc. (London)* **A247**, 505 (1958).
[98] M. Eigen, *Z. Elektrochem.* **64**, 115 (1960).
[99] G. Ertel and H. Gerischer, *Z. Elektrochem.* **66**, 560 (1962).
[100] M. Eigen, *Angew. Chem.* **75**, 489 (1963).
[101] M. Eigen, L. De Maeyer, and H. C. Spatz, *Ber. Bunsenges. Physik. Chem.* **68**, 19 (1964).
[102] E. F. Caldin, "Fast Reactions in Solution." Blackwell, Oxford, 1964.
[103] P. Ducros and M. Dupont, *Compt. Rend.* **254**, 1409 (1962).
[104] G. Zundel and G.-M. Schwab, *J. Phys. Chem.* **67**, 771 (1963).

Figure 87 shows spectra of polystyrenesulfonic acid at 292° and at 85°K. These membranes were quenched so rapidly that the dissociation equilibrium was not altered.

Result 97: The continuous absorption is not temperature dependent in the range of 85–292°K. It follows that, on the movement of the excess proton between the H_2O molecules of $H_5O_2^+$, no activation energy has to be provided. The excess proton then tunnels in $H_5O_2^+$. Thus between the oxygen atoms in this group a more or less symmetrical double minimum potential with a low barrier is present, i.e., a potential as shown in Fig. 63(b) (for further details see Section V.11). The continuous absorption can generally serve in what follows as a criterion for the presence of protons tunneling in hydrogen bridges.

In this case it appears at first perhaps surprising that continuous absorption is not found in the IR spectra of crystals containing such groups, but, rather, the bands of these groups are found. We shall see, however, that this continuity of energy levels only occurs when the mutual distances and orientations of the hydrogen bridges with tunneling protons vary (Section V.11.5.4). In crystals, however, these quantities have discrete values. Further, the steric constraint exerted by the crystal structure deforms the potential well, in general, to such an extent that the tunnel frequency of the proton is greatly decreased.

V.6.2.2. Proton Boundary Structures

How can we describe the tunneling proton in its hydrate structures? The answer is obtained by considering the "boundary positions" in which the excess proton can be present in these structures. Figure 88 shows the "boundary positions" of the excess proton in the $H_5O_2^+$ group. These "boundary positions" we term proton boundary structures.[105] The hydrate structures with tunneling protons are then always described by the superposition of all proton boundary structures (also see footnote 106). As in the case of mesomeric boundary structures, the real condition, averaged over a period of time, is to be regarded as the superposition of all proton boundary structures. The proton can, in its individual boundary structures, be described by condition functions ψ_i. If these be multiplied by factors c_i, the probability amplitudes, and then added, the condition function $\psi = \sum c_i \psi_i$ of the excess proton in the hydrate structures is obtained.

V.6.2.3. Proton Tunneling and Individuality of Water Molecules in $H_5O_2^+$

Do the water molecules retain their individuality as vibrating groups when the proton tunnels between them? Figure 89 shows the relation between

[105] G. Zundel, H. Noller, and G.-M. Schwab, *Z. Elektrochem.* **66**, 129 (1962).
[106] H. Zimmermann, *Angew. Chem. Intern.* Ed. *English* **3**, 1 (1964).

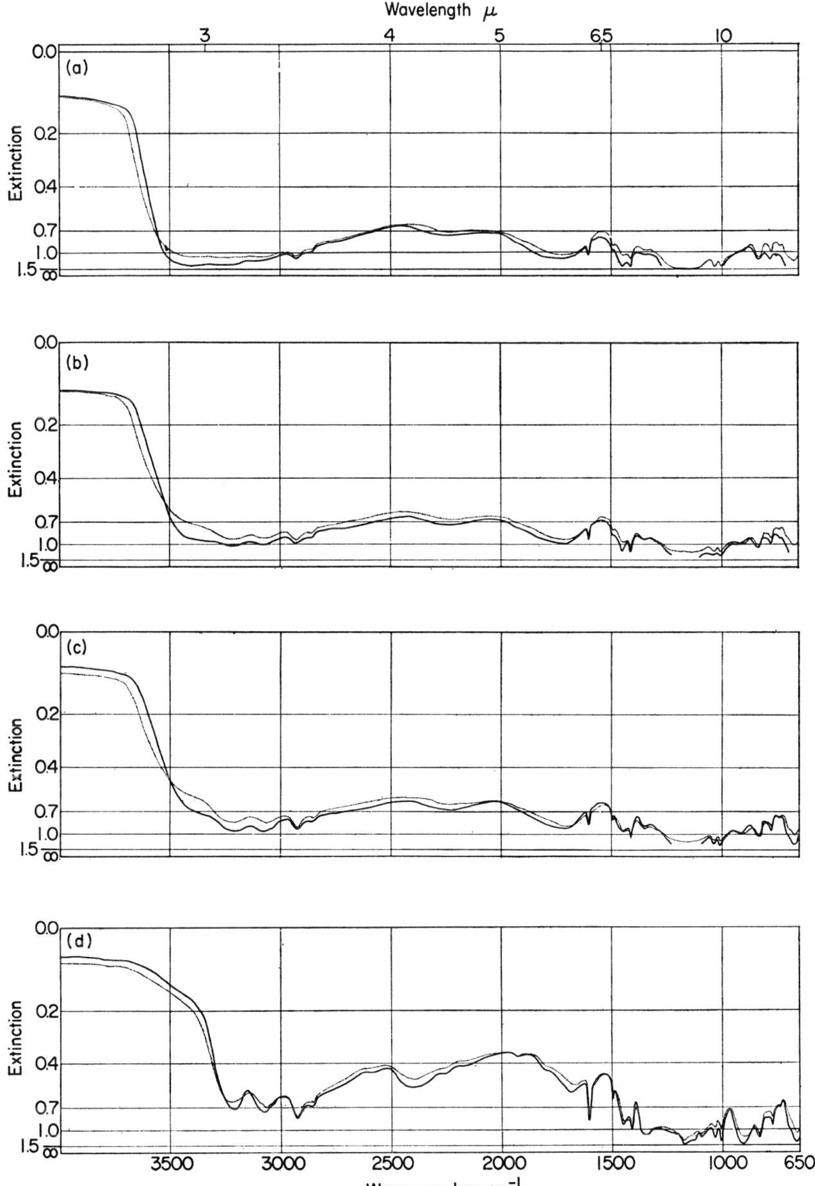

FIG. 87. Polystyrenesulfonic acid, 5% cross-linked, 5 days of sulfonation: · · · ·, 292°K; ——, 85°K (suddenly quenched). H_2O hydrated: (a) 33% relative atmospheric humidity; (b) 11% relative atmospheric humidity; (c) 7% relative atmospheric humidity; (d) approximately 1% relative atmospheric humidity.

FIG. 88. The proton boundary structures of the $H_5O_2^+$ group.

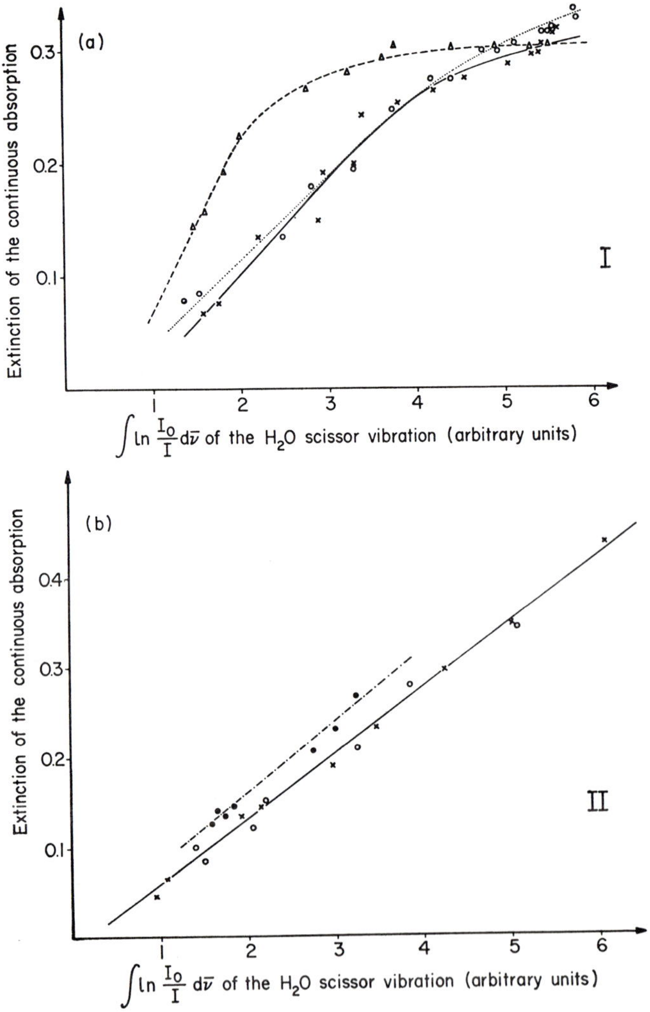

FIG. 89. The extinction of the continuous absorption at about 2000 cm^{-1} as a function of the integral extinction of the H_2O scissor vibration (polystyrenesulfonic acid), (a) in the case of the slowly dried membranes, (b) in the case of the rapidly dried membranes.

the extinction of the continuous absorption and the integral extinction of the scissor vibration. We see:

Result 98: At low degrees of hydration, the extinction of the continuous absorption increases linearly with the integral extinction of the scissor vibration of the water. In series II, these curves almost pass through the origin. This is not the case in series I, since water molecules are still present after thorough drying (Section V.5).

In the region of low degrees of hydration, the changes of the spectrum as a function of the degree of hydration are essentially determined by the formation of $H_5O_2^+$ groups. If the water molecules of $H_5O_2^+$ were to lose their individuality as vibrating groups through the tunneling of the proton, then at low degrees of hydration only the extinction of the continuous absorption would increase with the degree of hydration. The integral extinction of the scissor vibration would increase essentially only when larger groups were formed. This contradicts Result 98.

Hence it follows that the water molecules in the $H_5O_2^+$ group retain their individuality as vibrating groups. That means that, while the proton tunnels between the two water molecules in the $H_5O_2^+$ group, these two water molecules are preserved as individual vibrating groups and execute the scissor vibration. Thus, the three hydrogen nuclei on an oxygen atom do not exchange their roles rapidly in the $H_5O_2^+$ (see Section V.17).

In summary, we see that the excess proton in the hydrogen bridge of the $H_5O_2^+$ group is mobile through tunneling. The two water molecules of this group preserve their individuality as vibrating groups. The tunneling of the excess proton in the $H_5O_2^+$ group is closely connected with the appearance of an intense continuous absorption.

V.7. Homogeneous Liquid Acid and Base Solutions

Now that we have become familiar with the continuous absorption and the nature of the $H_5O_2^+$ group, we consider the observations on homogeneous liquid acid and base solutions. Later we will return to the $H_5O_2^+$ group in connection with the hydration of polystyrenesulfonic acid.

Investigations of aqueous acid solutions in the IR overtone vibration region have been carried out by Suhrmann and Breyer,[93] Biermann and Gilmour,[107] and Meerlender.[108] Corresponding investigations of acid solutions in the IR fundamental vibration region are also available (see footnotes

[107] W. J. Biermann and J. B. Gilmour, *Can. J. Chem.* **37**, 1249 (1959).

[108] G. Meerlender, Ph.D. Thesis, under R. Suhrmann, Technische Hochschule, Braunschweig, 1959.

81, 87, 90, and 109–115). A Raman-spectroscopic investigation has appeared from Busing and Hornig.[116]

Bands of H_3O^+? Falk and Giguere,[90,110] in IR investigation of aqueous acid solutions, found bands at about 2900, 1700, and 1200 cm^{-1}. They assigned these to the H_3O^+ group, following the results of the above-mentioned studies of crystals containing H_3O^+. Ackermann,[112] however, found in sodium hydroxide solution exactly the same bands as those found by Falk and Giguere in the investigation of acid solutions. This is shown by Fig. 90, taken from Ackermann's work. This author writes[112]: "The influence of the dissolved alkali hydroxides on the IR spectrum of water is not to be distinguished from the changes observed in solutions of hydrogen halides. Obviously the effect involved here is equally strongly caused by H^+ and OH^- ions." Hence, it should be assumed that the bands observed in acid solutions are not caused by the H_3O^+ group.

How are these bands then to be assigned? First, the bands at about 2900 and 1700 cm^{-1}: Grahn,[117-121] by the use of molecular orbital treatment, calculated properties of the hydrate structures which are formed about the excess proton.* He writes[119] on the nature of the external water molecules in the group $H_9O_4^+$ (see Section V.9): "The outer hydrogen atoms are ca. 20% more positive than in ordinary water . . .". Hence these external water molecules in $H_9O_4^+$ are able to form hydrogen bridges with other water molecules which are stronger than are those formed between two normal water molecules. The band at about 2900 cm^{-1} is thus that of the stretching vibration of the OH groups in the hydrogen bridges which the external water molecules in $H_9O_4^+$ form with corresponding acceptor groups—oxygen

* Grahn has reported a calculation for the hydrated OH^- ion (in footnote 122). (See also footnote 123.)

[109] G. E. Walrafen and D. M. Dodd, *Trans. Faraday Soc.* **57**, 1286 (1961).

[110] M. Falk and P. A. Giguere, *Chem. Eng. News* **35** (40), 59 (1957).

[111] P. A. Giguere and R. Savoie, *Can. J. Chem.* **38**, 2467 (1960).

[112] T. Ackermann, *Z. Physik. Chem.* (*Frankfurt*) **27**, 253 (1961).

[113] T. Ackermann, *Z. Physik. Chem.* (*Frankfurt*) **41**, 113 (1964).

[114] G. E. Walrafen, *J. Chem. Phys.* **40**, 2326 (1964).

[115] K. Stopperka, *Z. Anorg. Allgem. Chem.* **345**, 277 (1966).

[116] W. R. Busing and D. F. Hornig, *J. Phys. Chem.* **65**, 284 (1961).

[117] R. Grahn, *Arkiv Fysik* **19**, 147 (1961).

[118] R. Grahn, *Arkiv Fysik* **21**, 1 (1962).

[119] R. Grahn, *Arkiv Fysik* **21**, 13 (1962).

[120] R. Grahn, *Arkiv Fysik* **21**, 81 (1962).

[121] R. Grahn, On the hydration of the proton. Ph.D. Thesis, Uppsala University, Sweden, 1962.

[122] R. Grahn, *Arkiv Fysik* **28**, 85 (1964).

[123] R. Grahn, *Acta Chim. Scand.* **19**, 153 (1965).

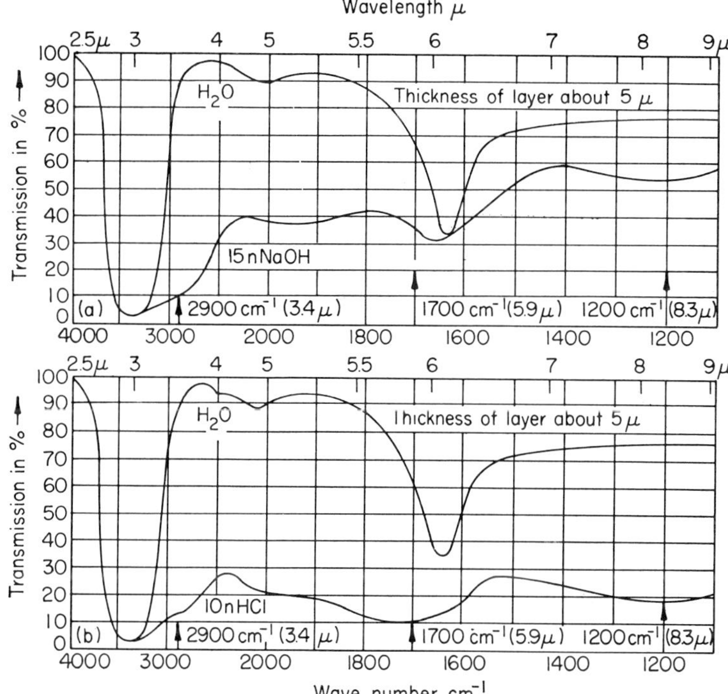

FIG. 90. IR spectra. (Sincerest thanks are due to Dr. T. Ackermann of the Physico-Chemical Institute, Muenster, Westfalen, West-Germany, for providing this figure.) (a) 15 N NaOH solution and pure water, (b) 10 N HCl solution and pure water.

atoms of anions or oxygen atoms of other water molecules. The scissor vibration of these external water molecules then causes the band at about 1700 cm^{-1}. The most surprising observation, however, is that of the weak band at about 1200 cm^{-1} in acids and alkalis. The water molecule has no band of a fundamental vibration at this wave number, but the band of the first overtone of the restricted torsion of the water molecule bound by hydrogen bridges does lie there.[124] The observed band is of low intensity, so it is conceivable that it could be ascribed to the overtone of the restricted torsion.

All this is in good agreement with the interpretation which Ackermann[112] gives of his findings. He writes: "Obviously this is primarily not a definite spectrum of the hydronium ion, but rather a solvation effect determined to an equal extent by the OH^- and H^+ ion in the solvent.*"

* Biermann and Gilmour[107] come to the same conclusion as a result of a new investigation in the spectral region from 1.4 to 1.8μ.

[124] E. Hartert and O. Glemser, *Z. Elektrochem.* **60**, 746 (1956).

Continuous absorption: In the studies of homogeneous liquid acid and alkali solutions, however, a corresponding continuous absorption was found. This continuous absorption was already observed in the overtone region of the spectra of acid solutions in 1933 by Suhrmann and Breyer.[93] This investigation was continued by Meerlender.[108] The same continuous absorption was found by Ackermann[112,113] in the fundamental region in a study of solutions of HCl and NaOH (on this point see Fig. 90) in H_2O, as well as of DCl and NaOD in D_2O.[113]

In summary, we see that no band of the H_3O^+ group is found in the spectroscopic investigation of homogeneous liquid acid solutions. The bands actually found are ascribed to the water molecules surrounding the excess proton. Continuous absorption was observed both in the spectra of aqueous HCl and aqueous NaOH solutions.

V.8. The $H_5O_2^+$ Group in Polystyrenesulfonic Acid

With increasing degree of hydration, $H_5O_2^+$ groups are built up; we have discussed their nature in Section V.6.

Are the hydrate structures surrounding the excess proton still bound to the anions in polystyrenesulfonic acid? The band of the antisymmetric stretching vibration of the $—SO_3^-$ ion and the consideration of the bands of the OH stretching vibrations both give information on this point. From Fig. 57, we obtain the following result.

Result 99: The band of the antisymmetric stretching vibration of the $—SO_3^-$ ion—which is found at about 1200 cm^{-1} when the proton has been removed from the anion—is always split into two broad bands. Also the degeneracy of the antisymmetric stretching vibration is removed in the acid. Thus the C_{3v} symmetry of the $—S\left(\begin{smallmatrix} O^- \\ O \\ O \end{smallmatrix}\right)$ ion is disturbed. This splitting of the band is so pronounced that it certainly cannot arise from the bond linking the $—SO_3^-$ ion to the benzene ring, as is shown by comparison with the results in Section IV.2.

Hence, it follows that the removal of the degeneracy arises from the difference in the extent to which the various oxygen atoms of the $—S\left(\begin{smallmatrix} O^- \\ O \\ O \end{smallmatrix}\right)$ ion function as acceptors of hydrogen bridges. The hydrate structures surrounding the hydrated excess protons are bound by hydrogen bridges to oxygen atoms of the $—S\left(\begin{smallmatrix} O^- \\ O \\ O \end{smallmatrix}\right)$ ions.

We now consider Figs. 68 and 69:

Result 100: If a shoulder is found at all in the spectra of polystyrenesulfonic acid with increasing degree of hydration at about 3615 cm^{-1}, it is very faint. Thus there are only a very few free OH groups of molecules of water of hydration.

This confirms the result that the hydrate structures surrounding the proton removed from the anion, i.e., in particular the $H_5O_2^+$ group, are bound by hydrogen bridges to oxygen atoms of the $\left(-\mathrm{S}\begin{matrix}\mathrm{O^-}\\ \mathrm{O}\\ \mathrm{O}\end{matrix}\right)$ ions. This is shown in Fig. 91.

We have seen in Section V.6.2.3 that the water molecules in $H_5O_2^+$ retain their individuality as vibrating groups. Hence it follows that if the hydrogen nuclei of these water molecules tunnel at all in the hydrogen bridges which they form to the oxygen atoms of the $\left(-\mathrm{S}\begin{matrix}\mathrm{O^-}\\ \mathrm{O}\\ \mathrm{O}\end{matrix}\right)$ ions, the tunneling frequency is very much smaller than that of the excess proton in the bridge in $H_5O_2^+$.

The bands of the anion point to the same conclusion. If the excess proton were to tunnel in these bridges with the same frequency as that of the excess proton in the single bridge in $H_5O_2^+$, it would then often be present briefly at the oxygen atom of an anion.

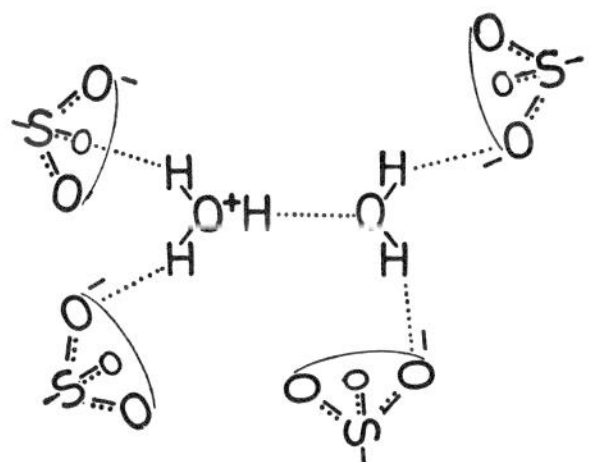

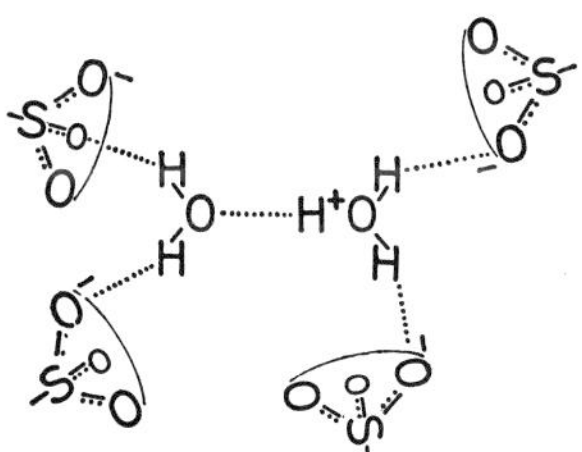

FIG. 91. The $H_5O_2^+$ group in polystyrenesulfonic acid (for clarity, only one excess proton is shown).

But since the bonding electrons in an anion are spontaneously rearranged on the approach of the excess proton, the electrons in the anions should be frequently rearranged for this reason. That is to say, the bands at 1350, 1172, and 907 cm^{-1} should still be found after dissociation, but should be considerably broadened. We now examine Fig. 57.

Result 101: The half widths of the bands at 1350 and 907 cm^{-1} do not change in dependence on the degree of hydration.

This confirms the result obtained above that, if the excess proton does tunnel between the oxygen atoms of the anions and of the water molecules, its frequency is very small, compared with that of the excess proton in the bridge in $H_5O_2^+$.

Further, from the fact that the bands of the $-S(=O)(=O)OH$ group vanish almost entirely with increasing degree of hydration (Result 59, Section V.1.2), it follows that the excess proton has only a small location probability near the anion under these conditions, so that only the two boundary structures shown in Fig. 91 have appreciable weight.

Where does the band of the OH or OD stretching vibration of the hydrogen bridge by which the $H_5O_2^+$ or $D_5O_2^+$ group is bound to the oxygen atoms of the anions occur? We consider Figs. 70, 71, and 92. The last shows spectra of polystyrenesulfonic acid with a double layer sample. These membranes have been hydrated with D_2O to a degree at which the presence of $D_5O_2^+$ groups is favored.

Result 102: The intensity of the broad intense band at about 2900 cm^{-1} does not essentially change with increasing degree of hydration (Figs. 70 and 71). However, if the band at 2405 cm^{-1} were to vanish with increasing

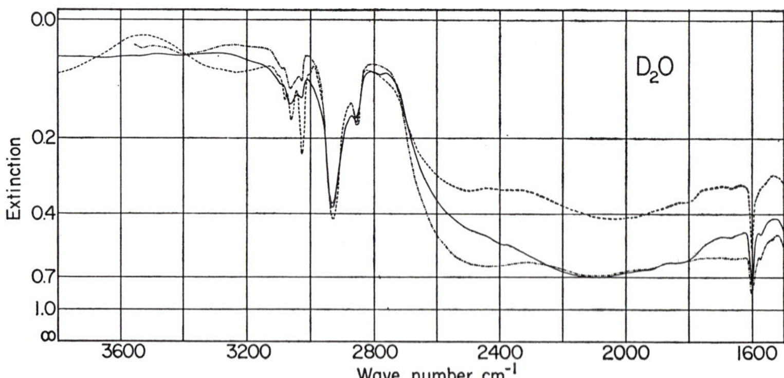

FIG. 92. Polystyrenesulfonic acid (two membranes one over the other). All membranes rapidly and thoroughly dried after hydration, to a high degree, with D_2O; ----, 10% cross-linked, 0.48 $-SO_3D$ groups per benzene ring; ····, 1% cross-linked, 1.50 $-SO_3D$ groups per benzene ring; -·-·, 10% cross-linked, 1.06 $-SO_3D$ groups per benzene ring.

degree of hydration, the band near 2950 cm^{-1} should also vanish, since we know that both bands are to be ascribed to vibrations of the OH group in the hydrogen bridges between $-S(=O)(=O)OH$ groups (Sections V.2 and V.3). A new broad intense band arises at about 2900 cm^{-1} with increasing degree of hydration, which takes the place of the disappearing 2950 cm^{-1} band. The corresponding broad band of the OD stretching vibration is found at about 2100 cm^{-1}. These bands are extremely broad. Corresponding bands are observed at higher degrees of hydration, i.e., when the $-S(=O)(=O)OH$ groups are no longer associated, so that these bands can no longer arise from the acid OH or OD groups (see Figs. 68, 69, and, particularly, 72).

With respect to the dependence on degree of hydration and the position of this band, it can only be that of the stretching vibration of the OH or OD groups in the bridges linking the water molecules surrounding the tunneling proton with the oxygen atoms in their neighborhood. At low degrees of hydration, the hydrogen bridges preferentially involved are those linking the $H_5O_2^+$ group here discussed to neighboring oxygen atoms of the $-SO_3^-$ ions; at higher degrees of hydration, bridges are preferentially those of the $H_9O_4^+$ group bound either to the neighboring oxygen atom of the $-SO_3^-$ ions or to other water molecules. As already mentioned in Section V.7, a corresponding band was found at about 2900 cm^{-1} by Falk and Giguere[90] and Ackermann[112] in the spectra of aqueous HCl and NaOH solutions.

That these bands are so extremely broad is not surprising if one considers the nature of the $H_5O_2^+$ (Section V.6) or $H_9O_4^+$ (Section V.9) group. The broadening is caused by the fluctuating electromagnetic field associated with the tunneling of the excess proton.

Figure 75 yields further interesting information on the processes occurring in the breakdown of the $H_5O_2^+$ group.

Result 103: The true degree of dissociation increases linearly with increasing association of the $-S(=O)(=O)OH$ groups through hydrogen bridges, if this association occurs on progressive drying.

From this it follows that, when the breakdown of the hydrate structures is so far advanced that the $-S(=O)(=O)OH$ groups can associate, the two water molecules in the $H_5O_2^+$ desorb in quick succession.

If we compare Fig. 2, with Fig. 57, we see the following result.

Result 104: If the membrane is dried rapidly, the band of the symmetric stretching vibration of the $\left(-S\begin{smallmatrix}O\\O\\O\end{smallmatrix}\right)^-$ ion at 1034 cm^{-1} disappears almost completely (Fig. 2). At comparable drying conditions, however, the band does not disappear completely if the membrane is dried slowly (Fig. 57). If we compare Figs. 70 and 71 with Figs. 68 and 69, we see that the continuous absorption at about 2000 cm^{-1} is more strongly reduced in the first pair of figures than in the second. This is the case, even though the first pair of figures shows spectra obtained on a double layer sample. Thus, on thorough drying, the continuous absorption disappears much more completely with rapid than with slow drying. This is quite clearly demonstrated in Fig. 80, on comparing the dotted and continuous curves.

It follows that, if drying proceeds slowly, then even on thorough drying, not all $\left(-S\begin{smallmatrix}O\\O\\O\end{smallmatrix}\right)^-$ ions are rearranged into $-S\begin{smallmatrix}=O\\=O\\-OH\end{smallmatrix}$ groups. Further, $H_5O_2^+$ groups are left, and these are not broken down by thorough drying. We can understand this, since we know that—assuming drying has been carried out slowly—water molecules are still present after thorough drying. They are bonded to the oxygen atoms of the $-S\begin{smallmatrix}=O\\=O\\-OH\end{smallmatrix}$ groups through hydrogen bridges. These oxygen atoms are thus blocked as acceptors for the OH groups of neighboring acid groups. The breakdown of the H_5O_2 group is thereby hindered.

In summary, we see that, if two water molecules are available for the acid proton in polystyrenesulfonic acid, it is removed from the anion. $H_5O_2^+$ groups are then found. The excess proton moves by tunneling in the hydrogen bridge in this group. The water molecules of this group are bound by strong hydrogen bridges to oxygen atoms of the $\left(-S\begin{smallmatrix}O\\O\\O\end{smallmatrix}\right)^-$ ions. These water molecules retain their individuality as vibrating groups. The band of the OH stretching vibration of these hydrogen bridges is found at about 2900 cm^{-1}, that of the OD stretching vibration at about 2200 cm^{-1}. Figure 91 shows the proton boundary structures.

V.9. Nature of The $H_9O_4^+$ Group

How does the excess proton behave when the hydrate structure is enlarged? Wicke *et al.*[94] showed, from measurements of heat capacity, that the $H_9O_4^+$ group is a particularly stable group in the network of the hydrate structures surrounding the dissociated proton. This was subsequently confirmed by numerous authors[119,125–141] using varying methods. Comprehensive surveys of this work are given by Eigen[100] and Clever.[142] The individual findings are collected in Table 15.

Do our experimental results demonstrate the special character which the $H_9O_4^+$ group assumes in the network of hydrate structures? To answer this question, we studied the dependence of the integral extinction of the scissor vibration of the molecule of water of hydration (a) on the number of H_2O molecules present per $—SO_3H$ group [Fig. 93(a)], (b) on $n^{\cdot}$, the number of water molecules attached to the excess proton removed from the anion [Fig. 93(b)] (for determination of $n^{\cdot}$ see p. 151). If we bear in mind that for $n^{\cdot} = 4$, α_t is always a little less than 1, then Fig. 93(b) indicates:

Result 105: The integral extinction of the band of the scissor vibration increases strongly with the attachment of the first four water molecules to the proton removed from the anion, i.e., with the building up of $H_9O_4^+$ groups. It increases only very slightly with the attachment of additional water molecules.

It follows that the extinction coefficient of the scissor vibration of the external water molecules, in the group formed by four water molecules surrounding the excess proton, is very large (see Section V.10.1). Accordingly, the $H_9O_4^+$ group is of special importance in the network of the hydrate structures. This is exactly what we should expect on the basis of published results.

Wicke *et al.*[94] made the assumption, based on reconsideration of the IR measurements of Suhrmann and Breyer,[93] that the excess proton in the $H_9O_4^+$

[125] H. Ulich, *Z. Elcktrochcm.* **36**, 497 (1930).
[126] A. M. Azzam, *Z. Elektrochem.* **58**, 889 (1954).
[127] A. M. Azzam, *Z. Physik. Chem. (Frankfurt)* **32**, 309 (1962).
[128] E. Glueckauf, *Trans. Faraday Soc.* **51**, 1235 (1955).
[129] D. G. Tuck and R. M. Diamond, *Proc. Chem. Soc.* p. 236 (1958).
[130] R. M. Diamond, *J. Phys. Chem.* **63**, 659 (1959).
[131] D. G. Tuck and R. M. Diamond, *J. Phys. Chem.* **65**, 193 (1961).
[132] A. H. Laurene, D. E. Campbell, S. E. Wiberley, and H. M. Clark, *J. Phys. Chem.* **60**, 901 (1956).
[133] W. H. Baldwin, C. E. Higgins, and B. A. Soldano, *J. Phys. Chem.* **63**, 118 (1959).
[134] K. N. Bascombe and R. P. Bell, *Discussions Faraday Soc.* **24**, 158 (1957).
[135] R. P. Bell, "The Proton in Chemistry." Methuen, London, 1959.
[136] M. Eigen and L. De Maeyer, *in* "The Structure of Electrolytic Solutions" (W. Hamer, ed.), p. 64. Wiley, New York, 1959.
[137] H. D. Beckey, *Intern. Kongr. Elektronenmikroskopie 4, Berlin, 1958, Verhandl.* (1960).
[138] H. D. Beckey, *Z. Naturforsch.* **14a**, 712 (1959).
[139] H. D. Beckey, *Z. Naturforsch.* **15a**, 822 (1960).
[140] P. F. Knewstubb and A. W. Tickner, *J. Chem. Phys.* **38**, 464 (1963).
[141] G. Perrault, *Compt. Rend.* **256**, 4203 (1963).
[142] H. L. Clever, *J. Chem. Educ.* **40**, 637 (1963).

TABLE 15

EVIDENCE OF THE SPECIAL POSITION OF THE $H_9O_4^+$ GROUP WITHIN THE HYDRATE STRUCTURES

Method	Medium	Authors	Reference
Entropy measurement	aqueous solution	Ulich	125
Temperature dependence of specific heat	aqueous solution	Wicke, Eigen, and Ackermann	94
Calculation by statistical mechanics	aqueous solution	Azzam	126, 127
Consideration of the activity coefficient	aqueous solution	Glueckauf	128
Extraction	solvent mixtures	Tuck and Diamond	129,131
		Diamond	130
		Laurene, Campbell, Wiberley, and Clark	132
		Baldwin, Higgins, and Soldano	133
Discussion of Hammet's acid function	aqueous solution	Bascombe and Bell	134
		Bell	135
Effective cross section in protolytic reactions	aqueous solution	Eigen and De Maeyer	97, 136
Mass spectrum:			
(a) field emission	water vapor	Beckey	137, 138, 139
(b) glow discharge	water vapor	Knewstubb and Tickner	140
Molecular orbit process[a]	—	Grahn	119
IR spectroscopy	polyelectrolyte	Zundel and Metzger	72

[a] Grahn[119] not only shows that the $H_9O_4^+$ group is exceptionally stable, but also, in particular, that the attachment of another water molecule directly on the "H_3O^+" group is unfavorable from the energy point of view.

group moves fluctuatingly.* We know that the excess proton in the hydrogen bridge of the $H_5O_2^+$ group tunnels. If now two additional water molecules are attached to this group, the extinction of the continuous absorption per excess proton does not change, and, furthermore, is temperature independent over the interval 292–85°K, even at high degrees of hydration [Fig. 87(a)]. Regarding the tunneling of the excess proton and its consequences, the $H_9O_4^+$ group shows no real difference in its behavior from the $H_5O_2^+$ group. However,

* It may also be pointed out that Darmois-Sutra and Darmois[143,144] in their attempt to calculate the anomalous proton conductivity on the basis of the Trude-Lorentz theory of electron conductivity, made the assumption that the excess proton is mobile in a cage of four water molecules. They assumed thermal motion.

[143] G. Sutra and E. Darmois, *Compt. Rend.* **222**, 1286 (1946).

[144] G. Darmois-Sutra and E. Darmois, *Z. Elektrochem.* **59**, 659 (1955).

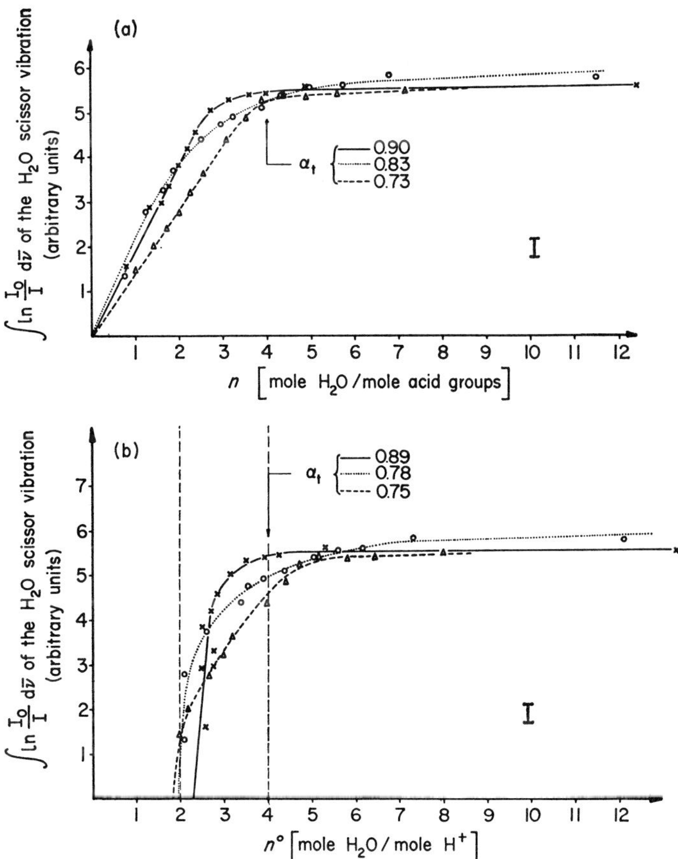

FIG. 93. The integral extinction of the H_2O scissor vibration (polystyrenesulfonic acid, slowly dried), (a) as a function of the number of H_2O molecules per acid group, (b) as a function of the number of H_2O molecules present at the proton removed from the anion.

in view of the symmetry of the $H_9O_4^+$ group, none of the three hydrogen bridges can be favored, i.e., in all three hydrogen bridges there must exist a potential with a low barrier which is symmetrical enough so that the protons can tunnel. Thus the three inner protons in this group play the part of the excess proton, while the three outer water molecules retain their individuality as vibrating groups; this also follows from Result 105. Thus the $H_9O_4^+$ group in the hydrate structure network is to be represented by the proton boundary structures shown in Fig. 94 (see also Section V.17).

In summary, we see that the $H_9O_4^+$ group occupies a special position in the network of the hydrate structures. The three internal protons in this group play the part of the excess proton. They tunnel in their hydrogen bridges.

Hence the $H_9O_4^+$ group must be represented by the four proton boundary structures shown in Fig. 94. The external water molecules retain their individuality as vibrating groups.

V.10. Hydrate Structure Network in Polystyrenesulfonic Acid at a High Degree of Hydration

Figure 77 indicates the following:

Result 106: The number of water molecules present at the proton removed from the anion increases slowly at first with increasing degree of dissociation. This changes abruptly when the majority of the $-S(=O)(=O)OH$ groups is dissociated.

It follows that the $H_5O_2^+$ group is strongly favored over the formation of larger groups (see second footnote p. 163). At first $H_5O_2^+$ groups are preferentially formed, and then $H_9O_4^+$ groups, while the $-S(=O)(=O)OH$ groups progressively dissociate.

Even when the degree of hydration increases, the groups surrounding the hydrated excess proton remain bound by hydrogen bridges to the oxygen atoms of the $\left(-SO_3\right)^-$ ions, as is shown by Result 99. The band of the stretching vibration of the OH groups in these hydrogen bridges lies—as already mentioned in Section V.8—at 2900 cm^{-1} (OH) or at 2100 cm^{-1} (OD), and is very broad. Further, the network of hydrate structures is only slightly disintegrated by an increase of the degree of hydration, since the number of free OH groups is small, as shown by Result 100. This is still the case when 10 to 12 water molecules per excess proton are present, as shown by comparison with the water adsorption isotherms in Fig. 73.

Attachment of Further Water Molecules to the $H_9O_4^+$ Group

From Figs. 68, 69, and 72 we obtain Result 107.

Result 107: A strongly marked band arises at about 3400 cm^{-1} with increasing degree of hydration with H_2O. A corresponding band arises at about 2500 cm^{-1} on D_2O hydration.

We have seen in Chapter IV that the stretching vibrations of the OH groups in the hydrogen bridges binding the molecules of water of hydration to the oxygen atoms of $\left(-SO_3\right)^-$ ions are found in the region 3460–3400 cm^{-1}, insofar as the hydrogen bridge donor property of these groups is not too

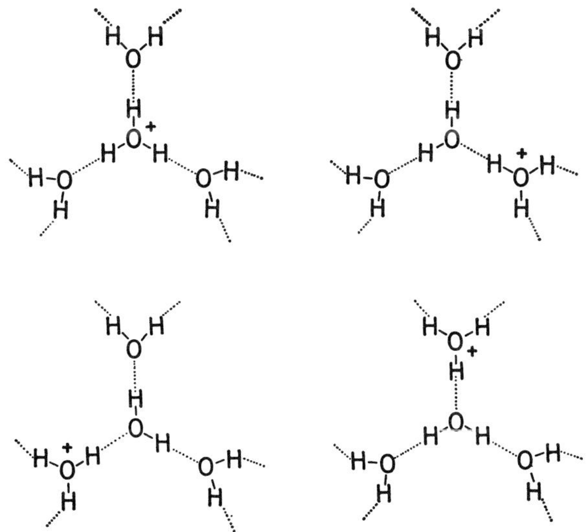

FIG. 94. The proton boundary structures of the $H_9O_4^+$ group.

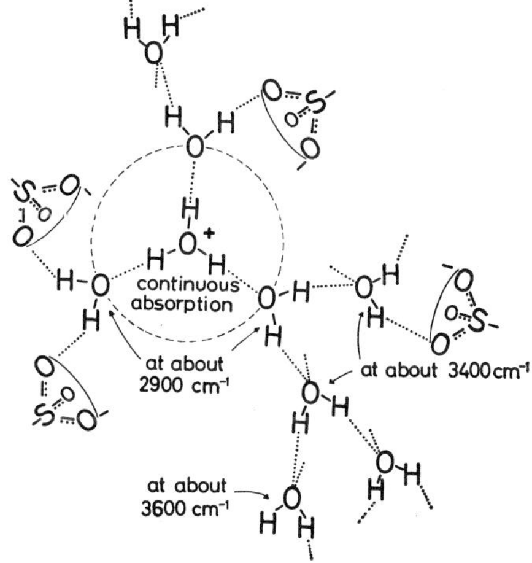

FIG. 95. The network of hydrate structures of hydrated polystyrenesulfonic acid (for clarity, only one excess proton, and of this only one boundary structure, is shown).

strongly enhanced by the cations. Further, we have seen in Chapter III that the OH stretching vibration of the hydrogen bridges in pure liquid water lies at 3420 cm^{-1} (see also Fig. 11). Hence the stretching vibration of the OH groups in the hydrogen bridges of the water molecules which do not themselves belong to the $H_9O_4^+$ group also must be sought in the region 3460–3400 cm^{-1}.

The position of the observed band, and its dependence on the degree of hydration, show that the band at about 3400 cm^{-1} (OH) or at about 2500 cm^{-1} (OD) is that of the stretching vibration of the OH or OD group, respectively, in the hydrogen bridges of the water molecules which are attached to the $H_9O_4^+$ or $D_9O_4^+$ group with increasing degree of hydration. These water molecules are bound by hydrogen bridges partly to the oxygen atoms of the $\left(-S\begin{matrix}O^-\\O\\O\end{matrix}\right)$ ions and partly to the oxygen atoms of other water molecules. The band at about 3400 cm^{-1} (OH) or 2500 cm^{-1} (OD) is thus caused by the OH or OD stretching vibration of two different hydrogen bridges in the network of hydrate structures. The contribution of the bridges between water molecules is increased with increasing degree of hydration, for in the 1% cross-linked, 4-day sulfonated membrane, if hydrated at 98% relative atmospheric humidity, more than 11 water molecules per acid group are present (see isotherm, Fig. 73).

Thus a network of hydrate structures is present, as shown in schematic representation in Fig. 95. For clarity, only one excess proton is shown. The nature of the excess proton in this network is described by the superposition of the four proton boundary structures, as discussed in Section V.9 (see Fig. 94).

In summary, we see that only a slightly disintegrated network of hydrate structure is built up around the excess proton (Fig. 95). The nature of the $H_9O_4^+$ group in this network of hydrate structures has already been established in Section V.9. The stretching vibration of the OH or OD groups in the hydrogen bridges of the water molecules attached to the $H_9O_4^+$ group is observed independently of whether the acceptors are oxygen atoms of $\left(-S\begin{matrix}O^-\\O\\O\end{matrix}\right)$ ions or lone electron pairs of other water molecules at about 3400 cm^{-1} (OH) or at about 2500 cm^{-1} (OD).

V.10.1. Dependence of the Extinction Coefficients of the Water Molecules on the Degree of Hydration

From Fig. 93 we have seen that the integral extinction of the band of the scissor vibration at a high degree of hydration hardly increases at all with the attachment of further water molecules, although the number of water molecules present, e.g., in the 1% cross-linked membrane, increases almost threefold after the attachment of four water molecules per acid group, as is shown by the isotherm (Fig. 73). Hence the increase of the integral extinction which the additional water molecules cause must be compensated by a decrease of the extinction coefficient of the external water of the $H_9O_4^+$ group with an increasing degree of hydration. The extinction coefficient of the scissor vibration of the external water molecules in the $H_9O_4^+$ group is considerably greater when this group is bound directly to the oxygen atoms of $-S(O^-)(O)(O)$ ions than when further water molecules are present between these groups and the $-SO_3^-$ ions. This is understandable, for an electric field extends from the excess proton to the anions. It is weaker at the OH groups of the external water molecules of the $H_9O_4^+$ group, the greater the distance between the excess proton and the anions; the weaker this field, the smaller the dipole moment induced in the OH groups of these water molecules. In general, however, IR bands of molecules become more intensive, the more polar these molecules are. It is therefore plausible that the extinction coefficient of the scissor vibration decreases with decreasing field, i.e., decreasing induced dipole moment. It must be borne in mind, however, that it is not the dipole moment but the transition moment which is effective for the intensity and that no strict relation has as yet been put forward for the relation between these two moments. (Concerning the connection between intensity changes and their dependence on changes in the dipole moments of groups in molecules see also the publications of Schmid *et al.*[145-149])

The situation is similar with the band of the OH stretching vibration at about 3400 cm^{-1}. We consider Figs. 68 and 69 and the isotherms in Fig. 73.

Result 108: In passing from 71 to 98% relative atmospheric humidity surrounding the membrane, the number of water molecules present per acid group increases in the one case by more than four, in the other case by more than eight. However, the increase of the integral extinction of the band at about 3400 cm^{-1} is not proportionately large. Hence the extinction coefficient in the case of the band at 3400 cm^{-1} also decreases with an increasing degree of hydration.

V.11. Continuous Absorption and Proton Dispersion Force

V.11.1. Systems Showing Continuous Absorption

The Continuous Absorption in Spectra of Acids

We have studied the continuous absorption more closely in the polystyrenesulfonic acid. The same continuous absorption was also observed by Ackermann[112,113] in aqueous solutions of HCl (Section V.7). In what follows,

[145] E. D. Schmid, V. Hoffmann, R. Joeckle, and F. Langenbucher, *Spectrochim. Acta* **22**, 1615 (1966).
[146] E. D. Schmid and V. Hoffmann, *Spectrochim. Acta* **22**, 1621 (1966).
[147] E. D. Schmid and V. Hoffmann, *Spectrochim. Acta* **22**, 1633 (1966).
[148] E. D. Schmid and R. Joeckle, *Spectrochim. Acta* **22**, 1645 (1966).
[149] E. D. Schmid, *Spectrochim. Acta* **22**, 1659 (1966).

we shall also find the continuous absorption in the spectra of hydrated polystyreneselenonic acid (Result 123) and of hydrated polystyrenethiophosphonic acid (Result 129). The spectrum of saturated aqueous *p*-toluenesulfonic acid solution is shown in Fig. 96.

Result 109: In the spectrum of saturated aqueous *p*-toluenesulfonic acid solution, an intense continuous absorption is found. This spectrum is quite similar to that of polystyrenesulfonic acid at a high degree of hydration.

Spectra of sulfuric acid are given in Fig. 97(a).

Result 110: Sulfuric acid shows a continuous absorption. This becomes more intense with progressive dilution in the range of concentration 100–80% w/w H_2SO_4. As in the other cases, it begins at the bands of the OH stretching vibration of the hydrogen bridges and extends toward smaller wave numbers.

The Continuous Absorption in Nonaqueous Systems

Spectra of saturated solutions of *p*-toluenesulfonic acid monohydrate in methanol and dimethylsulfoxide are shown in Fig. 97(b, c).*

Result 111: Both the *p*-toluenesulfonic acid monohydrate solution in methanol and that in dimethylsulfoxide show a powerful continuous absorption which begins at the bands of the OH stretching vibration of the hydrogen bridges and extends toward smaller wave numbers. It follows that the excess proton tunnels.

The bands of the $-S\left(\begin{smallmatrix}O\\O\\O\end{smallmatrix}\right)^{-}$ ion show that the acid protons have been removed from the anions [Fig. 97(b, c)]. In the case of dimethylsulfoxide, we further observe from Fig. 97(c), along with the continuous absorption, a band of the OH stretching vibration of relatively weak hydrogen bridges at 3425 cm^{-1}. This indicates that the single water molecule of the monohydrate has no further interaction with the excess proton. Accordingly it must be assumed that the excess proton links the oxygen atoms of two dimethylsulfoxide molecules through a hydrogen bridge, in which it tunnels. The structure can then be described by the two proton boundary structures shown in Fig. 98.*

Result 112†: The continuous absorption is also found in an anhydrous

* For IR spectroscopic investigations and band assignments of sulfoxides see footnotes 149a, 150, 151 and for the acidity of acids in dimethylsulfoxide see 152, 153.

† Private communication from Dr. T. Ackermann, University of Muenster, Westfalen, West Germany.

[149a] W. D. Horrocks, Jr., and F. A. Cotton, *Spectrochim. Acta* **17**, 134 (1961).
[150] T. Cairns, G. Eglinton, and D. T. Gibson, *Spectrochim. Acta* **20**, 31 (1964).
[151] T. Cairns, G. Eglinton, and D. T. Gibson, *Spectrochim. Acta* **20**, 159 (1964).
[152] E. C. Steiner and J. M. Gilbert, *J. Am. Chem. Soc.* **85**, 3054 (1963).
[153] E. C. Steiner and J. M. Gilbert, *J. Am. Chem. Soc.* **87**, 382 (1965).

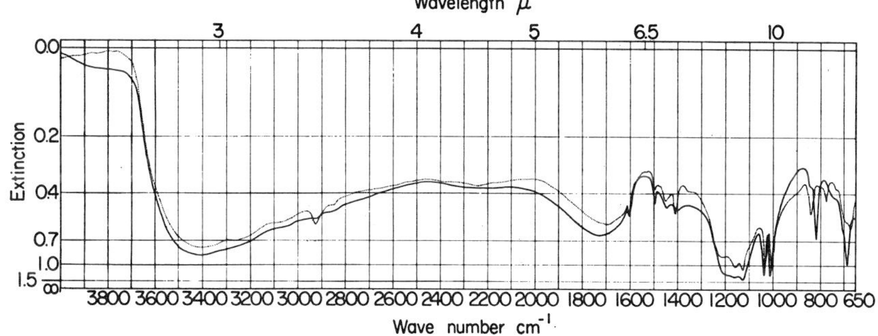

FIG. 96. ——, *p*-Toluenesulfonic acid saturated aqueous solution; · · · ·, polystyrenesulfonic acid. 5% cross-linked, 3 days sulfonation, hydrated with H_2O at 98% relative atmospheric humidity:

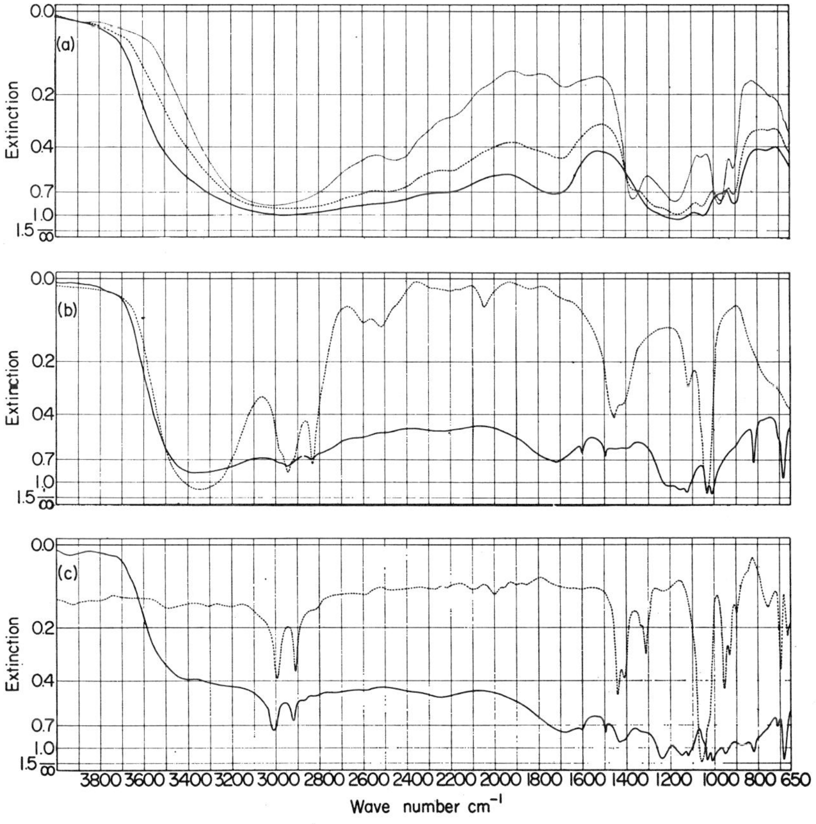

FIG. 97. IR spectra. (a) Sulfuric acid: · · · ·, 92.0 %w/w H_2SO_4; – - – -, 88.0 %w/w H_2SO_4; ——, 80.6 %w/w H_2SO_4. (b) · · · ·, CH_3OH; ——, saturated solution of *p*-toluenesulfonic acid monohydrate in CH_3OH. (c) · · · ·, $(CH_3)_2SO$; ——, saturated solution of *p*-toluenesulfonic acid monohydrate in $(CH_3)_2SO$.

solution of HCl in methanol. It follows that the excess proton also tunnels in the hydrogen bridge between two methanol molecules.

Note added in proof: Investigations of solutions of *p*-toluenesulfonic acid in dimethylsulfoxide (DMSO), water, and methanol showed[153a]:

(a) In the spectra of anhydrous solutions of *p*-toluenesulfonic acid in DMSO and methanol, an intensive continuous absorption is found. This indicates that the acid proton tunnels between the oxygen atoms of the solvent molecules. Thus, proton dispersion forces act between these bridges with tunneling protons (Section V,11.5.2). In the DMSO solution the band of the antisymmetric stretching vibration of the $—SO_3^-$ ions shows no doublet structure, which confirms that the acid proton is not in the neighborhood of the $—SO_3^-$ ion, but is present in hydrogen bridges between the DMSO molecules.

(b) For anhydrous solutions of *p*-toluenesulfonic acid in methanol and for water solutions, the extinction of the continuous absorption per excess proton is, within experimental error, constant in the investigated range (2500–650 cm^{-1}). In contrast, the continuous adsorption in the spectra of anhydrous *p*-toluenesulfonic acid–DMSO solution shows a marked structure. This is clear when the solvate structure for water and DMSO are compared. In order that a continuous absorption may be found, it is necessary that the distances and orientations of the hydrogen bridges with tunneling protons have a statistical random distribution (Section V.11.5.4). The proton dispersion forces promote the orientation among these hydrogen bridges with the tunneling protons. In water solutions, this orientation of the bridges with tunneling protons is counteracted by the constraint which the whole network of the hydrate structure brings to bear on the groups with the tunneling protons. In DMSO solutions the $[(CH_3)_2SOH^+ \cdots OS(CH_3)_2]$ groups are not linked to one another by hydrogen bridges and therefore the orientation caused by the proton dispersion forces is unhampered. Thereby, however, the statistical random distribution of the orientations and distances of the hydrogen bridges with tunneling protons that is present in aqueous solutions changes to give an accumulation value of distances and orientations. This leads to the observed structure of the continuous absorption.

(c) In the temperature range of 20–90°C, the extinction of the continuous absorption per tunneling proton is strongly temperature dependent in aqueous solutions, only somewhat so in methanol (anhydrous solution), and almost completely temperature-independent in DMSO (anhydrous solutions). In the water solution and in that of methanol, the extinction per tunneling proton increases with increasing temperature This is easy to understand if it is borne in mind that the hydrogen bridges of the network of the hydrate structure become weaker with increasing temperature and the structure network becomes more flexible. Thus, the orientation of the bridges of the tunneling protons to one another is less hampered. With this orientation the proton dispersion forces between the bridges become stronger. This increase in interaction causes an increase in the virtual dipole moment **p** in the bridges with tunneling protons, and thus an increase in the extinction coefficient, and hence of the continuous absorption. This interpretation is confirmed by the following results: first, in DMSO in which there is already an orientation among the hydrogen bridges with tunneling protons at 25°C (b), no increase of extinction of the continuous absorption with rising temperature is observed. Second, in the spectrum of the aqueous solution with rising temperature a similar structure of the continuous absorption as in DMSO solution starts to form.

[153a] I. Kampschulte-Scheuing and G. Zundel, in preparation.

In summary, it follows that the continuous absorption can be found, not only when excess protons tunnel in the hydrogen bridge in $H_5O_2^+$, i.e., between two water molecules, but also when protons tunnel in the hydrogen bridges between other molecules.

The Continuous Absorption in the Spectra of Alkalis

Ackermann[112] also observed the continuous absorption in a study of aqueous NaOH solutions (Section V.7).

Figure 99(a) shows the spectrum of a membrane of poly(*p*-dimethylamino)styrene [$—N(CH_3)_2$ groups]. In Fig. 99(b) the broken curve is the spectrum of a membrane of hydrated poly(*p*-trimethylammonium)styrene iodide [$—N(CH_3)_3^+ \ I^-$ groups], and the continuous curve is that of a membrane of hydrated poly(*p*-trimethylammonium)styrene hydroxide [$—N(CH_3)_3^+ \ OH^-$ groups]. The last-mentioned is a strongly basic ion exchanger.

Result 113[154,154a]*:* In the spectrum of the membrane with dissociated hydrated $—N(CH_3)_3^+ \ OH^-$ groups, a continuous absorption is found. It begins at the bands of the OH stretching vibrations of the hydrogen bridges and extends toward smaller wave numbers.

The motion of the "negative" charge takes place in the hydrate structures surrounding the OH^- in such a way that a proton tunnels to the OH^- group, which leads to an apparent displacement of the negative charge in the opposite direction. In what follows, we will use the term "defect proton." According to this, the defect proton is described by the two proton boundary structures shown in Fig. 100.

V.11.2. Continuous Absorption Extinction per Tunneling Proton and Tunneling Proton Concentration

Membranes of polystyrenesulfonic acid, alike in thickness and degrees of cross-linking and sulfonation, were treated with mixtures of aqueous solutions of HCl and NaCl in various proportions (Section VI.3). In this way, membranes containing hydrated $—SO_2OH$ and hydrated $—SO_3^-Na^+$ groups in different ratios were readily obtained. The concentration of acid groups was determined as described in Section V.4. The true degree of dissociation is given by the integral extinction of the band of the stretching vibration of the SO single bond in the $—S(=O)(=O)—OH$ group. From this the concentration of

[154] T. Ackermann, G. Zundel, and K. Zwernemann, *Z. Physik. Chem.* (*Frankfurt*) **49**, 331 (1966).

[154a] T. Ackermann, G. Zundel, and K. Zwernemann, *Ber. Bunsenges. Physik. Chem.*, in press.

$(CH_3)_2SO^+H \cdots OS(CH_3)_2 \qquad (CH_3)_2SO \cdots H^+OS(CH_3)_2$

FIG. 98. The proton boundary structures of the acid proton in the hydrogen bridge between two dimethylsulfoxide molecules.

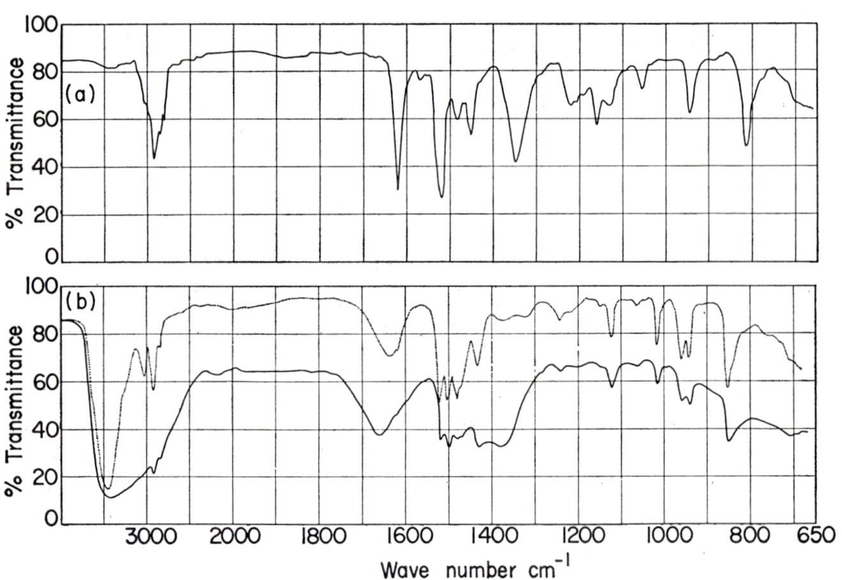

FIG. 99. (a) ——, Poly(*p*-dimethylamino)styrene (10 μ thick membrane, 10% cross-linked). (b) · · · ·, Poly(*p*-trimethylammonium)styrene iodide (10 μ thick, 10% cross-linked, H_2O-hydrated, 90% relative atmospheric humidity); ——, poly(*p*-trimethylammonium)styrene hydroxide (10 μ thick membrane, 10% cross-linked, H_2O hydrated, 90% relative atmospheric humidity). (Sincere thanks are due Dr. T. Ackermann and Dr. K. Zwernemann, of the Physico-Chemical Institute at the University of Muenster, Westfalen, West Germany, for supplying these spectra. They were taken with the spectrophotometer of the firm of Leitz, at Wetzlar.)

FIG. 100. The proton boundary structure of the defect proton in the $H_3O_2^-$ group (alkalis).

tunneling protons is obtained.* The spectra of the membranes were plotted with the membranes at 71% relative atmospheric humidity. In Fig. 101† the extinction of the continuous absorption is plotted against the concentration of tunneling protons.* We see:

Result 114[155]: If the concentration of tunneling protons is less than 1.5 mmole H^+/gm, the extinction of the continuous absorption per tunneling proton increases linearly with it. If it is greater than this value, the extinction of the continuous absorption does not essentially increase further if the concentration of tunneling protons is increased. The extinction of the continuous absorption per tunneling proton decreases at higher concentrations of the latter, approximately in proportion as the concentration of the latter increases. The extinction of the continuous absorption exhibits saturation. Ackermann[112] found a corresponding effect in a study of aqueous HCl solutions. Concerning the continuous absorption, he writes (see footnote 112, p. 272): ". . . the concentration dependence is less marked at high concentrations than with dilute or moderately concentrated solutions."

The same result, though less pronounced, is shown by the following experiment in which membranes of different degrees of sulfonation were prepared (Section VI.2.1.3). The concentration of the tunneling protons was determined as described above. The spectra were taken with the membranes at 43% relative atmospheric humidity. In comparing the results with those of the foregoing experiment, it must be borne in mind that the membranes used here were somewhat thicker. In Fig. 102, the extinction of the continuous absorption is again plotted against the concentration of the tunneling protons. We see that this curve is likewise bent. However, the bend is certainly not so pronounced, because the $—SO_3^-$ ions are not uniformly distributed in the membrane at low degrees of sulfonation. This is shown by the kinetics of the sulfonation reaction discussed in Section VI.2.1.4, Result 133.[156]

We were led to make these experiments by the following observation: In the polystyrenesulfonic acid membranes studied in Section V.4, the concentration of tunneling protons is greater than 1.5 mmole H^+/gm, provided a sufficient number of acid groups is dissociated. Hence, the saturation effect becomes perceptible if the extinction of the continuous absorption is followed as a function of the dissociation. In Fig. 103(a) the extinction of the continuous absorption is plotted as a function of the integral extinction of the band of the SO single bond of the $—S(=O)(=O)—OH$ groups and in Fig. 103(b) against α_t, the true degree of dissociation.

* Always calculated per gram of dry material.

† The three points (0, Δ, x) are taken from the spectral series in Section V.4.

[155] G. Zundel and H. Metzger, *Z. Physik. Chem.* (*Leipzig*) **235**, 333 (1967).

[156] G. Zundel and H. Metzger, *Z. Physik. Chem.* (*Leipzig*) **240**, 90 (1969).

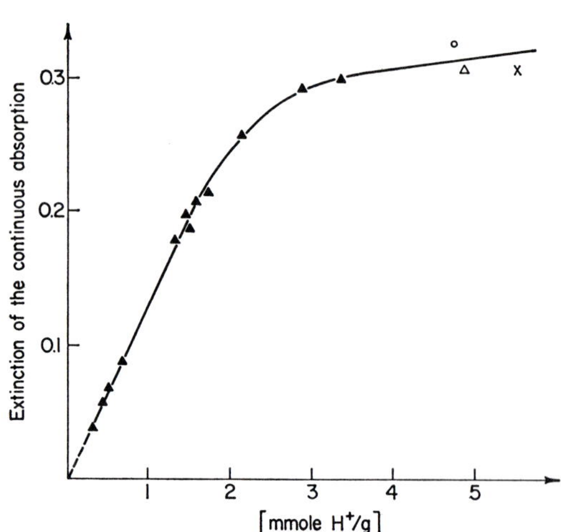

FIG. 101. The extinction of the continuous absorption as a function of the concentration of the tunneling protons. Results in polystyrenesulfonic acid membranes containing hydrated H^+ and Na^+ ions together.

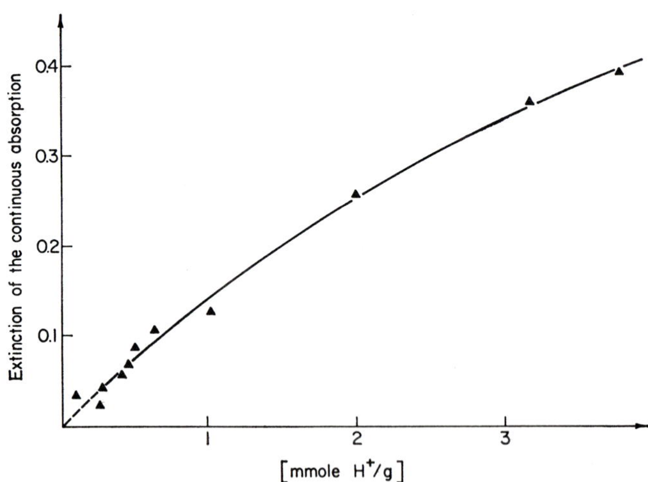

FIG. 102. The extinction of the continuous absorption as a function of the concentration of the tunneling protons. Results on membranes of polystyrenesulfonic acid sulfonated to different extents.

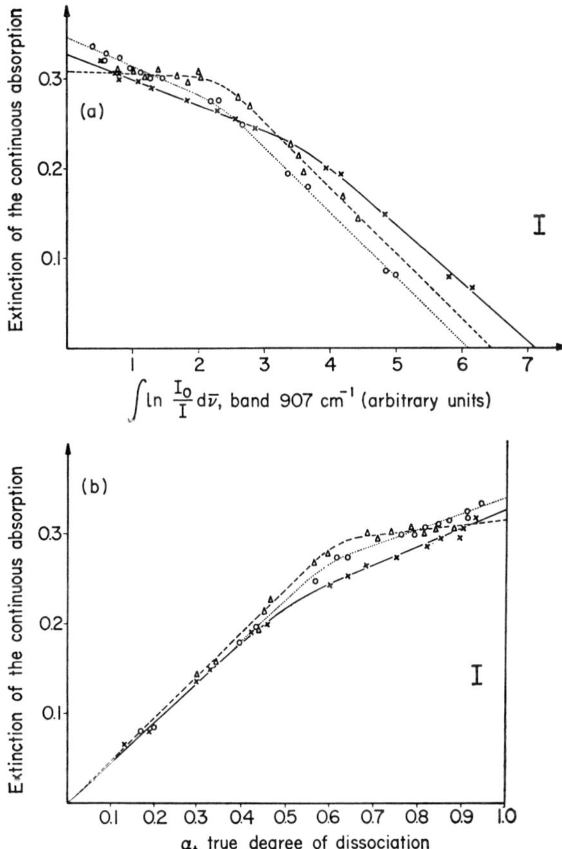

FIG. 103. Extinction of the continuous absorption: (a) as a function of the intergal extinction of the band of the stretching vibration of the SO single bond in the —S(=O)(=O)—OH group; (b) as a function of the true degree of dissociation.

Result 115: At a low concentration of tunneling protons, the extinction of the continuous absorption increases linearly with the true degree of dissociation. At a definite true degree of dissociation, the concentration of tunneling protons attains a critical value and the slope of the curves is strongly reduced.

In summary, the extinction of the continuous absorption shows a saturation effect depending on the concentration of tunneling protons. This effect is perceptible when the extinction of the continuous absorption is followed as a function of the true degree of dissociation, provided the concentration of the acid groups in the membrane is sufficiently large.

V.11.3. Summary: Continuous Absorption

(a) The continuous absorption begins at the bands of the OH or OD stretching vibrations of the water molecules, i.e., at about 3500 or 2600 cm^{-1}, respectively, and extends toward smaller wave numbers (Result 92).

(b) It is not dependent on temperature in the range 292–85°K (Result 97). This eliminates all the causes for the continuous absorption in which an activation energy would play a role.

(c) In acids, the continuous absorption arises when $H_5O_2^+$ groups are formed (Result 94). Numerous observations (Results 93–96) have led us to the hypothesis that the excess proton moves in the hydrogen bridge of this group and that the continuous absorption is closely connected with this fact. We have seen that the excess proton tunnels in this hydrogen bridge (Result 97). Hence, conversely, the continuous absorption is a criterion for the presence of tunneling protons. Thus a symmetrical potential well with a double minimum having a relatively low barrier is present in the hydrogen bridge of the $H_5O_2^+$ group. The two water molecules of this group retain their individuality as vibrating groups even when the excess proton tunnels in the hydrogen bridge (Result 98). An enlargement of the hydrate structure surrounding the excess proton has no effect on the continuous absorption.

(d) The continuous absorption is observed when protons tunnel in hydrogen bridges, not only between water molecules, but also between other molecules, e.g., $(CH_3)_2SO$ or CH_3OH (Results 111 and 112 and notes added in proof p. 186 and p. 207).

(e) This continuous absorption is also encountered in the spectra of dissociated alkalis (Result 113). Thus it is also caused by the tunneling of defect protons in hydrogen bridges.

(f) In all these cases the behavior of the proton can be described by proton boundary structures (Section V.6.2.2 and Figs. 88, 94, 98 and 100).

(g) The extinction of the continuous absorption increases with the concentration of tunneling protons, at first linearly (Result 93), but then tending to a saturation value (Results 114 and 115).

V.11.3.1. Energy Bands of the Tunneling Protons

The observation of a continuous absorption indicates, considering the energy levels of the absorbing particles, that either the ground state or the excited state, or both, lies in a continuous part of the energy term scheme. The continuous absorption thus indicates that a continuous energy distribution of the excess or defect protons occurs in solutions of acids, alkalis, and the corresponding polyelectrolytes. Hence energy bands of the excess or defect protons do exist.

In the following section, we concern ourselves with the question of how this comes about.[157]

V.11.4. Earlier Attempt at an Explanation

We assumed that the position of the energy levels of the tunneling proton in the double potential minimum is mainly influenced by the steric conditions, i.e., by the variation in length and angle of bending of the hydrogen bridges, for this will alter the potential well, and, in particular, the barrier in the double potential minimum. Bearing this in mind, we calculated the positions of the energy levels as a function of the form of the potential well in the case of potential curves such as that shown in Fig. 104.

The energy levels are plotted in Fig. 105(a) as a function of the height of the barrier ($1 \text{eV} \triangleq 8065.7 \text{cm}^{-1}$). The thickness $2b$ of the barrier in the individual cases is given in the figure. The breadth of the potential well is $2(b + \Delta)$, where Δ is always taken as 0.36 Å. These curves give approximately the positions of the levels, when the hydrogen bridges are more or less strongly bent, in the case that a double potential minimum with a low barrier is present. Such bending results even when the molecules execute thermally excited rotational movements.

Figure 105(b) also shows the energy levels as a function of the height of the barrier, when the height, βU_0, and thickness, $2\beta b_0$, of the barrier and the breadth of the potential well $2(\beta b_0 + \Delta)$ are varied simultaneously. This was carried out by varying the parameter β. U_0 is taken as 0.5 eV, $2b_0$ as 0.44 Å, and Δ as 0.36 Å. These curves give the approximate positions, when the length of the hydrogen bridge is varied within the bounds of the steric conditions, when a double potential minimum with a low barrier is present. Such a variation of length results when the molecules execute thermally excited restricted translation vibrations.

Figure 105 shows:

Result 116: The positions of the energy level of the proton in a double potential minimum are strongly dependent on the height and the thickness of the barrier in the potential well, but only when this barrier is comparatively low.

Brickmann and Zimmermann[66] calculated the tunnel frequency, on the one hand, according to the distance between the two potential minima, and, on the other hand, according to whether or not the hydrogen bridges were slightly unsymmetrical, i.e., whether one minimum is slightly raised in relation to the other. These authors also found that the tunnel frequency changes by several orders of magnitude as a function of these parameters.

In this way, a continuous distribution of energy levels of the tunneling protons could arise, if many hydrogen bridges were present under various steric conditions. The necessary condition for this is, nevertheless, that the potential barrier be very low. It was already stated, above, that the variation of the barrier is determined by the thermal motions—namely, by the restricted torsion and translation of the water molecules. In this case, however, if the variation of the barrier is the principal cause, it would be expected that the continuity of the energy levels is temperature dependent. The continuous absorption, however, shows no temperature dependence in the region 292–85°K (Fig. 87 and Result 97). It is then impossible that the already-discussed variation of the barrier in the double potential minimum should be the cause of the continuous absorption. We

[157] E. G. Weidemann and G. Zundel, *Z. Physik.* **198**, 288 (1967).
[158] E. G. Weidemann and G. Zundel, Unpublished calculations, 1963.

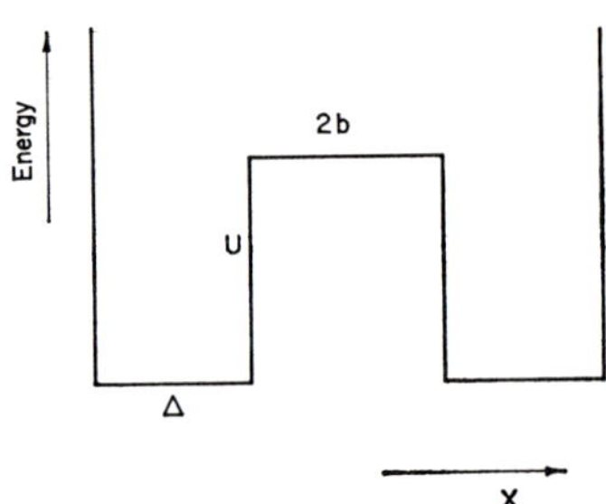

FIG. 104. Square potential well. The position coordinate X is given in arbitrary units.

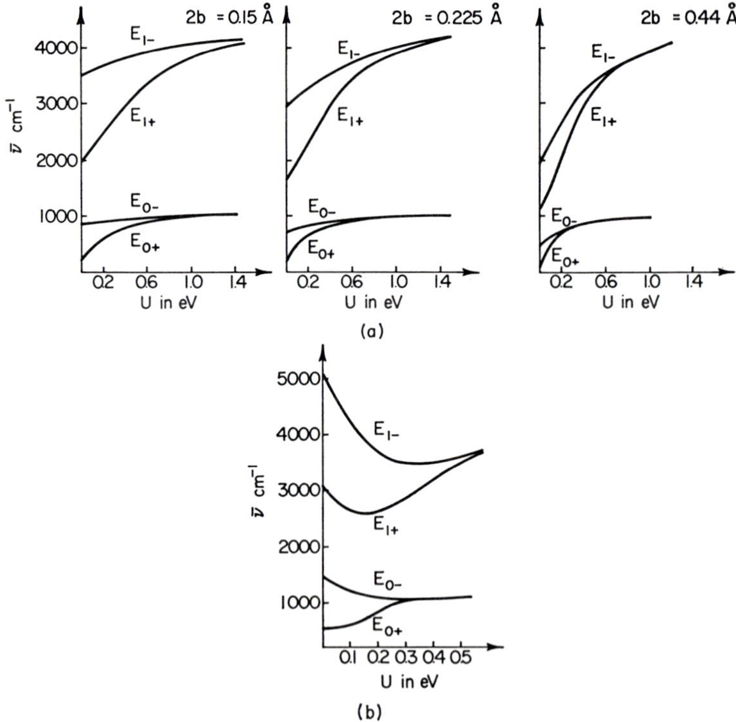

FIG. 105. The energy terms as a function of the potential well[158], for potential wells as shown in Fig. 104. (a) The individual diagrams give the energy terms for potential wells with different barrier thicknesses. (b) The energy terms, when not only the height but the breadth of the barrier and the breadth of the potential well itself are varied.

therefore consider whether there is another cause which could lead to a continuity of energy levels of the tunneling protons.

V.11.5. Explanation of the Continuous Absorption

We consider first a single proton isolated from its surroundings, tunneling in a hydrogen bridge. We know that the energy levels of this proton are always split by the tunnel effect, but never broadened [see Fig. 63(b)]. A single excess proton tunneling between two water molecules accordingly always has discrete energy levels, if it is considered in isolation from its surroundings. Therefore, another mechanism must cause the continuous absorption. The tunneling protons are in fact not isolated from their surroundings, and especially not from each other.

Let us now ask: What is the nature of the interaction between the tunneling protons, and what is its magnitude? Consider Fig. 106. It shows a single proton boundary structure for two tunneling excess protons. A variation of the electric field in the neighborhood is associated with the transition of one excess proton from one boundary position to the other. This variation of field changes, on the one hand, directly, and, on the other hand, via the electron system of the anion, the potential at the position of the other excess proton. Conversely, if now the other excess proton goes from one boundary position to the other, the same holds true. In this connection it is essential that the center of charge is displaced more than is the tunneling hydrogen nucleus, since the motion of the proton causes a considerable shift of the negative charge in the opposite direction. The field fluctuation is, however, determined by the displacement of the charge center and not by that of the hydrogen nucleus. Neighboring tunneling excess protons are then coupled by these fluctuating fields. The strength of this interaction depends on the distance between the hydrogen bridges in which the protons tunnel, and on their mutual orientations. If these quantities are randomly distributed, many neighboring energy levels arise and, provided the interaction is sufficiently

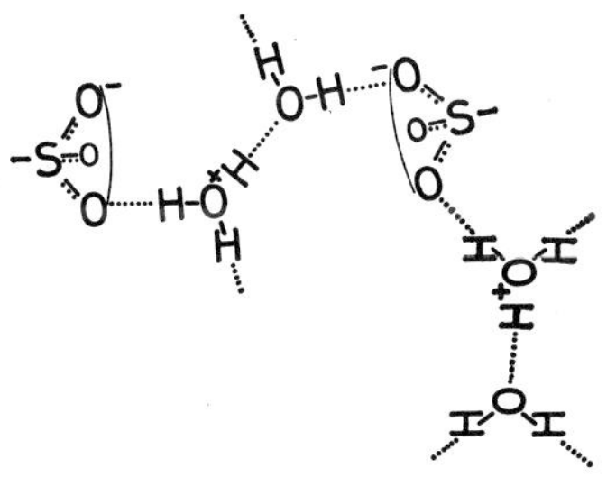

FIG. 106. A proton boundary structure for two excess protons for polystyrenesulfonic acid.

strong, an energy band. These coupling forces, arising through the tunneling of the excess proton, are the cause of the continuous absorption—assuming that they are sufficiently strong. We shall now proceed to estimate the strength of the coupling forces by means of a perturbation calculation.[157]

Consider two protons tunneling in their hydrogen bridges (Fig. 106). We wish to establish the energy term scheme of this system, assuming, at first, no interaction between the bridges. The energy of the whole system is then the sum of the energies of the two protons in the individual bridges. We assume that the tunnel effect in the individual bridges leads only to a splitting, which is smaller than the distance between the unperturbed levels E_{1+} and E_{0-}

$$\nu_0 \text{ and } \nu_1 \ll \frac{E_{1+} - E_{0-}}{h} \tag{1}$$

ν_0 and ν_1 being the tunnel frequencies [see Fig. 63(b)]. From this assumption we obtain groups of energy levels as shown in the diagram in Fig. 107. In this diagram, $0+0-$, indicates that one proton is in the ψ_{0+} state and the other in the ψ_{0-} state; $1+1+$ indicates that both protons are in an excited state with the quantum number $1+$. This is extremely unlikely under present conditions.

In the energy term scheme in Fig. 107, the 10 energy terms of the whole system of the two hydrogen bridges are obtained by considering the protons in the ground state and in the first excited state. There are 10 terms but in all $4 \times 4 = 16$ states. Some terms are doubly degenerate. In the following, we must always bear in mind that this degeneracy is removed by the interaction.

The appropriate wave functions are the products of the wave functions of the individual protons, i.e., for the lower group of states:

$$\begin{gathered}\psi_{0+}(\mathbf{r}_1)\psi_{0+}(\mathbf{r}_2), \quad \psi_{0-}(\mathbf{r}_1)\psi_{0-}(\mathbf{r}_2),\\ \psi_{0+}(\mathbf{r}_1)\psi_{0-}(\mathbf{r}_2), \quad \text{and} \quad \psi_{0-}(\mathbf{r}_1)\psi_{0+}(\mathbf{r}_2)\end{gathered} \tag{2}$$

Here $\mathbf{r}_1$ and $\mathbf{r}_2$ are the position vectors of the protons (see Fig. 108). The wave functions need not be made antisymmetric, as they do not overlap under prevailing conditions.

In the following we shall see how large is the influence which the interaction between the bridges exerts on the energy terms. Consider neutral groups, as shown in Fig. 106. The direct Coulombic repulsion would only be superimposed as an additive term.

We must determine first the interaction operator W. For this, we expand the Coulombic interaction between the charge distributions of the two bridges as an interaction between multipoles.[159] Since the groups containing

[159] L. D. Landau and E. M. Lifshitz, "The Classical Theory of Fields," p. 107. Addison-Wesley, Cambridge, Mass., 1951.

the bridges with tunneling protons are electrically overall neutral, the first term not vanishing in Eq. (3) is the dipole–dipole interaction. If we neglect the higher multipoles, we obtain:

$$W = \frac{1}{\varepsilon}\frac{e^2}{R^3}[\mathbf{r}_1\mathbf{r}_2 - 3(\mathbf{r}_1\mathbf{n})(\mathbf{r}_2\mathbf{n})] \tag{3}$$

n is a unit vector in the direction of the line joining the midpoints of the centers of the two bridges (see Fig. 108), e is the elementary charge, and ε is the dielectric constant of the medium between the tunneling protons.

For calculating the eigenvalues of the energy, we treat the interaction as a perturbation and calculate the eigenvalues according to perturbation theory. In this, the integrals

$$\iint \psi_\alpha^*(\mathbf{r}_1)\psi_\beta^*(\mathbf{r}_2)W(\mathbf{r}_1, \mathbf{r}_2)\psi_\gamma(\mathbf{r}_1)\psi_\delta(\mathbf{r}_2)d^3\mathbf{r}_1 d^3\mathbf{r}_2 = \langle\alpha\beta|\mathrm{W}|\gamma\delta\rangle \tag{4}$$

occur.[160] We have for brevity written $\alpha\beta$ or $\gamma\delta$, for example, for the quantum numbers $0+0-$. These integrals are the matrix elements W_{ik} (with $i = \alpha\beta$ and $k = \gamma\delta$) of the interaction operator W [Eq. (3)] between the states of the unperturbed system (2).

The calculation of the eigenvalues of the perturbed system is considerably simplified as many matrix elements vanish. From the symmetry property of the perturbation operator:

$$W(\mathbf{r}_1, \mathbf{r}_2) = -W(-\mathbf{r}_1, \mathbf{r}_2) = -W(\mathbf{r}_1, -\mathbf{r}_2) \tag{5}$$

follows that

$$\langle\alpha\beta|W|\gamma\delta\rangle = 0 \qquad \text{if} \qquad P_\alpha = P_\gamma \text{ or } P_\beta = P_\delta \tag{6}$$

i.e., when the parities of the states α and γ or β and δ, respectively, coincide.

This is because:

$$\langle\alpha\beta|W|\gamma\delta\rangle = \iint \psi_\alpha^*(\mathbf{r}_1)\psi_\beta^*(\mathbf{r}_2)W(\mathbf{r}_1, \mathbf{r}_2)\psi_\gamma(\mathbf{r}_1)\psi_\delta(\mathbf{r}_2)d^3\mathbf{r}_1 d^3\mathbf{r}_2$$

If the variable of integration $\mathbf{r}_1$ is replaced by $-\mathbf{r}_1$, there results:

$$\langle\alpha\beta|W|\gamma\delta\rangle = \iint \psi_\alpha^*(-\mathbf{r}_1)\psi_\beta^*(\mathbf{r}_2)W(-\mathbf{r}_1, \mathbf{r}_2)\psi_\gamma(-\mathbf{r}_1)\psi_\delta(\mathbf{r}_2)d^3\mathbf{r}_1 d^3\mathbf{r}_2$$

According to the definition of parity and from Eq. (5):

$$\langle\alpha\beta|W|\gamma\delta\rangle = -P_\alpha P_\gamma\langle\alpha\beta|W|\gamma\delta\rangle$$

from which, since $P_\alpha P_\gamma$ is always $+1$ when $P_\alpha = P_\gamma$, the above result follows.

[160] L. D. Landau and E. M. Lifshitz, "Quantum Mechanics." Macmillan (Pergamon), New York, 1958.

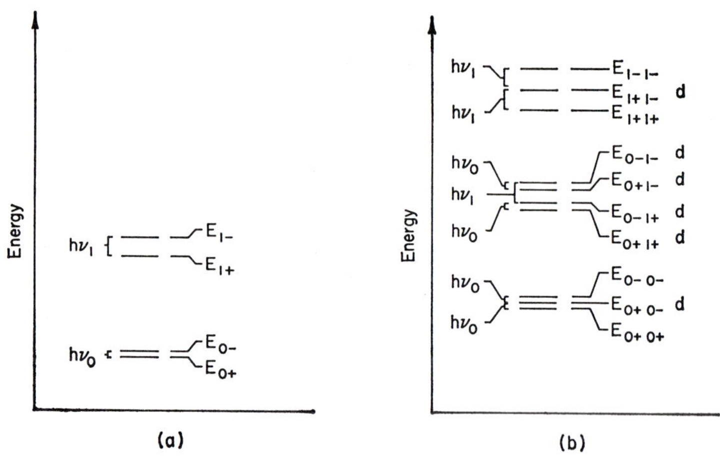

FIG. 107. Energy term scheme: (a) of a symmetrical hydrogen bridge with tunneling proton; (b) of a system of two symmetrical hydrogen bridges with tunneling protons, assuming no interaction between the latter.

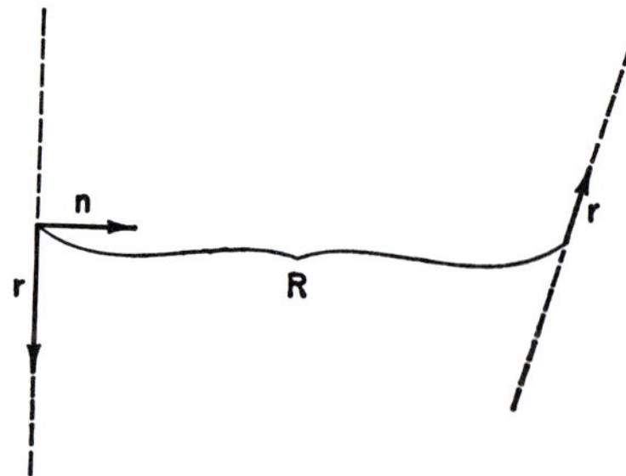

FIG. 108. Geometry of a pair of hydrogen bridges: ----, positions of the bridges; **r**, position vectors of the protons in the bridges (midpoints of the bridges as origins); **n**, unit vector in the direction of the joint of the midpoints of the bridges; R distance between the midpoints of the bridges.

It follows further from Eq. (5), that:

$$\langle\alpha\beta|W|\gamma\delta\rangle = 0 \qquad \text{if} \qquad P_\alpha P_\beta \neq P_\gamma P_\delta \tag{7}$$

i.e., the matrix elements of the perturbation operator vanish between states of different parities.

The usual equation of the Rayleigh-Schroedinger perturbation theory for nondegenerate states (see footnote 160, p. 134 ff.) cannot be directly applied in this case, as the series does not converge sufficiently rapidly for the distances between the bridges of interest here. The convergence is bad, because many levels are at only a short distance from one another (Fig. 107). We therefore employ the perturbation theory for nearly degenerate states.[161] In this, the nondegenerate case is reduced to the degenerate by splitting off a part of the perturbation operator, so that the groups of terms are reduced to degenerate levels. This can be attained by adding to the eigenvalues E_i° of the unperturbed system values D_i such that for each group of terms $E_i^\circ + D_i$ becomes equal to E°, a constant independent of i. In order not to alter the total Hamiltonian operator, the same values D_i are subtracted from the diagonal elements W_{ii} of the perturbation matrix. In this way a new perturbation matrix is obtained with

$$W_{ii}' = W_{ii} - D_i \qquad \text{and} \qquad W_{ik}' = W_{ik}, \quad i \neq k \tag{8}$$

If now the perturbation calculation for degenerate states is carried out to a first approximation, for each individual group of terms, one obtains a secular equation with an $m \times m$ determinant, m being the degree of degeneracy, i.e., in this case the number of states in each term group (see footnote 160, p. 137).

$$\begin{vmatrix} E^\circ + W_{11}' - E & W_{12}' \cdots\cdots & W_{1m}' \\ W_{21}' & E^\circ + W_{22}' - E \cdots & W_{2m}' \\ \vdots & \vdots & \\ W_{m1}' & W_{m2}' \cdots\cdots & E^\circ + W_{mm}' - E \end{vmatrix} = 0 \tag{9}$$

According to $E_i^\circ + D_i = E^\circ$ and Eq. (8), this is

$$\begin{vmatrix} E_1^\circ + W_{11} - E & W_{12} \cdots\cdots & W_{1m} \\ W_{21} & E_2^\circ + W_{22} - E \cdots & W_{2m} \\ \vdots & \vdots & \vdots \\ W_{m1} & W_{m2} \cdots\cdots & E_m^\circ + W_{mm} - E \end{vmatrix} = 0 \tag{10}$$

This procedure has the effect that the determinant of the complete secular

[161] A. Dalgarno, *in* "Quantum Theory" (D. R. Bates, ed.), Pt. I, p. 202. Academic Press, New York, 1961.

equation, whose solutions are the exact energy eigenvalues, decomposes to a first approximation into subdeterminants as shown schematically:

$$\begin{vmatrix} \blacksquare & 0 & 0 \\ 0 & \blacksquare & 0 \\ 0 & 0 & \blacksquare \end{vmatrix} = 0 \tag{11}$$

The accuracy of the approximation is determined by the magnitude of the perturbation relative to the distance between the term groups.

We shall see that, because of the symmetry of the perturbation operator, always only 2×2 determinants of the form

$$\begin{vmatrix} E_a^\circ - E + W_{aa} & W_{ab} \\ W_{ba} & E_b^\circ - E + W_{bb} \end{vmatrix} = 0 \tag{12}$$

remain. These in each case have the two solutions:

$$E_{a,b} = \frac{E_a^\circ + E_b^\circ}{2} + \frac{W_{aa} + W_{bb}}{2} \pm \sqrt{\left(\frac{E_a^\circ - E_b^\circ + W_{aa} - W_{bb}}{2}\right)^2 + |W_{ab}|^2} \tag{13}$$

As the limiting case $W = 0$ shows, the upper sign holds for E_a, the lower for E_b. In the special case

$$W_{aa} = W_{bb} \tag{14}$$

the shift of the levels as a result of the perturbation is obtained as:

$$\Delta E_a = E_a - E_a^\circ = -\frac{E_a^\circ - E_b^\circ}{2} + W_{aa} + \sqrt{\left(\frac{E_a^\circ - E_b^\circ}{2}\right)^2 + |W_{ab}|^2} \tag{15}$$

$$\Delta E_b = E_b - E_b^\circ = \frac{E_a^\circ - E_b^\circ}{2} + W_{aa} - \sqrt{\left(\frac{E_a^\circ - E_b^\circ}{2}\right)^2 + |W_{ab}|^2} \tag{16}$$

V.11.5.1. The Shift and Splitting of Lower Group Levels

To the levels of the lower group belong the four states given in (2). The secular determinant of this group is thus 4×4. Since the first two states in (2) have positive and the other two negative parity, then by Eq. (7) all matrix elements between these states with different parities will be zero, so that the determinant decomposes into two 2×2 determinants, as shown in Eq. (12). These can then be diagonalized separately, i.e., the shift of the terms E_{0+0+} and E_{0-0-} can be calculated independently of the splitting of the term E_{0+0-}.

Behavior of the terms E_{0+0+} and E_{0-0-}: from Eqs. (15) and (16) we obtain

$$\Delta E_{0+0+} = h\nu_0 - \sqrt{(h\nu_0)^2 + |\langle 0+0+|W|0-0-\rangle|^2} \tag{17}$$

$$\Delta E_{0-0-} = -h\nu_0 + \sqrt{(h\nu_0)^2 + |\langle 0+0+|W|0-0-\rangle|^2} \tag{18}$$

since $E_a^\circ - E_b^\circ = E_{0-0-}^\circ - E_{0+0+}^\circ = 2h\nu_0$ (Fig. 107).
Further

$$W_{aa} = \langle 0-0-|W|0-0-\rangle$$

$$W_{bb} = \langle 0+0+|W|0+0+\rangle$$

These matrix elements vanish according to Eq. (6), so that the condition shown in Eq. (14) is also fulfilled.

It is now necessary to calculate the matrix element $\langle 0+0+|W|0-0-\rangle$ substituting the interaction operator W from Eq. (3). In this way there results:

$$\langle 0+0+|W|0-0-\rangle = \frac{1}{\varepsilon}\frac{e^2}{R^3}\iint \psi_{0+}^*(\mathbf{r}_1)\psi_{0+}^*(\mathbf{r}_2)[\mathbf{r}_1\mathbf{r}_2 - 3(\mathbf{r}_1\mathbf{n})(\mathbf{r}_2\mathbf{n})] \psi_{0-}(\mathbf{r}_1)\psi_{0-}(\mathbf{r}_2)d^3\mathbf{r}_1 d^3\mathbf{r}_2 \tag{19}$$

In this equation we choose real values for the wave functions and put:

$$e\psi_{0+}(\mathbf{r})\psi_{0-}(\mathbf{r}) = \rho(\mathbf{r}) \tag{20}$$

with which the matrix element takes the following form:

$$\langle 0+0+|W|0-0-\rangle = \frac{1}{\varepsilon R^3}\iint \rho(\mathbf{r}_1)\rho(\mathbf{r}_2)[\mathbf{r}_1\mathbf{r}_2 - 3(\mathbf{r}_1\mathbf{n})(\mathbf{r}_2\mathbf{n})]d^3\mathbf{r}_1 d^3\mathbf{r}_2 \tag{21}$$

The quantity $\rho(\mathbf{r})$ given in Eq. (20) can be regarded as charge density. The matrix element thus represents the interaction of charge density distributions $\rho(\mathbf{r}_1)$ and $\rho(\mathbf{r}_2)$. These charge density distributions have the form shown in Fig. 109 in the direction of the axis of the hydrogen bridge. The quantities of charge $-\frac{e}{2}$ and $+\frac{e}{2}$ are situated on the two sides of the potential barrier, respectively. For estimation of the value of the matrix element, we can replace the continuous charge distribution by point charges $-\frac{e}{2}$ and $+\frac{e}{2}$ at the respective charge centers, which are situated at $x = -\frac{d}{2}$ and $x = +\frac{d}{2}$, $\frac{d}{2}$ being

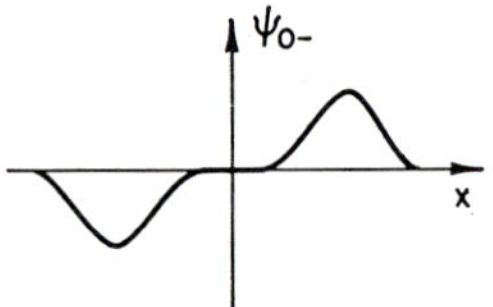

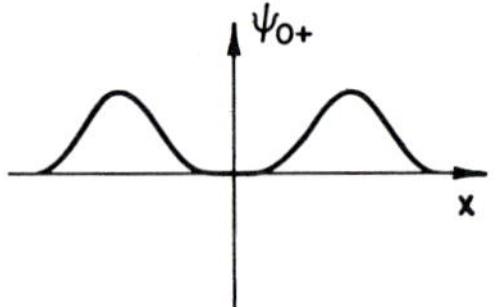

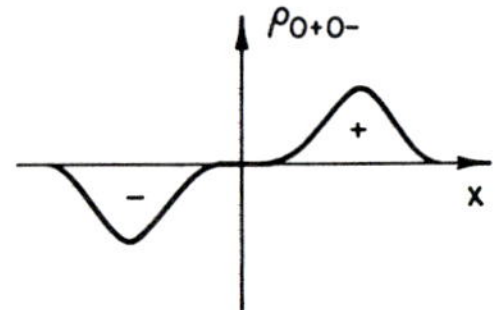

FIG. 109. The wave functions ψ_{0+} and ψ_{0-} of the proton in the hydrogen bridges and the charge density ρ_{0+0-} for the state $\psi_{0+}\psi_{0-}$ (schematic representation).

the distances of the charge centers from the middle of the bridge. We can then approximate the matrix element by the dipole–dipole interaction of these equivalent point charges. The interaction thus corresponds to that of two dipoles $\mathbf{p}_1$ and $\mathbf{p}_2$ of magnitude $p = ed/2$ lying in the direction taken by the axis of the hydrogen bridges at these centers. We then obtain:

$$\langle 0 + 0 + | W | 0 - 0 - \rangle = \frac{1}{\varepsilon R^3} [\mathbf{p}_1\mathbf{p}_2 - 3(\mathbf{p}_1\mathbf{n})(\mathbf{p}_2\mathbf{n})] \tag{22}$$

The magnitude of the matrix element accordingly depends on the positions of the charge centers in the individual proton boundary structures (Figs. 88, 98, and 100) and on the internal orientation of the hydrogen bridges. If we represent the orientation dependence of Eq. (22) by a factor:

$$g = \frac{1}{p^2} [\mathbf{p}_1\mathbf{p}_2 - 3(\mathbf{p}_1\mathbf{n})(\mathbf{p}_2\mathbf{n})] \tag{23}$$

we obtain:

$$\langle 0 + 0 + | W | 0 - 0 - \rangle = \frac{p^2}{\varepsilon} \frac{g}{R^3} \tag{24}$$

with

$$-2 < g < +2 \tag{25}$$

By substitution of Eq. (24) in Eqs. (17–18), we obtain, for the shifts of the terms considered as a consequence of the interaction:

$$\Delta E_{0+0+} = h\nu_0\left[1 - \sqrt{1 + \left(\frac{p^2}{\varepsilon h\nu_0}\frac{g}{R^3}\right)^2}\right] \tag{26}$$

$$\Delta E_{0-0-} = -h\nu_0\left[1 - \sqrt{1 + \left(\frac{p^2}{\varepsilon h\nu_0}\frac{g}{R^3}\right)^2}\right] \tag{27}$$

We now have to estimate p. In this, it is important to bear in mind that the charge center is much more strongly displaced than the proton when the latter tunncls, since the electrons move in the opposite direction. In proton transition in $H_5O_2^+$, the charge center is displaced from the oxygen atom of one "H_3O^+" group almost to that of the other, i.e., approximately 2Å.

We put for ε the value of the dielectric constant of water. The value used is that in the visible frequency region, since the field oscillations associated with tunneling are very rapid, so that the water dipoles cannot follow them. This is the more pronounced because the rearrangement frequency of the water molecules in very concentrated acid solutions is much smaller than in pure liquid water. This is because the outer water molecules in $H_5O_2^+$ and $H_9O_4^+$ form by far stronger hydrogen bridges (Sections V.7, V.8, and V.10). Hence, according to Pauling[164] (see also footnotes 162 and 163) the dielectric constant used is 1.78. The tunneling frequency cannot be much smaller than the frequencies of the bending vibrations of the "H_3O^+" group and of H_2O, since the bands of the "H_3O^+" group are not observed and the H_2O molecules preserve their individuality as vibrating groups (see Results 96 and 98). We accordingly carry out the calculation for a tunneling frequency ν_0 of 3×10^{12} Hz (i.e., $\tilde{\nu}_0 = 100\ \mathrm{cm}^{-1}$). In Fig. 110, the values thus obtained for E_{0+0+} and E_{0-0-} are plotted as a function of R. In this calculation, g is given its maximum value of 2.

Result 116: The energy level E_{0+0+} of the tunneling protons is lowered by this interaction. The E_{0-0-} level is raised by the same amount. For a distance of 6 Å between the tunneling protons, the shift amounts to about 500 cm^{-1}.

Behavior of the degenerate term E_{0+0-}:

Because of the degeneracy, instead of the simple products $\psi_{0+}(\mathbf{r}_1)\psi_{0-}(\mathbf{r}_2)$ and $\psi_{0-}(\mathbf{r}_1)\psi_{0+}(\mathbf{r}_2)$, orthogonal linear combinations of these states must be

[162] J. W. Smith, "Electric Dipole Moments." Butterworths, London, 1955.

[163] C. P. Smyth, "Dielectric Behavior and Structure." McGraw-Hill, New York, 1955.

[164] L. Pauling, "The Nature of the Chemical Bond," 3rd Ed. Cornell University Press, Ithaca, New York, 1960.

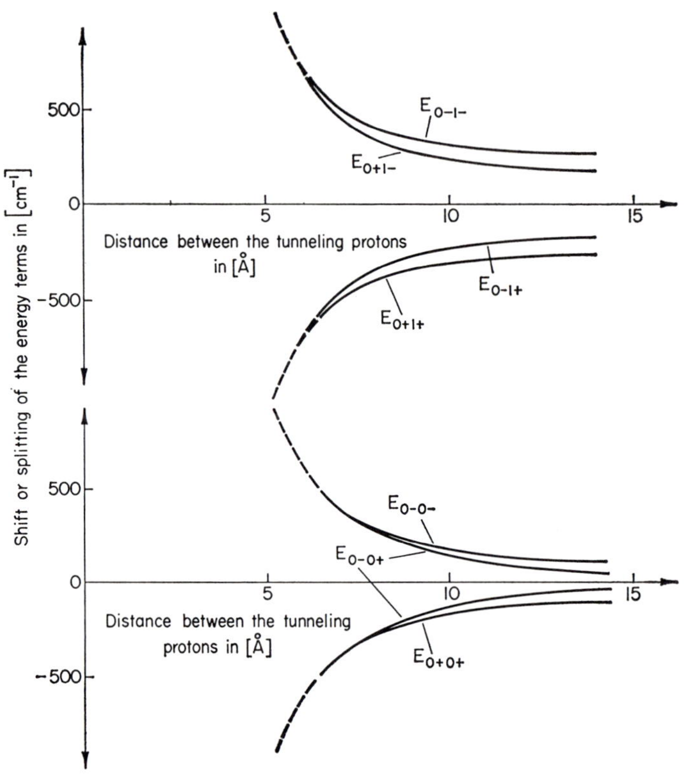

FIG. 110. The shift or splitting ΔE of the energy terms of the tunneling protons as a function of the distance between the midpoints of the hydrogen bridges.

used (see footnote 160, p. 137). If the symmetric and antisymmetric linear combinations:

$$\psi_{0+0-,s}(\mathbf{r}_1, \mathbf{r}_2) = \frac{1}{\sqrt{2}} [\psi_{0+}(\mathbf{r}_1)\psi_{0-}(\mathbf{r}_2) + \psi_{0-}(\mathbf{r}_1)\psi_{0+}(\mathbf{r}_2)]$$
$$\psi_{0+0-,a}(\mathbf{r}_1, \mathbf{r}_2) = \frac{1}{\sqrt{2}} [\psi_{0+}(\mathbf{r}_1)\psi_{0-}(\mathbf{r}_2) - \psi_{0-}(\mathbf{r}_1)\psi_{0+}(\mathbf{r}_2)] \tag{28}$$

are chosen, all nondiagonal elements of the determinants in the secular equation are zero. This follows from the symmetry of the perturbation operator [see Eq. (3)]

$$W(\mathbf{r}_1, \mathbf{r}_2) = W(\mathbf{r}_2, \mathbf{r}_1) \tag{29}$$

Hence, the matrix elements of W vanish between states of different symmetry:

$$\langle \alpha\, \beta, s | W | \gamma\, \delta, a \rangle = 0 \tag{30}$$

The proof is as follows:

$$W_{sa} = \iint \psi_s(\mathbf{r}_1, \mathbf{r}_2) W(\mathbf{r}_1, \mathbf{r}_2) \psi_a(\mathbf{r}_1, \mathbf{r}_2) d^3\mathbf{r}_1 d^3\mathbf{r}_2$$

is an arbitrarily chosen matrix element between a symmetric and an antisymmetric state, and the variables of integration may be redesignated:

$$W_{sa} = \iint \psi_s(\mathbf{r}_2, \mathbf{r}_1) W(\mathbf{r}_2, \mathbf{r}_1) \psi_a(\mathbf{r}_2, \mathbf{r}_1) d^3\mathbf{r}_1 d^3\mathbf{r}_2$$

However, because of the symmetry of ψ_s and W [Eq. (29)] and the antisymmetry of ψ_a the whole expression changes its sign by interchanging $\mathbf{r}_1$ and $\mathbf{r}_2$. Thus $W_{sa} = - W_{sa} = 0$.

In particular:

$$W_{ab} = \langle 0 + 0 -, s | W | 0 + 0 -, a \rangle = 0$$

Thus, according to Eq. (12), the splitting of the term E_{0+0-} into two terms gives the shifts:

$$\begin{aligned} \Delta E_{0+0-,\frac{s}{a}} &= \langle 0 + 0 -, \tfrac{s}{a} | W | 0 + 0 -, \tfrac{s}{a} \rangle \\ &= \langle 0 + 0 - | W | 0 + 0 - \rangle \pm \langle 0 + 0 - | W | 0 - 0 + \rangle \end{aligned} \tag{31}$$

In this transformation the relationship:

$$\langle \alpha\beta, \tfrac{s}{a} | W | \gamma\delta, \tfrac{s}{a} \rangle = \langle \alpha\beta | W | \gamma\delta \rangle \pm \langle \alpha\beta | W | \delta\gamma \rangle \tag{32}$$

which follows from Eq. (28), is used.

Also used is

$$\langle 0 + 0 - | W | 0 + 0 - \rangle = 0$$

which follows from Eq. (6). Hence the splitting of the level E_{0+0-} is:

$$\Delta E_{0+0-} = \pm \langle 0 + 0 - | W | 0 - 0 + \rangle \tag{33}$$

With the estimated value for the matrix element W from Eq. (24), Eq. (33) gives:

$$\Delta E_{0+0-} = \pm \frac{p^2}{\varepsilon} \frac{g}{R^3} = \pm 8.1 \frac{g}{R^3} \text{ [eV]} \tag{34}$$

in which R is in Å, and p and ε are chosen as in the numerical evaluation of Eqs. (26) and (27). The energy levels arising through the splitting are plotted in Fig. 110.

Result 117: The degeneracy of the energy level E_{0+0-} of the tunneling protons is removed by this interaction. With a decrease in the distance between the tunneling protons, this level is split. The extent of the split is about twice as great as the shift of the levels E_{0+0+} and E_{0-0-}.

V.11.5.2. Proton and Defect-Proton Dispersion Forces between Tunneling Protons

The E_{0+0+} level, which the tunneling protons preferentially occupy, is lowered by this interaction, as Result 116 shows. Hence, there follows Result 118:

Result 118: The groups with tunneling protons attract one another. This attractive force between the tunneling protons, which we term proton or defect-proton dispersion force, is similar in nature to van der Waals forces.

To obtain an idea of the magnitude of the proton dispersion force, we compare it with the dispersion force between mercury atoms.*

Result 119: When the distance between the mercury atoms is 3.5 Å, the dispersion force, according to London,[169] is 0.15 eV/Å. If we assume the distance between the centers of the hydrogen bridges to be 5 Å, the proton dispersion force is 0.08 eV/Å. Even at a distance of 5 Å, it is thus not markedly smaller than the dispersion force between mercury atoms. The proton dispersion forces are comparable in strength with other intermolecular forces. This makes it very probable that the anion hydrated-H^+ groups in acid solutions are arranged in pairs. The same holds for the cation hydrated-OH^- groups in alkalis.

Theory of Kirkwood and Shumaker[176]

Kirkwood and Shumaker write: "We shall investigate here another type of electrostatic interaction between protein molecules, which arises from fluctuations in charge and

* On van der Waals forces, see the papers by London,[167–169,171] Kallmann and London,[165] and Eisenschitz and London.[166] Also see the reviews of Margenau[170] and Dalgarno and Davison.[170a] Cf. also the papers of McLachlan,[172] Lifshitz,[173] and Linder,[175] and particularly those of Blinc and Svetina.[174]

[165] H. Kallmann and F. London, *J. Phys. Chem.* **B2**, 207 (1929).
[166] R. Eisenschitz and F. London, *Z. Physik.* **60**, 491 (1930).
[167] F. London, *Z. Physik* **63**, 245 (1930).
[168] F. London, *Trans. Faraday Soc.* **33**, 8 (1937).
[169] F. London, *Z, Physik. Chem. (Leipzig)* **B11**, 222 (1930).
[170] H. Margenau, *Rev. Mod. Phys.* **11**, 1 (1939).
[170a] A. Dalgarno and W. D. Davison, *Advan. Atomic Molecular Phys.* **2**, 1 (1966).
[171] F. London, *J. Phys. Chem.* **46**, 305 (1942).
[172] A. D. McLachlan, *Mol. Phys.* **6**, 423 (1963).
[173] E. M. Lifshitz, *Soviet Phys. JETP (English Transl.)* **2**, 73 (1956).
[174] R. Blinc and S. Svetina, *Phys. Rev.* **147**, 423, 430 (1966).
[175] B. Linder, *J. Chem. Phys.* **40**, 2003 (1964).
[176] J. G. Kirkwood and J. B. Shumaker, *Proc. Natl. Acad. Sci.* U.S. **38**, 863 (1952).

charge distribution associated with fluctuations in number and configuration of the protons bound to the molecules. Proteins, considered as ampholytes, contain a large number of neutral and negatively charged basic groups, such as $—NH_2$ and $—COO^-$, to which protons are attached. Except in highly concentrated acid solutions, the number of basic sites generally exceeds the average number of protons bound to the molecules, so that there exist many possible configurations of the protons, differing little in free energy, among which fluctuations may occur. Fluctuations in the number and configuration of the mobile protons impart to the molecules fluctuating charges and fluctuating electric multipole moments. We have demonstrated in a previous investigation[177] that the dipole moment fluctuations, arising from configurational fluctuations of the protons, are sufficient to account for the dielectric constant increments of many proteins without postulating the existence of permanent dipole moments." (Also see footnotes 178 and 179.)

There is, however, one essential difference between the force proposed by Kirkwood and Shumaker and the proton or defect-proton dispersion force. Kirkwood and Shumaker assume thermal motions of the protons, while the charge displacement in the case of the proton dispersion force is produced by tunneling of protons.

In biological processes, hydrogen bridge systems are of great importance. In consequence it appears probable that the proton dispersion force is also sometimes of considerable significance in biological processes. We are investigating this problem further.

Note added in proof: In the spectra of imidazole (aqueous solution), *N*-methylimidazole (anhydrous liquid), pyrazole (aqueous solution), and, especially poly-L-histidine (film), continuous absorption arises and increases as excess protons are added to the nitrogen atom of the imidazole rings. The intensity of this continuous absorption is greatest when there is one excess proton present for two imidazole rings. Pyrazole excepted, it decreases with further addition of excess protons. Thus, the addition of excess protons links the nitrogen atoms of two imidazole rings by a hydrogen bridge. $>NH^+\cdots N<$ bridges are formed. The continuous absorption indicates that there is a highly symmetrical double minimum potential well in this hydrogen bridge and that the proton tunnels in this bridge. The proton dispersion forces which cause the continuity of the energy levels act between these bridges.

If with decreasing pH the addition of excess protons increases further, the continuous absorption vanishes because the $>NH^+\cdots N<$ hydrogen bridges are broken up since $>NH^+\cdots Cl^-$ groups are formed. In the $>NH^+\cdots Cl^-$ groups there is no symmetrical double minimum potential well. Therefore, the proton cannot tunnel and instead of the continuous absorption the broad

[177] J. G. Kirkwood and J. B. Shumaker, *Proc. Natl. Acad. Sci.* U.S. **38**, 855 (1952).
[178] S. Takashima, *J. Phys. Chem.* **69**, 2281 (1965).
[179] W. Scheider, *Biophys. J.* **5**, 617 (1965).

intense band of the NH stretching vibration in the $\rangle NH^{+}\cdots Cl^{-}$ groups in the range 3000–2500 cm^{-1} is observed. The formation of this band can be particularly well observed with *N*-methylimidazole, because this can be investigated as an anhydrous system.[179a]

According to NMR investigations, it is probable that there is a stacking effect in aqueous solutions of histidine when one excess proton is present for every two histidine molecules[179b] (see also the work of Shinitzky[179c]). According to present results, this stacking effect is caused essentially by the proton dispersion forces.

In summary, we see that groups with tunneling protons mutually attract one another. This proton or defect-proton dispersion force is comparable in order of magnitude to other intermolecular forces. It presumably leads to the pairing of anion hydrated-H^{+} groups. It is assumed that these forces play a significant role in biological processes.

V.11.5.3. Energy Level Shifts of the Upper Group

The upper group (see Fig. 107) consists of the terms E_{0+1+}, E_{0-1+}, E_{0+1-}, and E_{0-1-}. These are characterized by one proton being in each of the two states designated by the quantum numbers 0 and 1, respectively. All these states are doubly degenerate, since either of the two protons can be excited. The determinant in the secular equation of the terms of this group is accordingly 8 × 8, and decomposes into two 4 × 4 determinants, one for the terms of positive parity, E_{0+1+} and E_{0-1-}, and one for the terms of negative parity, E_{0-1+} and E_{0+1-}. Each of these 4 × 4 determinants decomposes, because of the symmetry of the perturbation operator [(Eq. 29)], into two 2 × 2 determinants, if the symmetric and antisymmetric linear combinations of the eigenfunctions are chosen, analogously to Eq. (28). For these 2 × 2 determinants, we can again use Eq. (15) and (16) if the condition [(Eq. (14)] is fulfilled. In Table 16, the quantities necessary for evaluation of these formulas are collected. The values in column 4 are taken from Fig. 107. In the calculation of the values in the last three columns, we have used Eqs. (30) and (6).

We note, in the states of positive parity, the condition [Eq. (14)] is fulfilled. For the states of negative parity the condition is approximately fulfilled, as we shall see in the estimation of the magnitudes of the matrix elements in the paragraph following Eq. (37). Hence we obtain for the shift and splitting ΔE of the levels of the upper group:

[179a] G. Zundel and J. Muehlinghaus, in preparation.
[179b] H. Rueterjans, University of Muenster, Westfalen, private communication.
[179c] M. Shinitzky, *Israel J. Chem.* **6**, 525 (1968).

TABLE 16

VALUES FOR THE EVALUATION OF EQS. (15) AND (16)

a	b	Parity	$E_a^\circ - E_b^\circ$	W_{aa}	W_{bb}	W_{ab}
1	2	3	4	5	6	7
$0-1-, s$	$0+1+, s$	+	$h(\nu_1+\nu_0)$	0	0	$\langle 0-1-\|W\|0+1+\rangle + \langle 0-1-\|W\|1+0+\rangle$
$0-1-, a$	$0+1+, a$	+	$h(\nu_1+\nu_0)$	0	0	$\langle 0-1-\|W\|0+1+\rangle - \langle 0-1-\|W\|1+0+\rangle$
$0+1-, s$	$0-1+, s$	−	$h(\nu_1-\nu_0)$	$\langle 0+1-\|W\|1-0+\rangle$	$\langle 0-1+\|W\|1+0-\rangle$	$\langle 0+1-\|W\|0-1+\rangle$
$0+1-, a$	$0-1+, a$	−	$h(\nu_1-\nu_0)$	$-\langle 0+1-\|W\|1-0+\rangle$	$-\langle 0-1+\|W\|1+0-\rangle$	$\langle 0+1-\|W\|0-1+\rangle$

$$
\begin{aligned}
\Delta E_{0+1+} &= +\frac{h(\nu_0+\nu_1)}{2} \\
&\quad - \sqrt{\left(\frac{h(\nu_0+\nu_1)}{2}\right)^2 + |\langle 0-1-|W|0+1+\rangle \pm \langle 0-1-|W|1+0+\rangle|^2} \\
\Delta E_{0-1-} &= -\frac{h(\nu_0+\nu_1)}{2} \\
&\quad + \sqrt{\left(\frac{h(\nu_0+\nu_1)}{2}\right)^2 + |\langle 0-1-|W|0+1+\rangle \pm \langle 0-1-|W|1+0+\rangle|^2} \\
\Delta E_{0-1+} &= +\frac{h(\nu_1-\nu_0)}{2} \\
&\quad \pm \langle 0+1-|W|1-0+\rangle - \sqrt{\left(\frac{h(\nu_1-\nu_0)}{2}\right)^2 + |\langle 0+1-|W|0-1+\rangle|^2} \\
\Delta E_{0+1-} &= -\frac{h(\nu_1-\nu_0)}{2} \\
&\quad \pm \langle 0+1-|W|1-0+\rangle + \sqrt{\left(\frac{h(\nu_1-\nu_0)}{2}\right)^2 + |\langle 0+1-|W|0-1+\rangle|^2}
\end{aligned}
\tag{35}
$$

For the energy terms of the upper group, we must again estimate the matrix elements. We put:

$$
\begin{aligned}
e\psi_\alpha(\mathbf{r}_1)\psi_\gamma(\mathbf{r}_1) &= \rho_{\alpha\gamma}(\mathbf{r}_1) \\
e\psi_\beta(\mathbf{r}_2)\psi_\delta(\mathbf{r}_2) &= \rho_{\beta\delta}(\mathbf{r}_2)
\end{aligned}
\tag{36}
$$

Then the matrix elements assume the following form:

$$
\langle\alpha\beta|W|\gamma\delta\rangle = \frac{1}{\varepsilon e^2}\iint \rho_{\alpha\gamma}(\mathbf{r}_1)\rho_{\beta\delta}(\mathbf{r}_2)W(\mathbf{r}_1, \mathbf{r}_2)d^3\mathbf{r}_1 d^3\mathbf{r}_2 \tag{37}
$$

The quantities given in Eq. (36) can again be regarded as charge density distributions. The matrix element then again represents the dipole–dipole interaction of charge density distributions $\rho(\mathbf{r}_1)$ and $\rho(\mathbf{r}_2)$. The qualitative form of the wave functions and charge density distributions in the hydrogen bridges are shown in Fig. 111. From this figure we see that the dipole moment of the charge density distribution ρ_{1+1-} is approximately equal to the dipole moment of the charge density distribution ρ_{0+0-}, and that the dipole moments of ρ_{0+1-} and ρ_{0-1+} are considerably smaller than p. Hence the dipole moment of the two last-named charge density distributions can be neglected. This is permissible since, in calculating the shift of the terms of the lower group, it should be borne in mind that the charge center is much more strongly

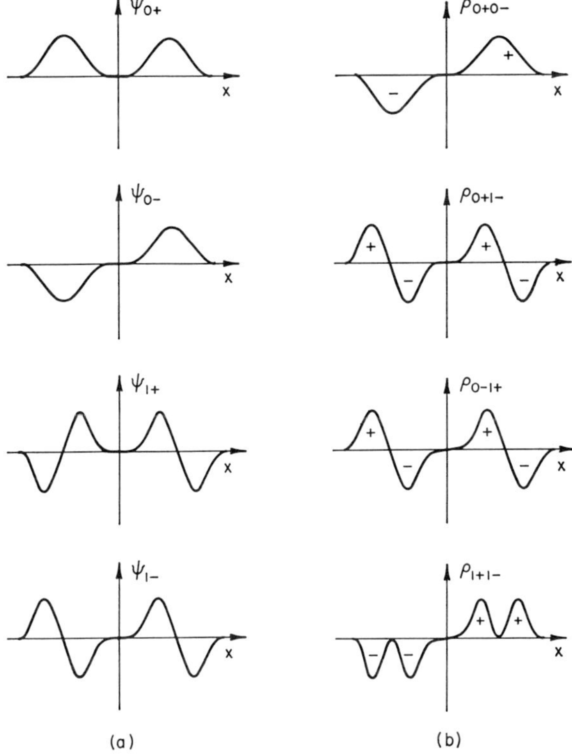

FIG. 111. (a) The wave functions of the protons in the hydrogen bridges (ψ_α, ψ_β, ψ_γ, and ψ_δ) (schematic representation). (b) The charge densities $\rho_{\alpha\gamma}$ and $\rho_{\beta\delta}$ for states $\psi_{\alpha\gamma}$ or $\psi_{\beta\delta}$, respectively (schematic representation).

shifted than the proton when the proton tunnels. At this degree of approximation, the splitting of the terms thus is small compared with the shift.

$$\begin{aligned}
\Delta E_{0+1+} &= + \frac{h(\nu_0 + \nu_1)}{2} - \sqrt{\left(\frac{h(\nu_0 + \nu_1)}{2}\right)^2 + \frac{p^4 g^2}{\varepsilon^2 R^6}} \\
\Delta E_{0-1-} &= - \frac{h(\nu_0 + \nu_1)}{2} + \sqrt{\left(\frac{h(\nu_0 + \nu_1)}{2}\right)^2 + \frac{p^4 g^2}{\varepsilon^2 R^6}} \\
\Delta E_{0-1+} &= + \frac{h(\nu_1 - \nu_0)}{2} - \sqrt{\left(\frac{h(\nu_1 - \nu_0)}{2}\right)^2 + \frac{p^4 g^2}{\varepsilon^2 R^6}} \\
\Delta E_{0+1-} &= - \frac{h(\nu_1 - \nu_0)}{2} + \sqrt{\left(\frac{h(\nu_1 - \nu_0)}{2}\right)^2 + \frac{p^4 g^2}{\varepsilon^2 R^6}}
\end{aligned} \tag{38}$$

We have put $\nu_1 = 10^{13}$ Hz, i.e., $\bar{\nu}_1 = 333$ cm^{-1}, calculated the shift ΔE as a function of R under the assumptions we used for the lower energy term group, and plotted it in Fig. 110.

Result 120: The energy levels of the upper group are in part raised and in part lowered by this interaction. These shifts amount to about 500 cm^{-1} at a distance between the tunneling protons of 6 Å.

V.11.5.4. Tunneling Proton Energy Bands—The Explanation

We shall now consider many pairs of hydrogen bridges with tunneling protons. A random distribution of distances and orientations of the hydrogen bridges is present. Equations (26), (27), (34), and (38) show that a random distribution of the energy levels of the tunneling protons then arises. This is due both to R and g, of which the latter takes on all values from -2 to $+2$. Further, Results 116, 117, and 120 show that the shift or splitting of the energy levels by this interaction is so large that, if many pairs of tunneling protons are considered, broad energy bands occur. Finally, as we have seen in Section V.11.4, the tunneling frequency changes, on the one hand, as a function of the form of the barrier in the double minimum potential, and, on the other hand, as a function of the symmetry of the potential well in the hydrogen bridge. The potential well in the hydrogen bridges is somewhat deformed by the steric conditions in the liquid, mentioned in Section IV.13, so that the tunneling frequency is also changed. If we consider the above equations from this standpoint, they show that the levels also form a continuous distribution of values as a function of the steric conditions. The variation of the tunneling frequency, however, cannot be the determining cause of the continuous distribution of energy levels, since, if that were the case, the continuous absorption would be expected to be temperature dependent.

Result 121: The energy levels of the tunneling protons form a continuous distribution of values, first as a function of the distance between the tunneling protons, second as a function of the mutual orientation of the bridges, and third as a function of the tunneling frequency. Under prevailing conditions, the coupling of the protons is strong enough for energy bands to arise from the levels for many pairs of tunneling protons.

The observed energy bands differ in one respect essentially from the energy bands of the electrons in metals. The latter are caused by the overlapping of the ψ functions of the electrons, whereas the energy bands of the tunneling protons are caused by the field interaction.

The result that the continuous absorption begins at 3500 cm^{-1} (protons) or 2600 cm^{-1} (deuterons) is understandable, since in very strong hydrogen bridges the distance between the levels E_0 and E_1 is about 2000 cm^{-1} (on this point see footnotes 23, 89, and 180). As a result of interaction through the proton dispersion forces there arises a continuous distribution of levels, particularly of those strongly lowered in comparison with the E_0 level or strongly raised in comparison with E_1. This makes the energy difference between the lower edge of the lower band of the terms of the lower group and the upper edge of the terms of the upper group about 3500 cm^{-1} in the case of tunneling protons. The observation that the continuous absorption begins at smaller wave numbers when tunneling deuterons are involved is understandable, as the distance between the levels E_0 and E_1 is less in this case.

The proton dispersion forces are so large that the anion hydrated-H^+ groups are probably arranged in pairs (Section V.11.5.2) This explains why the extinction of the continuous absorption increases linearly with the concentration of the tunneling protons, even when the latter is very small (see Fig. 84).

Do these results also explain the saturation effect of the extinction of the continuous absorption (Results 114 and 115)? The present calculations have been performed under the assumption that the interaction occurs only within pairs of tunneling protons. If the concentration of tunneling protons becomes greater, these protons will not be simply coupled in pairs. The pairwise coupling passes progressively into a coupling of all tunneling protons. In this case, however, the medium surrounding the tunneling protons can no longer be regarded as pure water. Rather the following must be borne in mind. Figure 112 is taken from an investigation of the dielectric constant of polystyrenesulfonic acid by Dickel and Bunzl.[181] We see that the dielectric constant increases sharply when the excess protons are removed from the anions, i.e., when an increasing number of excess protons tunnels. If now between a pair of tunneling protons there is another tunneling proton and not only water, the dielectric constant of the medium between these protons is considerably greater, so that, according to Eqs. (26), (27), (34), and (38), the interaction between them is considerably smaller. It is very probable that the saturation effect of the intensity of the continuous absorption is closely related to this, and hence shows the transition from pairwise coupling to a coupling of all tunneling protons.

Result 122: In conclusion, we see that all results relating to the continuous absorption are explained by the proton dispersion force, and the energy bands of the tunneling protons are caused by this force. Conversely, the continuous absorption is evidence for the proton dispersion force and the energy bands.

[180] D. Hadži and A. Novak, *Spectrochim. Acta* **18**, 1059 (1962).
[181] G. Dickel and K. Bunzl, *Makromol. Chem.* **79**, 54 (1964).

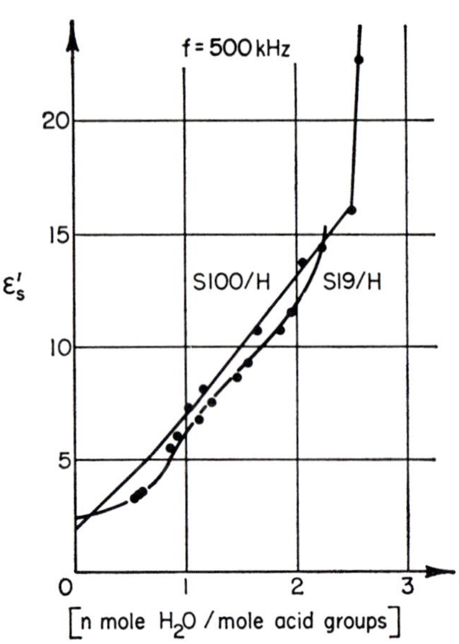

FIG. 112. The dielectric constant of polystyrenesulfonic acid as a function of the degree of hydration. ε' is the real part of the complex quantity $\varepsilon = \varepsilon' - i\varepsilon''$. The ion exchangers involved are numbers S 100/H and S 19/H of the firm of Bayer, Leverkusen, West Germany. (Sincere thanks are due to Dr. G. Dickel of the Physico-Chemical Institute of the University of Munich for providing this figure.)

V.12. Polystyreneselenonic Acid

Figures 113 and 114 show spectra of a membrane of polystyreneselenonic acid.[182] We compare these spectra with those of polystyrenesulfonic acid in Figs. 68, 69, and 72 and see the next result.

Result 123: The intensity of the band of the OH groups in the hydrogen bridges between the $—Se(=O)(=O)(OH)$ groups at about 2385 cm^{-1} decreases with increasing degree of hydration, i.e., the $—Se(=O)(OH\cdots O)(O\cdots HO)Se(=O)—$ group is increasingly disassociated (Section V.2.1). If the membrane is dried slowly, the bands of the stretching vibrations of the OH or OD groups in the hydrogen bridges

[182] G. Zundel and H. Metzger, *Z. Physik Chem.* (*Leipzig*) **240**, 50 (1969).

are found which are connected with the attachment of the last water molecule to the hydrogen atom of the $-Se(=O)_2OH$ group, as with polystyrenesulfonic acid (compare Section V.5). In D_2O hydration, these bands are observed in Fig. 114 at 2350 and 2225 cm^{-1}. We further see from these figures that, if the membranes are dried rapidly, these bands are not found. (For an explanation of this dependence on the rate of drying, see Section V.5.) If these groups are dissociated i.e., if the $-Se(=O)_2OH$ groups are rearranged on removal of the acid proton (Section V.1.2), then continuous absorption occurs. It is nevertheless less intense, since the true degree of dissociation is smaller. The bands of the stretching vibrations of the OH or OD groups in the hydrogen bridges by which these groups are linked to oxygen atoms of $\left(-SeO_3\right)^-$ ions or of other water molecules should be found at about 2900 cm^{-1} (H_2O) or at about 2100 cm^{-1} (D_2O), as a comparison with polystyrenesulfonic acid shows (Section V.8). This, however, cannot be detected in the present case, because even when much water of hydration is present, some hydrogen bridges are still present between acid groups. Their OH or OD groups absorb in this region, yet it is not to be expected that these bands should have a very different position from that observed for polystyrenesulfonic acid. We further see from Fig. 113 that an intense band arises at about 3400 cm^{-1} (H_2O) or 2500 cm^{-1} (D_2O)* with increasing degree of hydration. Finally, we see that a band of free OH groups (i.e., not bound by hydrogen bridges) is hardly noticeable at about 3600 cm^{-1} at a high degree of hydration.

It follows that, on the whole, the hydration and the behavior of the acid proton in polystyreneselenonic acid are the same as in the polystyrenesulfonic acid. The essential differences have already been established in Section V.1.2. We have seen there that the true degree of dissociation of polystyreneselenonic acid is smaller than that of polystyrenesulfonic acid (Result 59); undissociated acid groups are still present, even when the membranes have been hydrated at 98% relative atmospheric humidity. Further, we have seen (Section V.2.1) that not all $-Se(=O)(\text{–OH}\cdots\text{O}=)(\text{–O}\cdots\text{HO–})Se(=O)-$ groups are dissociated, even when the membrane is hydrated at 98% relative atmospheric humidity (Table 12, p. 125, and Result 77). The dissociation equilibrium:

* With regard to the slight doublet structure of these bands see p. 219.

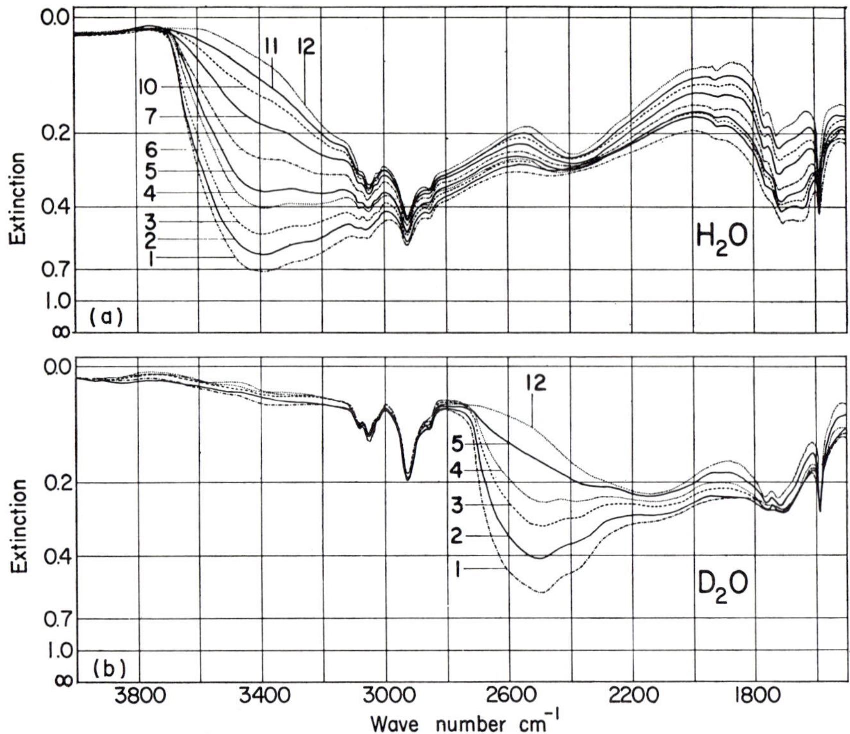

FIG. 113. Polystyreneselenonic acid, 7% cross-linked, 3 days selenonated: (a) H_2O hydrated, one drying stage every 2 days; 1, 98%; 2, 71%; 3, 53%; 4, 33%; 5, 22%; 6, 11%; 7, 5% relative atmospheric humidity; 10, 11, incompletely dried; 12, thoroughly dried membranes. (b) D_2O hydrated: 1, 100% relative atmospheric humidity; 2 → 5, decreasing atmospheric humidity; 12, thoroughly dried membrane.

```
            H
           /
    OH····O                                  H
   /       \                                /
—Se=O       H                  O-   ····H—O
   \\        ·.               .·'
    O          O—H ⇌ —Se····O )          H+
               |          ·.                ·.
               H            O                 O—H
                                              |
                                              H
```

is not as far displaced on the side toward dissociation as in the polystyrene-sulfonic acid. This is discussed in connection with the potential curves in the hydrogen bridges in Section V.16.

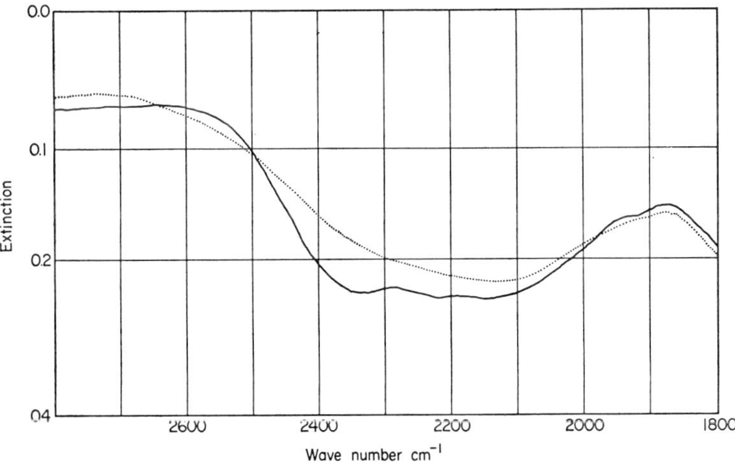

FIG. 114. Polystyreneselenonic acid, 7% cross-linked, 3 days selenonated, spectra taken after D_2O hydration of thoroughly dried membranes; ····, membrane rapidly dried; ——, membrane slowly dried.

V.13. Polystyreneseleninic and Polystyrenephosphinic Acids

Very different hydration behavior is found in polystyreneseleninic and -phosphinic acids.[182] Figure 115 shows spectra of a membrane of polystyreneseleninic acid. Figure 116 shows spectra of a membrane of polystyrenephosphinic acid. From these figures we obtain Result 124.

Result 124: Neither polystyreneseleninic acid nor polystyrenephosphinic acid yields a continuous absorption.

Here, even when the membranes have been hydrated at 98% relative atmospheric humidity, there are very few protons tunneling in the network of hydrate structures. This is exactly what would be expected from our results in Section V.1.2, for there the bands of the anions showed that these acid groups do not dissociate appreciably. With respect to our results in Section V.6, this shows that hardly any $H_5O_2^+$ groups are formed; the dissociation equilibrium is almost completely displaced away from dissociation. The cause of this difference is explained more fully in Section V.16.

We have further seen in Section V.2.2 that the acid groups in these acids still remain associated through powerful hydrogen bridges, even when the acids are hydrated at 98% relative atmospheric humidity. This must now be borne in mind when we consider the hydration. From Figs. 115 and 116 we obtain:

Result 125: With both polystyreneseleninic and polystyrenephosphinic acids, a band arises with an increasing degree of hydration at 3390 cm^{-1}

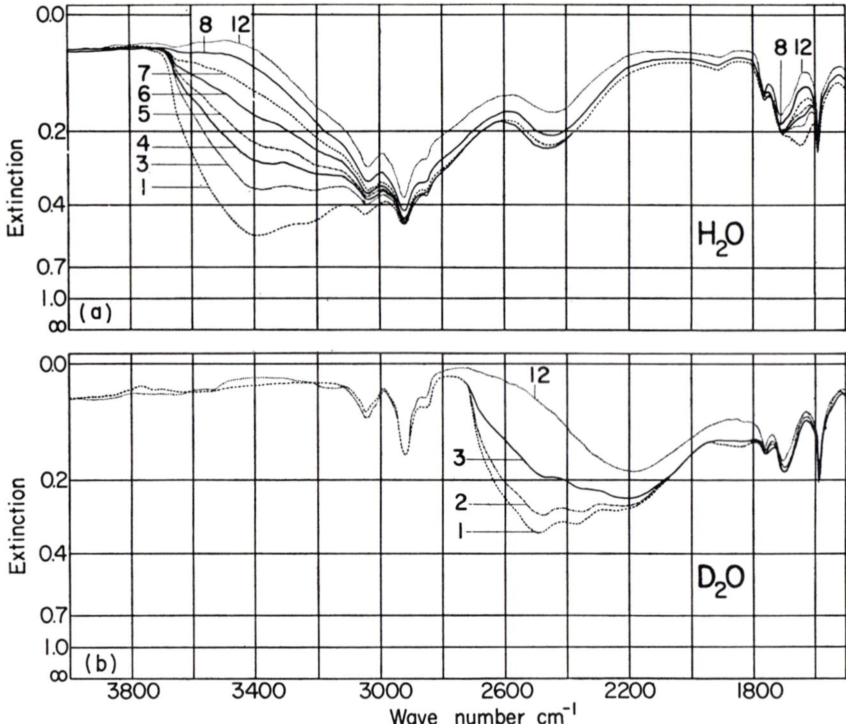

FIG. 115. Polystyreneseleninic acid, 7% cross-linked, 3 days selenonated: (a) H_2O hydrated: 1, 98%; 3, 53%; 4, 33%; 5, 22%; 6, 11%; 7, 5%; 8, 1% relative atmospheric humidity; 12, thoroughly dried membrane. (b) D_2O hydrated: 1, atmospheric humidity over a saturated solution of $BaCl_2$ in D_2O; 2, 3, decreasing atmospheric humidity; 12, thoroughly dried membrane.

(H_2O) or 2490 cm^{-1} (D_2O). The band at 2490 cm^{-1} for polystyreneseleninic acid has a broad shoulder on the side toward larger wave numbers. In addition, in both cases a shoulder is found at about 3200 cm^{-1} (H_2O) or 2365 cm^{-1} (D_2O).

To decide whether the shoulder at 3200 cm^{-1} (H_2O) or 2365 cm^{-1} (D_2O) is to be assigned to the overtone of the scissor vibration, we took spectra of polystyreneseleninic acid with H_2O and HDO hydration, respectively. They are shown in Fig. 117.

Result 126: The shoulder at about 3200 cm^{-1} is not observed in the HDO hydration. Hence it follows that the shoulder at about 3200 cm^{-1} is caused by the band of the overtone of the scissor vibration of the H_2O molecule (Section III.2.1). In the same way, the shoulder found at about 2365 cm^{-1} in the case

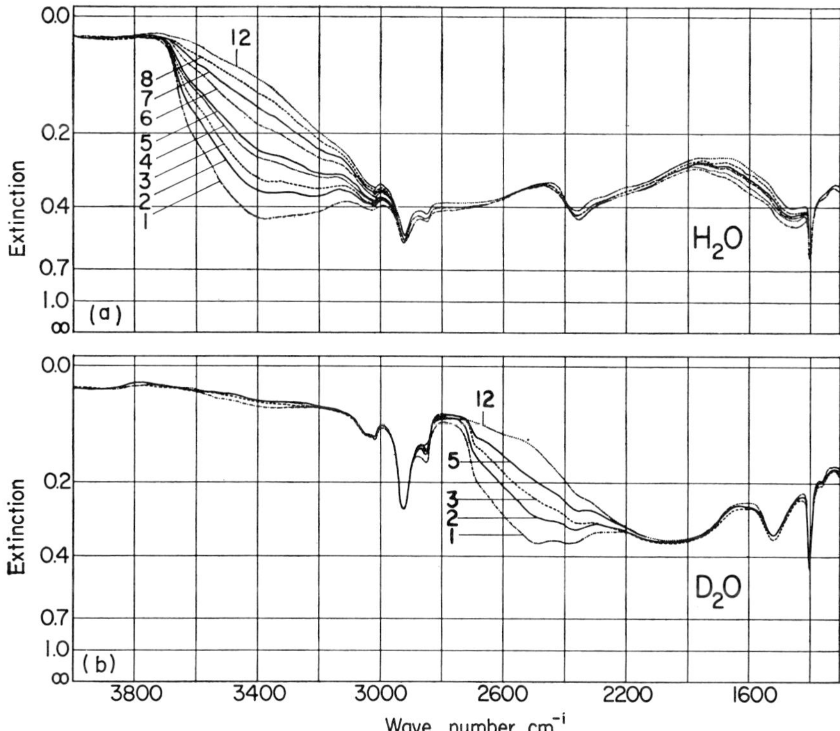

FIG. 116. Polystyrenephosphinic acid, 7% cross-linked, 3 days phosphinated: (a) H_2O hydrated: 1, 98%; 2, 71%; 3, 53%; 4, 33%; 5, 22%; 6, 11%; 7, 5%; 8, 1% relative atmospheric humidity; 12, thoroughly dried membrane. (b) D_2O hydrated: 1, the atmospheric humidity over a saturated solution of $BaCl_2$ in D_2O; 2 → 5, decreasing atmospheric humidity; 12, thoroughly dried membrane.

of D_2O hydration is to be ascribed to the overtone of the scissor vibration of the D_2O molecule.

On the other hand, the band at 3390 cm^{-1} (OH) or 2390 cm^{-1} (OD) is to be ascribed, with respect to its position and to the dependence of its intensity on degree of hydration, to the stretching vibration of OH or OD groups in hydrogen bridges. Since the acid protons fixed in hydrogen bridges cannot be hydrated, only such hydrogen bridges with water molecules bound either to the oxygen atoms of the anions, or to one another, can be the reason for this band. In this way we can understand the broad shoulder on the side of the OD band toward larger wave numbers. This splitting into a band and a shoulder* arises from the formation of hydrogen bridges to acceptors of

* In polystyreneselenonic acid, a splitting of the band at 2490 cm^{-1} is also found at high degrees of D_2O hydration (Fig. 113). This is, however, much less clearly marked, but could arise from the same cause.

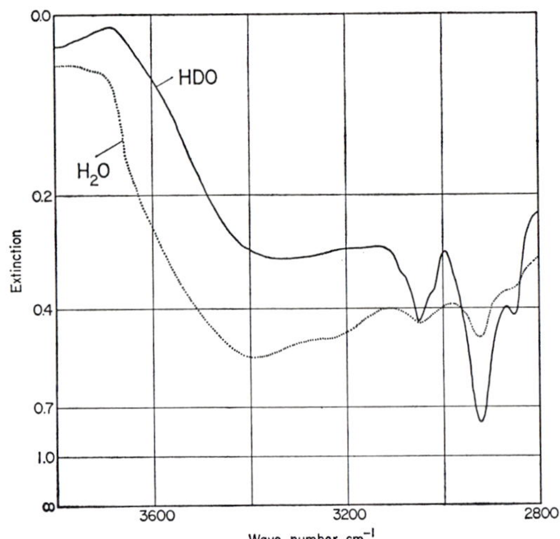

FIG. 117. Polystyreneseleninic acid, 7% cross-linked, 3 days selenonated: ——, HDO hydrated, atmospheric humidity over a saturated solution of $BaCl_2$ in H_2O–D_2O (mixture of about 7 molecular % H_2O and 93 molecular % D_2O); · · · ·, H_2O hydrated, 98% relative atmospheric humidity.

various strengths (i.e., to oxygen atoms of ions and of other water molecules), by the OD groups of the molecules of water of hydration. With H_2O, such a doublet structure is not found, because both bands are so broad that they can no longer be resolved.

Result 127: On hydration with H_2O a shoulder is found at about 3640 cm^{-1}, and on hydration with D_2O at about 2680 cm^{-1}. This shoulder is only faintly marked in polystyreneseleninic acid, but is clearly marked in polystyrene-phosphinic acid.

This shoulder lies in a region in which no stretching vibration of OH or OD groups in hydrogen bridges can be present. It follows that it results from the stretching vibration of free OH or OD groups of water molecules.

Figure 118 shows spectra of polystyreneseleninic acid, one membrane at 292°K and one at 85°K. The comparison yields:

Result 128: At 85°K exactly the same bands are observed as at 292°K, excepting only that the bands become sharper on cooling. Thus at low temperature the hydration is similar to that at room temperature.

The attachment of the molecules of water of hydration must be assumed to be as summarized in Fig. 119. In addition to the associated acid groups, the

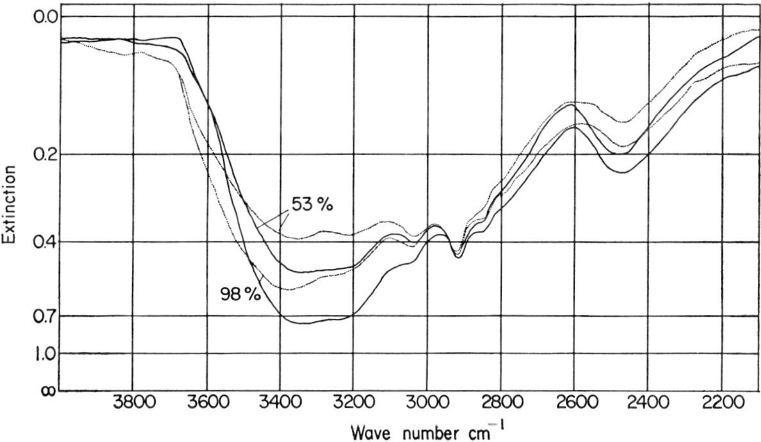

FIG. 118. Polystyreneseleninic acid, 7% cross-linked, 3 days selenonated, H_2O hydrated, 98 and 53% relative atmospheric humidity; ···· 292°K; ——, 85%K.

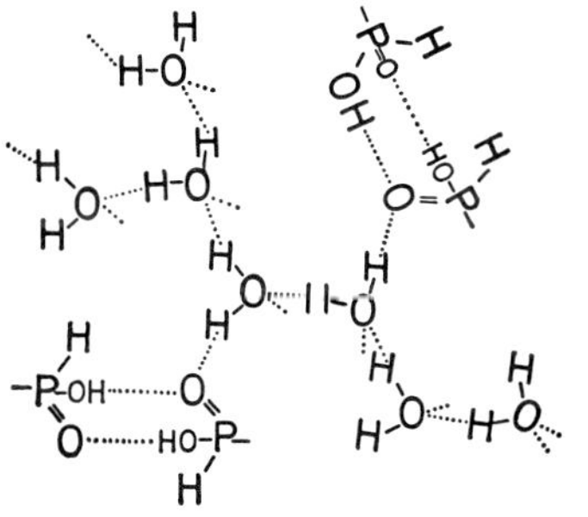

FIG. 119. The hydrate structures of polystyrenephosphinic acid (see footnote p. 67).

water molecules are connected one to the other and to the oxygen atoms of the acid groups by hydrogen bridges. Free OH groups of water molecules are to be found, i.e., this network of hydrogen bridges is considerably disintegrated. In particular, more free OH groups are found in polystyrenephosphinic acid than in pure liquid water (Fig. 11). The associated acid groups and the polymer network thus act to disintegrate the hydrate structure. This is understandable, for one of the lone electron pairs on the oxygen atom of this ion, which can serve for water molecules as a hydrogen bridge acceptor, is already occupied in the formation of the hydrogen bridges by which these acid groups are linked.

V.14. Polystyrenethiophosphonic Acid

Figure 120 shows spectra of a membrane of polystyrenethiophosphonic acid. In Chapter II we have already assumed, in assigning the bands of the PO stretching vibrations, that some $-\mathrm{P}(=\mathrm{S})(\mathrm{OH})-\mathrm{OH}$ groups were present in tautomeric equilibrium along with the $-\mathrm{P}(\mathrm{SH})(\mathrm{OH})=\mathrm{O}$ groups (also see footnote 183). This is confirmed below by the results which we obtain from Fig. 120. From

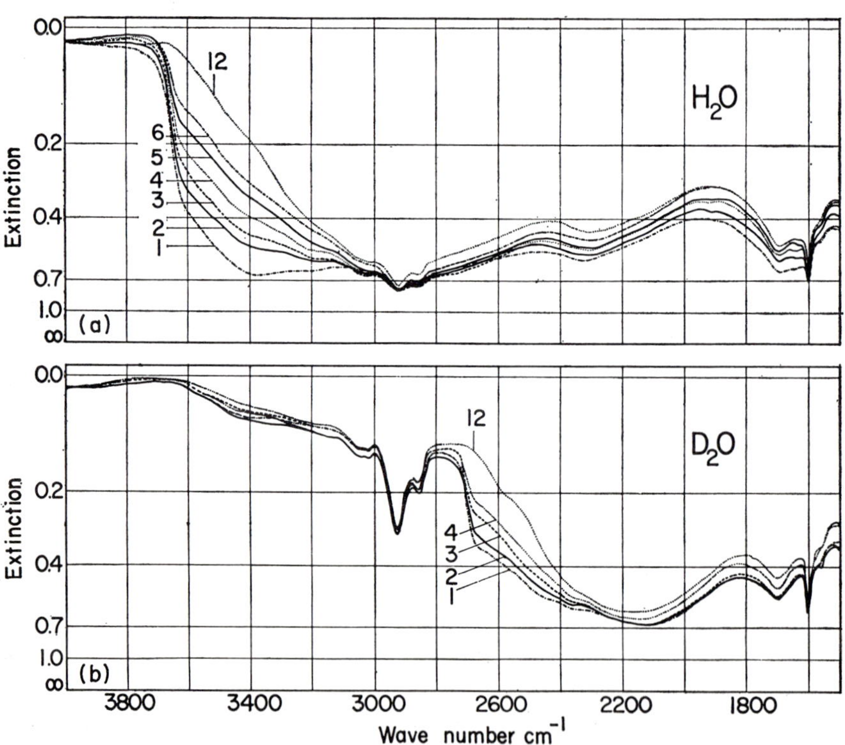

FIG. 120. Polystyrenethiophosphonic acid, 7% cross-linked, 3 days thiophosphonated. (a) H_2O hydrated: 1, 98%; 2, 71%; 3, 53%; 4, 33%; 5, 24%; 6, 11% relative atmospheric humidity; 12, thoroughly dried membrane. (b) D_2O hydrated: 1, 100% relative atmospheric humidity; 2 → 5, decreasing atmospheric humidity; 12, thoroughly dried membrane.

[183] W. I. Semljanski and L. K. Klimowskaja, *Zh. Obshch. Khim.* **30**, 4056 (1960).

Section V.2.2, we know that the —P(=O)(SH)(OH) groups are linked by strong hydrogen bridges to form
—P(SH)=O···HO—P(SH)=O··· (—P(SH)(=O···HO)(OH···O=P(HS)(—)) groups, even if the membrane has been hydrated at 98% relative atmospheric humidity.

Result 129: In hydrated polystyrenethiophosphonic acid there is a weak continuous absorption. This vanishes with progressive drying.

It follows that, with increasing degrees of hydration, acid protons are removed from the anions and become mobile in the hydrate structures by tunneling. No change is observed in the bands of the PO stretching vibration with the degree of hydration; hence the acid protons which are removed from the anions are those of the SH groups.

However, no band of a SH stretching vibration is found in the spectrum of the thoroughly dried membrane. According to Bellamy (footnote 184, p. 351 ff.), SH bands occur at about 2550 cm^{-1} (SH) or 1900 cm^{-1} (SD). A stronger shift of this band toward smaller wave numbers through formation of hydrogen bridges is not to be expected since, according to Bellamy,[184] the inclination of the SH group to form hydrogen bridges is not very large. We cannot observe any corresponding bands in the spectra. These bands are obviously masked by the broad and strong bands of the OH or OD stretching vibration of the POH···OP or POD···OP bridges. This is not surprising, since Bellamy (see footnote 184, p. 352), writes: "The SH absorption is not inherently strong, and whilst it is usually well defined, it is often difficult to detect in dilute solutions or in samples examined in very thin cells."

Result 130: When the membrane is dried after hydration with D_2O, a clearly marked shoulder is observed at about 2580 cm^{-1}. If it is dried after hydration with H_2O, a corresponding shoulder should be found at 3500 cm^{-1}. The form of the flank of the intense OH band in this region leads us to suppose that a weak band is actually present at about 3500 cm^{-1}.

It is probable that this band is that of the OD or OH stretching vibration of the free OD or OH groups in —P(=S)(OD)(OD) or —P(=S)(OH)(OH) groups.

Result 131: In polystyrenethiophosphonic acid hydrated with H_2O, a band is found at about 3400 cm^{-1}, and, when the acid is hydrated with D_2O, at about 2500 cm^{-1}.

Result 132: In the spectrum of polystyrenethiophosphonic acid hydrated with H_2O, a strongly marked shoulder is found at about 3640 cm^{-1}, and, when the acid is hydrated with D_2O, at about 2680 cm^{-1}.

[184] L. J. Bellamy, "The Infrared Spectra of Complex Molecules," 2nd Ed. Methuen, London, 1958.

The band at about 3400 cm^{-1} (OH) or about 2500 cm^{-1} (OD) arises from the OH or OD stretching vibration of the hydrogen bridges formed by the hydration water molecules. The molecules involved are presumably both the "outer" hydration water molecules surrounding the tunneling proton (Section V.10, Fig. 95) and the water molecules linked by hydrogen bridges to one another or to oxygen atoms of acid groups.

The shoulder at 3640 cm^{-1} (OH) or 2680 cm^{-1} (OD) shows that there are many free OH or OD groups in the network of hydrate structures in this case, i.e., that the network of the hydrate structure is very strongly disintegrated in the polystyrenethiophosphonic acid. The result, that these acid groups have such a strong structure-disintegrating effect, is to be explained in the same way as the corresponding result in the foregoing section. In addition, it must be borne in mind that the sulfur atom has a smaller acceptor affinity for hydrogen bridges than have oxygen atoms.[185] The structure of the anions thus explains, in the present case also, their influence on the nature of the network of hydrate structures.

V.15. Survey of Hydration in the Acids

The behavior of the acid protons and the hydration behavior observed with the individual acids is summarized in Table 17.*

V.16. The Dissociation Process

The change in the electron configuration and, hence, the increase of mesomerism of the anions and the formation of the $H_5O_2^+$ groups are decisive for the changes of enthalpy and entropy in the dissociation process.

The molecular processes involved in dissociation are fairly well known. They are shown for polystyrenesulfonic acid in schematic representation in Fig. 121(a). We now discuss the dissociation process on the basis of this representation.

We begin with polystyrenesulfonic acid. For this, we apply the results discussed in Sections V.1, V.2, V.5, V.6, and V.8, without indicating this in every case. Figure 121a, *α* shows the potential well in the hydrogen bridge formed by the acid group when only one water molecule is attached to the acid proton.

* Concerning the hydrate structure in the hydrated poly(*p*-trimethylammonium)-styrene hydroxide [—$N(CH_3)_3^+OH^-$ groups] and their salts, see footnote 185a.

[185] G. C. Pimentel and A. L. McClellan, "The Hydrogen Bond," p. 201. Freeman, San Francisco, California, 1960.

[185a] T. Ackermann, G. Zundel and K. Zwernemann, *Ber. Bunsenges. Physik. Chem.*, in press.

TABLE 17

REVIEW OF HYDRATION RESULTS OF THE ACIDS STUDIED

Acid	Cross-linking of the acid groups by hydrogen bridges	True degree of dissociation α_t at 98% relative atmospheric humidity	Continuous absorption	Hydrate structures at 98% relative atmospheric humidity	Hydrate structure network disintegration
1	2	3	4	5	6
Polystyrenesulfonic acid	only after thorough drying, but all broken at 98% relative atmospheric humidity	almost all acid groups are dissociated (α_t nearly 1)	strong	hydrate structure network surrounding tunneling protons linked with oxygen atoms of the anions by hydrogen bridges (see Fig. 95)	extremely little
Polystyrene-selenonic acid	only after thorough drying, but not all broken at 98% relative atmospheric humidity	some undissociated acid groups are still present	medium	like polystyrenesulfonic acid, but some acid groups still associated at 98% relative atmospheric humidity	extremely little
Polystyreneseleninic acid and Polystyrene-phosphinic acid	still present at 98% relative atmospheric humidity	α_t very small	absent	the water molecules located near associated acid groups, linked with one another and with the oxygen atoms of the acid groups by hydrogen bridges (See Fig. 119)	little noticeably
Polystyrene-thiophosphonic acid	still present at 98% relative atmospheric humidity	α_t small	weak	hydrate structure network surrounding tunneling excess protons, independent of excess proton water molecules linked with one another and with the oxygen atoms of the acid groups by hydrogen bridges	very strongly

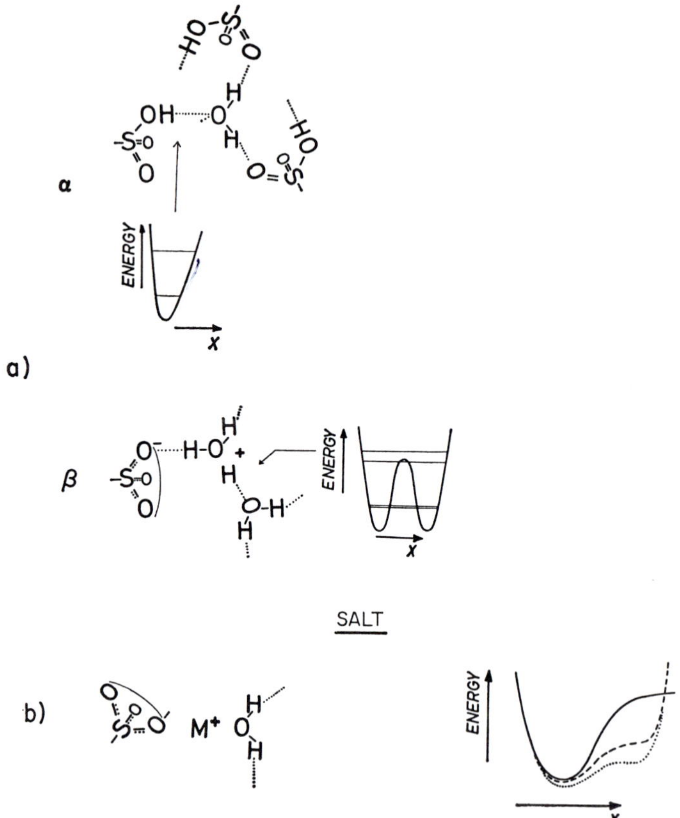

FIG. 121. (a) Schematic representations for discussion of the dissociation process of the acids; hydrate structures and potential curves in the hydrogen bridges. (b) Schematic representation for discussion of the dissociation process in the salts.

When the latter is in its ground state, it is present at the anion. If a second water molecule is attached, an essential change can only occur when the acid proton is removed from the anion. If the acid proton approaches the water molecule, the potential between the two water molecules becomes symmetrical (Fig. 121a, see β). The acid proton exchanges its function with the proton of the water molecule. An $H_5O_2^+$ group is formed, in which this proton now tunnels as an excess proton. When the excess proton is removed from the anion, the bonding electrons of the anion are rearranged. The potential well in the hydrogen bridge between the oxygen atom of the anion and the water molecule is altered. The proton in this bridge is now, in the ground state, preferentially located at the water molecule.

The true degree of dissociation of polystyreneselenonic acid is somewhat smaller, that is, the equilibrium is not so far displaced toward dissociation (Section V.1.2). We can understand this, since we know that the hydrogen bridge acceptor property of the oxygen atoms of the $\left(—Se{\overset{O^-}{\underset{O}{\lessgtr}}}O\right)$ ion is stronger than that of the oxygen atoms of the $\left(—S{\overset{O^-}{\underset{O}{\lessgtr}}}O\right)$ ion (Section IV.8). This is not surprising, since Paetzold[186] states that the SeO bond shows more polarity than the SO bond. It can be assumed that the proton acceptor property increases as well as the hydrogen bridge acceptor property. The potential curve in the hydrogen bridge between the oxygen atom of the anion and that of the water molecule in polystyreneselenonic acid thus differs from that in polystyrenesulfonic acid in that, for the former, it is more difficult for the acid proton to leave the anion and easier to return to it than in the latter. This explains why the equilibrium is not quite so strongly displaced toward dissociation.

We know that the true degree of dissociation is very small in polystyreneseleninic and polystyrenephosphinic acids (Section V.1.2). The reason why these acid groups remain almost completely undissociated must be sought in their nature, since the contribution to the enthalpy and entropy changes which the formation of the $H_5O_2^+$ group would make in the event of a dissociation in these acids, would be almost of the same magnitude as that in polystyrenesulfonic and -selenonic acids.

Gillespie[187] divides the inorganic oxyacids into various groups. These acids may be represented by the general formula $XO_n(OH)_m$. The larger n is, the more strongly dissociated is the acid. Kortuem[4] writes on the degree of dissociation and nature of the anion, p. 332: "The larger the number of double bonds, the better can the charge distribute itself over the whole anion by mesomerism, the greater is the resonance stabilization of the anion against the acid, and the less is the proton removal energy."

As we have seen, in the dissociation of $—Se{\overset{\displaystyle /\!\!/ O}{\underset{\displaystyle \backslash OH}{}}}$ or $—P{\overset{\displaystyle / H}{\underset{\displaystyle \backslash OH}{=}}}O$ group the charge is distributed through mesomerism almost exclusively between two bonds, for the PH and PS bonds take but little part in the mesomeric bond resonance in $—S{\overset{\displaystyle /\!\!/ O}{\underset{\displaystyle \backslash OH}{=}}}O$ or $Se{\overset{\displaystyle /\!\!/ O}{\underset{\displaystyle \backslash OH}{=}}}O$ groups it is distributed between three bonds. This explains the differences seen in the true degree of dissociation for these acids.

[186] R. Paetzold, *Z. Anorg. Allgem. Chem.* **325**, 47 (1963).
[187] R. J. Gillespie, *J. Chem. Soc.* (*London*) p. 2537 (1950).

In summary, we see that the processes in the dissociations of the acids can be understood on the basis of the rearrangement processes in the anions and the formation of the $H_5O_2^+$ group. The difference in the true degree of dissociation between $-S(=O)(=O)OH$ and $-Se(=O)(=O)OH$ is explained by the difference in the proton acceptor property of the oxygen atoms of these anions. The difference between $-S(=O)(=O)OH$ and $-Se(=O)(=O)OH$, on the one hand, and $-Se(=O)OH$ and $-P(H)(=O)OH$, on the other, is explained by the difference in the variations in mesomerism of these anions.

V.16.1. Comparison of Dissociation Processes in Acids and Salts

Now that we know the processes which occur in the dissociation of acids, we can consider the dissociation of salts, already discussed in Section IV.15, from a somewhat different point of view.

In the salts, the cation lies in a potential well at one of the oxygen atoms of the anion; this is represented by the potential curve shown as a continuous line in the schematic representation in Fig. 121(b). If now a water molecule is attached to the cation, the potential well is "bent down," as shown by the broken curve. It is bent down rather more when the second water molecule is attached—dotted curve—and so on. Hence the cation can be more easily removed, the more water molecules are attached.

Comparing the above with the discussion in the previous section, it can be seen that the processes in the dissociation of salts are basically different from those in the dissociation of acids.

V.17. Considerations on Anomalous Proton Conductivity

In Sections V.6–V.11, we have seen that the nature of the hydration of the excess proton in polyelectrolytes is not at all different from that in concentrated electrolytes. This is particularly clearly shown by comparison of the spectrum of concentrated *p*-toluenesulfonic acid with that of polystyrenesulfonic acid in Fig. 96. Using this as a basis for discussion, we now proceed to the processes of anomalous proton conductivity.

Numerous authors[96,99–101,188–191] have concerned themselves in recent years with the question of the rate-determining process in anomalous proton

[188] B. E. Conway, J. O'M. Bockris, and H. Linton, *J. Chem. Phys.* **24**, 834 (1956).
[189] R. A. Horne and R. A. Courant, *J. Phys. Chem.* **69**, 2224 (1956).
[190] R. A. Horne, R. A. Courant, and D. S. Johnson, *Electrochim. Acta* **11**, 987 (1966).

conductivity. Most came to the conclusion that this process, at 20°C, is not a motion of the excess proton in the hydrogen bridges, but rather structure migration.*

Thus Eigen[100] writes: "For the mechanism of proton excess and defect conduction in liquid H_2O there are two possibilities:

"(a) The rate determining step consists of the transition of the proton from H_3O^+ to an H_2O molecule of the secondary hydrate structure. The transformation must then take place within the tertiary structure (transformation to a secondary and formation of a new tertiary hydrate structure) before the proton or the proton defect jumps back into its previous central position.

"(b) The rate determining step is structure embedding (structure diffusion). In this case the proton will fluctuate to and fro between a primary and a secondary structure a large number of times before the center of the complex is displaced by the length of a hydrogen bridge.

"A choice between the two mechanisms is offered by a comparison of the drift mobility of the proton in water and ice. In the latter—at least at low proton concentrations where the disorientation of the lattice by proton jumps can be neglected—mechanism (a) occurs, the perfect tetrahedral hydrogen bridge structure offering an optimum of transmission possibilities for the proton.

"The proton mobility in ice crystals has been directly determined by the methods of reaction kinetics.[136,193] It is one or two orders of magnitude higher than the mobility in the liquid phase. From this it may be directly concluded that mechanism (b) occurs in water."

We now consider the schematic representation in Fig. 122. On the left hand side, two of the four proton boundary structures of an excess proton of $H_9O_4^+$ in polystyrenesulfonic acid are shown. The rate determining step is not a tunneling of the excess proton in the hydrogen bridges of the $H_9O_4^+$ group, but rather displacement of the charge center of this group, as shown by the schematic representation in Fig. 122. How does this structure diffusion of the $H_9O_4^+$ groups in the network of the hydrate structure come about? Consider the proton boundary structure 2 on the left side of Fig. 122. If the thermal motion of the molecules causes one of the hydrogen bridges (b) to linearize, a doubly degenerate condition of the excess proton prevails at this

* On the proton conduction in methanol, see the paper of Grunwald *et al.*[192]; on that in aqueous solutions of imidazole the paper of Ralph and Grunwald[192a]; and on that in glycerol the paper of Erdey-Grúz and Kugler.[192b]

[191] The Kinetics of Protons Transfer Processes, *Discussions Faraday Soc.* **39**, (1965).

[192] E. Grunwald, C. F. Jumper, and S. Meiboom, *J. Am. Chem. Soc.* **84**, 4664 (1962).

[192a] E. K. Ralph, III and E. Grunwald, *J. Am. Chem. Soc.* **90**, 517 (1968).

[192b] T. Erdey-Grúz and E. Kugler, *Acta Chim. Acad. Sci. Hungaricae* **57**, 301 (1968).

[193] M. Eigen and L. De Maeyer, *Z. Elektrochem.* **60**, 1037 (1956).

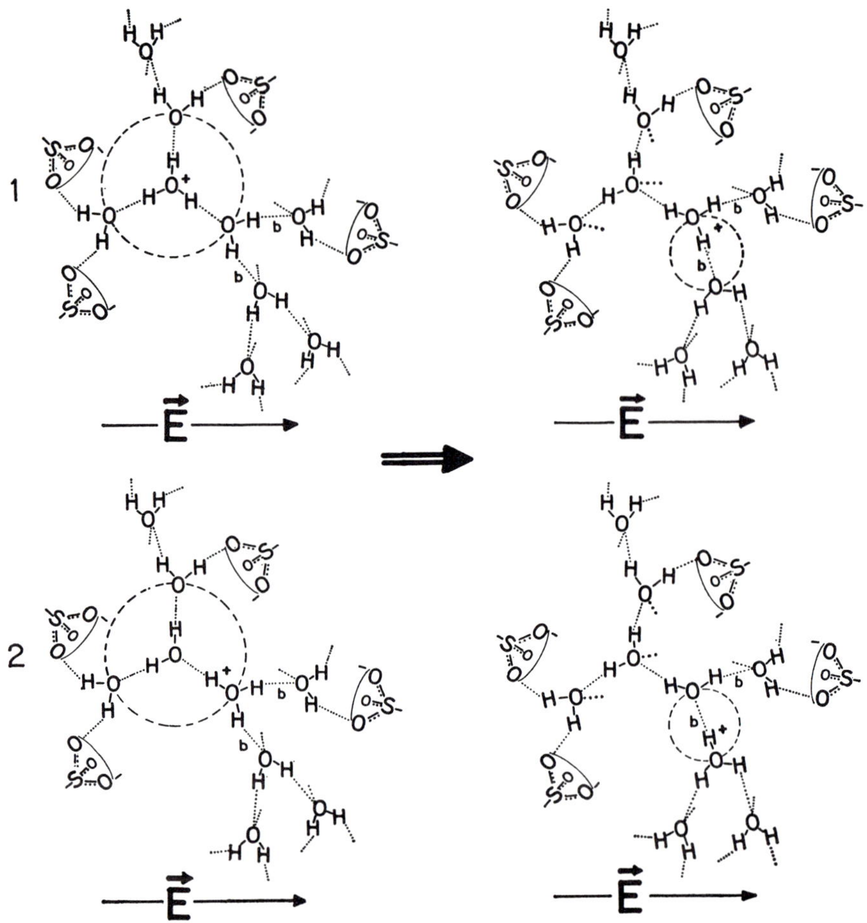

FIG. 122. Left side: two proton boundary structures of the excess proton in the $H_9O_4^+$ group in polystyrenesulfonic acid; right side: two proton boundary structures of the excess proton after one step of the structural diffusion has been carried out.

"H_3O^+" group, because the proton in the hydrogen bridge in the interior of the $H_9O_4^+$ and the proton in bridge (b) can now tunnel. If the proton tunnels in bridge (b), and if finally the bridges in which the excess proton previously tunneled bend, then the structure with the excess proton has migrated. A comparison of the proton boundary structures on the left-hand side with those on the right in Fig. 122 will illustrate this process.

Why has this molecular process a preferential direction when an outer electrical field is applied?[194] How does the anomalous proton conductivity, the Grotthus conductivity, come about? There may be two reasons for this:

(a) Because of the outer field, the weight of the proton boundary structure in which the proton lies in the direction of the field, i.e., the weight of the proton boundary structure 2 on the left side of Fig. 122, might be increased. This would increase the probability that the excess proton changes roles with the proton in one of the hydrogen bridges (b) as soon as this is linearized. Thus, the structure migration and, consequently, the charge transport would be preferentially in the field direction. However, this cannot be so, since calculation shows that for an increase in the weight of the proton boundary structure, a very large electrical field is necessary. Thus, the outer field cannot considerably increase the weight of a proton boundary structure.

(b) We have seen that the almost complete linearization of one of the bridges (b) (proton boundary structure 2 left-hand side in Fig. 122) and, therefore, the existence of a nearly doubly degenerate condition at "H_3O^+" is necessary for the structure migration. This degeneracy is eliminated by the outer field, since at the "H_3O^+" that boundary structure is energetically preferred for which the excess charge lies at the hydrogen nucleus in the bridge (b), i.e., in the direction of the field. Similarly, the migration of the structure with the excess proton in the direction opposite to the field is encumbered.

This discussion and the results in Section V.6 show: Anomalous proton conductivity occurs when $H_5O_2^+$ groups are present in the network of the hydrate structures, since the structure migration of groups of the size of $H_5O_2^+$ results in migration of the excess protons in the field. The presence of the larger group $H_9O_4^+$ is not necessary for anomalous proton conductivity.

In summary, the structure migration of the groups with the tunneling excess protons is closely connected with the linearization of the hydrogen bridges by which the groups $H_5O_2^+$ or $H_9O_4^+$ are cross-linked with the adjacent network of the hydrate structure. The field sensitivity of the molecular processes leading to anomalous proton conductivity is caused by the elimination of the double degeneracy of the condition of the excess proton at the "H_3O^+" by the outer field. This degeneracy occurs if one of the bridges (b) is highly linearized. The result of the elimination of the degeneracy is that a charge displacement in the field direction is favored.

[194] E. G. Weidemann and G. Zundel, in preparation.

CHAPTER VI

PREPARATION OF THE MEMBRANES

Two methods appear possible for the preparation of ion-exchange resins as membranes by means of the polymerization process. The first method involves the fixed ions when they are already present in the monomer[1]; the second method, the fixed ions when they are incorporated after polymerization.[2]

We have consistently used the second method.[3]

In this process, membranes are prepared in four steps.

1. Polymerization of a polystyrene membrane of definite thickness and degree of cross-linking. By "degree of cross-linking" we understand the quantity of divinylbenzene added to the styrene prior to polymerization. The figures are % w/w divinylbenzene.
2. Incorporation of the fixed ions into the polystyrene membranes.
3. Incorporation by exchange of the required counter anions, i.e., preparation of the various exchanger forms.
4. Preparation of the membranes for IR-spectroscopic measurement.

VI.1. Small-Scale Preparation of Polystyrene Membranes

Polystyrene membranes of thickness 5 μ or more, which may if required be cross-linked, can be polymerized by the process to be described. When preparing membranes for IR-spectroscopic purposes, we polymerize—when not otherwise stated—membranes of about 5–6 μ thickness.

VI.1.1. Experimental Technique

Monomeric styrene is polymerized as the principal component, with divinylbenzene as a bridging agent (quantity determined by degree of cross-linking

[1] H. Spinner, J. Ciric, and W. F. Graydon, *Can. J. Chem.* **32**, 143 (1954).
[2] R. Griessbach, "Austauschadsorption in Theorie und Praxis," p. 50 ff. Akademie Verlag, Berlin, 1957.
[3] G. Zundel, H. Noller and G.-M. Schwab, *Z. Naturforsch.* **16b**, 716 (1961).

required) and 0.8% benzoyl peroxide as initiator, for 5 days at 75°C and a further 2 days at 90°C in a special apparatus (Fig. 123).

The following points are important:

The styrene must be redistilled before polymerization, to remove the inhibitor and any dimeric and polymeric constituents present. The distillation is carried out *in vacuo* (b.p. about 34°C at 11 Torr). The distilled styrene can be stored for some weeks at −20°C without affecting the quality of the resulting membranes.

Commercial divinylbenzene generally is not pure, but contains ethylvinylbenzene. This must either be taken into account in calculating the degree of cross-linking, or it must be removed by the process given by Blasius and Beushausen.[4] Divinylbenzene has a very limited storage life. We always obtain fresh divinylbenzene in a Teflon bottle. Divinylbenzene should never be stored in a glass bottle.

Commercial benzoyl peroxide is mixed with water, to reduce the explosion risk. This water must be completely removed, as otherwise the membrane is cloudy. For this purpose a few grams of benzoyl peroxide are dried over P_2O_5 for 10 days at about 10^{-1} Torr.

The polymerization of the polystyrene membrane is carried out in the apparatus shown in Fig. 123.

The thickness of the aluminum frame (4), used as a spacer between the glass plates (3), determines the thickness of the polystyrene membrane produced and must be chosen accordingly. We cut these frames out of aluminum foil with a thickness of 5 μ or more.

The frames (2) consist of heat-resistant rubber, 1 mm thick. They must transmit the pressure elastically from the metal plates (1) (thickness 6 mm) to the glass plates (3) (thickness 8 mm). The outside dimensions of 1, 2, and 3 are 140 × 80 mm. The shape of the frames (2) is essentially the same as that of the aluminum frame.

The mixture of styrene, divinylbenzene, and benzoyl peroxide is transferred to the space defined by the glass plates (3) and the spacer frame (4), preferably with the aid of a syringe. The assembly of 1, 2, 3, and 4 is then compressed in the brass clamps (5) by means of the screws (6). The screws must be screwed in place tightly. However, it is advisable to tighten the screws more tightly in one clamp than in the other. This makes the membranes not quite uniform in thickness. The infrared spectrum of such a membrane then shows no interferences, which is desirable (see below).

The aluminum spacer foil does not provide a sufficiently tight seal to prevent the styrene from evaporating. For this reason the whole apparatus is immediately placed under water in a vessel. This vessel must be sealed, so that the water cannot evaporate. The vessel with the apparatus in it is left for 5 days at 75°C and a further 2 days at 90°C in a drying oven.

The following points must be observed in this procedure:

The glass plates (3) must be carefully cleaned, particularly when used more than once.

[4] E. Blasius and J. Beushausen, *Z. Anal. Chem.* **197**, 228 (1963).

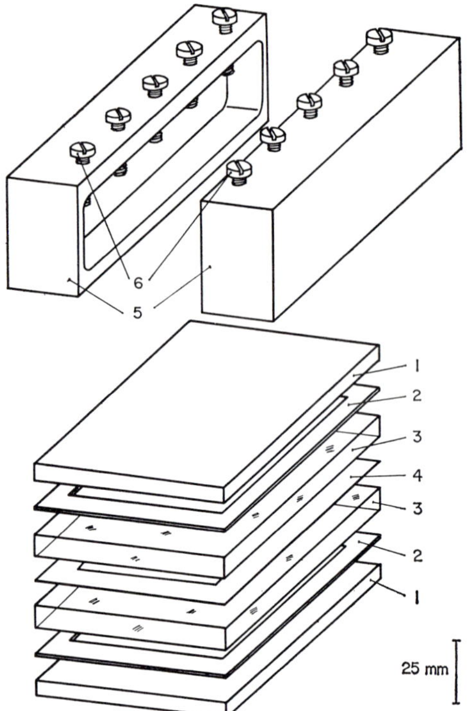

FIG. 123. Apparatus for polymerizing the membranes: 1, metal plates, 6 mm thick; 2, frames of temperature-resistant (100°C) rubber, 1 mm thick; 3, crystal glass plates, 8 mm thick; 4, frame of aluminum foil; 5, metal clamps; 6, screws.

It is best to place them in chromic–sulfuric acid solution for 2 to 3 days at 75°C. After being used once, the glass plates sometimes bend slightly, and can no longer be used. Care must be taken that the glass plates taken from the drying oven are completely cooled before the styrene is added, otherwise the styrene will be more rapidly evaporated than the divinylbenzene. The degree of cross-linking is then indefinite.

The aluminum foil (4) must never be touched with the fingers, but only with tweezers. The slightest contamination is noticeable and clouds the polystyrene membrane.

The rubber (2) should be in minimal contact with the styrene. Silicone rubber is preferable, as it is very resistant to styrene.

When adding the styrene, care must be taken that no air bubbles remain in the space.

The clamps (5) must be made of corrosion-resistant metal. The screws (6) may stick over a period of time. The pressure exerted by the clamps is then no longer uniform. This may be prevented by greasing the screws with silicone grease.

After seven days the apparatus is removed from the water bath, the clamps are unscrewed, and the glass plates, between which the aluminum foil and the polymerized polystyrene membrane are placed, are allowed to cool slightly and then placed in water. The plates are easily separated with a razor blade. The polystyrene membrane can then be separated from the plate by floating it on water. If the membrane is difficult to remove, it is probably not fully polymerized, which will lead to ill-defined conditions (see below). The polystyrene membrane is now fished out with a piece of paper and dried on it.

If the polystyrene membrane is cloudy, the cause is usually contamination, either of the aluminum foil or of the glass plates.

Caution: If membranes with "a lower degree of cross-linking" than that calculated from the quantity of added divinylbenzene are obtained, the effect is that, under the same reaction conditions (Section VI.2.1), a larger number of fixed ions is incorporated. It is then observed that the salts, in particular those with monovalent counterions, swell more strongly. This lower degree of cross-linking can have two causes. First: The divinylbenzene was already partially polymerized. Second: The polystyrene membrane was not fully polymerized. Incomplete polymerization has the same effect as a lower degree of cross-linking.

The IR spectrum of a polystyrene membrane prepared in this way is shown in Fig. 1(a).

Determination of the Thickness of the Polystyrene Membranes

The exact thickness of the membranes can easily be determined from the interference patterns in the IR spectrum [see Fig. 1(a)]. The membrane thickness d in microns is given by

$$d = \frac{1000\,(m + 1)}{2\, n_{IR}\,(\bar{\nu}_i - \bar{\nu}_k)}$$

where $\bar{\nu}_i$, $\bar{\nu}_k$ = wave number of ith and kth interference maxima, m = number of interference maxima between the measured ith and kth, and n_{IR} = refractive index of polystyrene in the infrared = 1.43.

It is preferable to select membranes whose spectra show no interference patterns because most interference patterns are still perceptible after incorporation of the fixed ions and can then cause considerable difficulty in the evaluation of the spectra, as they overlap with the bands. This is the case, for example, in the spectra reproduced in Figs. 37 and 38. If membranes showing no or weak interference are used, the thickness can be determined by measuring the integral extinction of a polystyrene band. The thickness of an interference-free membrane can then be read off from a calibration curve. The calibration curve is plotted from measurements of membranes showing interferences or the polystyrene membranes can be weighed [density ρ = 1.05 gm cm^{-3} (see footnote 5)].

[5] H. Staude, "Phys.-Chem. Taschenbuch," Vol. 1, p. 1008. Akad. Verlagsges. Leipzig, 1945.

VI.1.2. Membranes of Perdeutero-Polystyrene

Styrene-d_8 is not commercially available. Hence, we had to prepare it ourselves.[6] Two considerations decided the choice of the method of synthesis: First, the deuterated starting materials had to be commercially available. Second, the method used had to give a high yield. The reactions used were: acetyl chloride-d_3, acetophenone-d_8, α-methylbenzyl alcohol-d_{10}, α-phenyl-diethyl carbonic acid ester-d_9, leading to styrene-d_8 and finally to perdeutero-polystyrene. The spectra of these compounds are shown in Fig. 124* and the spectrum of perdeuterated polystyrene in Fig. 1(g).

Acetyl chloride-d_3. The yield of this compound from the reaction of acetic acid-d_4 with thionyl chloride was almost quantitative. It was secured by condensing SO_2 and letting the reaction proceed to completion in liquid SO_2. This strongly polar solvent accelerated the reaction while preserving extremely mild conditions.

Procedure: To 1.05 mole thionyl chloride, purified by distillation over quinoline and linseed oil,[7] 1 mole of acetic acid is added dropwise. The temperature of the mixture usually rises by about 1–2°C, but after a time it falls to −12°C with evolution of gas. The SO_2 evolved is cooled in a condenser (cooling liquid: methanol at −50°C) and the reaction is carried to completion in the liquid SO_2 produced. When the reaction is complete, the temperature in the reaction vessel is raised, and the SO_2 is evaporated. The acetyl chloride-d_3 is distilled; traces of $SOCl_2$ carried over do not disturb the subsequent reaction. The residue in the reaction vessel consists of a little acetic anhydride and decomposition products of the thionyl chloride produced by disproportionation. The overall yield is 98%, based on CD_3COOD (b.p. of acetyl chloride 54°C, of acetyl chloride-d_3 56°C).

Acetophenone-d_8. We were also able to optimize the Friedel-Crafts acetylation[8] by taking into account the effects of the individual reaction parameters. The best catalyst is aluminum chloride, as was already found by Calloway.[9] To initiate the reaction a little water must be added to the catalyst, since according to Nenitzesco[10] the active Lewis acid is $H(AlCl_3OH)$. The catalyst is present in excess (1.25 mole $AlCl_3$ to 1 mole acetyl chloride), so that the continued condensation of two molecules of acetophenone to dypnone may be prevented through complex formation with acetophenone.[11] As we were preparing a perdeuterated compound, we were naturally very

* For the assignment of the bands of these spectra see Mross and Zundel.[6]

[6] W. D. Mross and G. Zundel, *Chem. Ber*, **101**, 2865 (1968).

[6a] W. D. Mross and G. Zundel, *Spectrochim. Acta*, in press.

[7] "Organicum," Chapter F, p. 615, 6th Ed., VEB Verlag Wiss., Berlin, 1967.

[8] G. A. Olah, "Friedel-Crafts and Related Reactions," Vol. 1. Wiley (Interscience), New York, 1963.

[9] N. O. Calloway, *Chem. Rev.* **17**, 327 (1935).

[10] C. D. Nenitzesco, M. Auram, and E. Sliam, *Bull. Soc. Chim. France*, p. 1266 (1955).

[11] N. O. Calloway and L. D. Green, *J. Am. Chem. Soc.* **59**, 809 (1937).

restricted in the choice of a solvent, as we could only use one which would not yield protons under these conditions. Carbon disulfide was employed. The reaction proceeds very slowly in this medium, since both the catalyst and its acetyl chloride and acetophenone complexes are sparingly soluble in carbon disulfide. The reaction was accordingly carried out at the boiling point of the solvent. The maximum yield (85%) was attained after about 6 hours. After this period the yield falls off as more molecules are removed by condensation than are formed. Unexpectedly, the quantity of solvent affects the yield strongly: a pronounced maximum is obtained with 2 mole carbon disulfide to 1 mole acetyl chloride. The reaction temperature has an insignificant effect on the maximum yield and only changes the time at which it is attained.

With respect to these points, we chose the following conditions. 1.2 mole (anhydrous) aluminum chloride was suspended in 70 ml carbon disulfide to which a drop of D_2O was then added, and 1 mole acetyl chloride-d_3 was rapidly added dropwise. On heating, a part of the aluminum chloride dissolved. After boiling briefly, 1 mole benzene-d_6 in 50 ml carbon disulfide was added and the mixture was boiled gently. After six hours, the mixture was lemon-yellow in color. If the catalyst was contaminated with traces of ferric chloride, the reaction mixture was more or less blood red in color. The warm solution was placed in a dropping funnel and added dropwise into a cooled suspension of 6 mole D_2O (frozen) in 20 ml CS_2. Toward the end of the addition, coarse crystals of $Al(D_2O)_6Cl_x(OD)_3$ separated out. The carbon disulfide solution was removed with an immersion fritted disk and the crystals were washed several times with CS_2. To remove DCl from the solution, a few drops of pyridine were added and the precipitated pyridine deuterochloride was removed. The solution was dried over dehydrated sodium sulfate, the carbon disulfide evaporated, and the acetophenone vacuum distilled. Yield: 85%, based on the benzene-d_6 or acetyl chloride-d_3 used (b.p. of acetophenone 76.5°C at 11 Torr, ρ = 1.026 gm/ml; of acetophenone-d_8 78°C at 10 Torr, ρ = 1.1216 gm/ml).

α-Methylbenzyl alcohol-d_{10}. The carbinol was prepared according to a procedure in "Organicum" (see footnote 7, p. 481). Acetophenone-d_8 was reduced with $LiAlD_4$ in ether at 34°C. After completion of the reaction there was added a quantity of deuterium oxide exactly sufficient to form the mixed oxide quantitatively, i.e., 2 mole D_2O per mole $LiAlD_4$. This precipitated readily, and the remaining ethereal solution could readily be removed with the immersion fritted disk. Yield: 95% based on the acetophenone-d_8 (b.p. of carbinol 87°C at 11 Torr, ρ = 1.008 gm/ml, of carbinol-d_{10} 92.5°C at 11 Torr, ρ = 1.093 gm/ml).

Styrene-d_8. The direct method of dehydration[12] gave relatively poor yields and impure products (polymers and redox products), whether by acid, basic, or catalytic dehydration in the vapor phase. If, however, the ester pyrolysis reaction is chosen, very pure styrene is obtained. We used for pyrolysis the ethylcarbonic ester,[13,14] since the preparation of the ester from the carbinol

[12] D. V. Banthorpe, "Elimination Reactions." Elsevier, Amsterdam, 1963.
[13] G. L. O'Connor and H. R. Nace, *J. Am. Chem. Soc.* **75**, 2118 (1953).
[14] C. H. de Puy and R. W. King, *Chem. Rev.* **60**, 431 (1960).

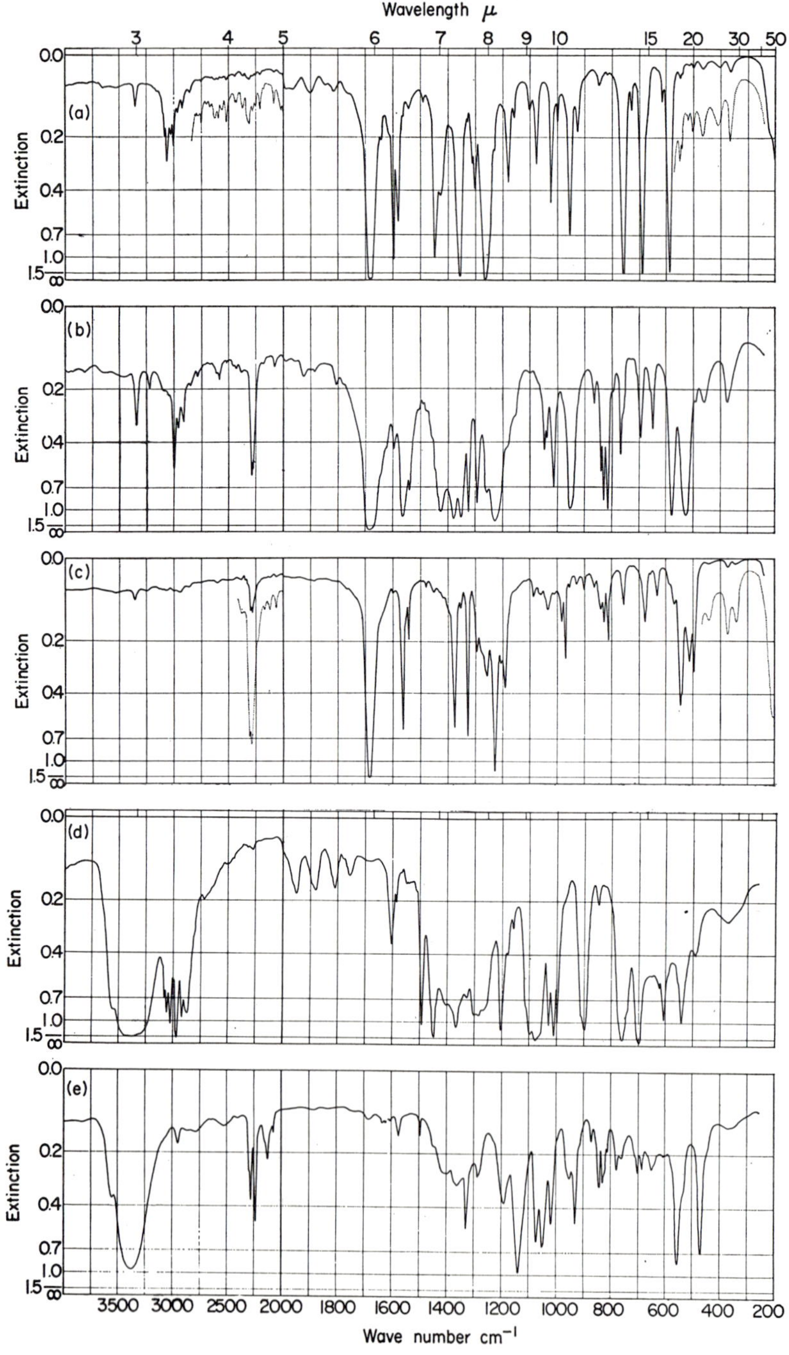
Wavelength μ
3
4
5
6
7
8
9
10
15
20
30
50
(a)
(b)
(c)
(d)
(e)
Extinction
0.0
0.2
0.4
0.7
1.0
1.5
∞
3500
3000
2500
2000
1800
1600
1400
1200
1000
800
600
400
200
Wave number cm⁻¹

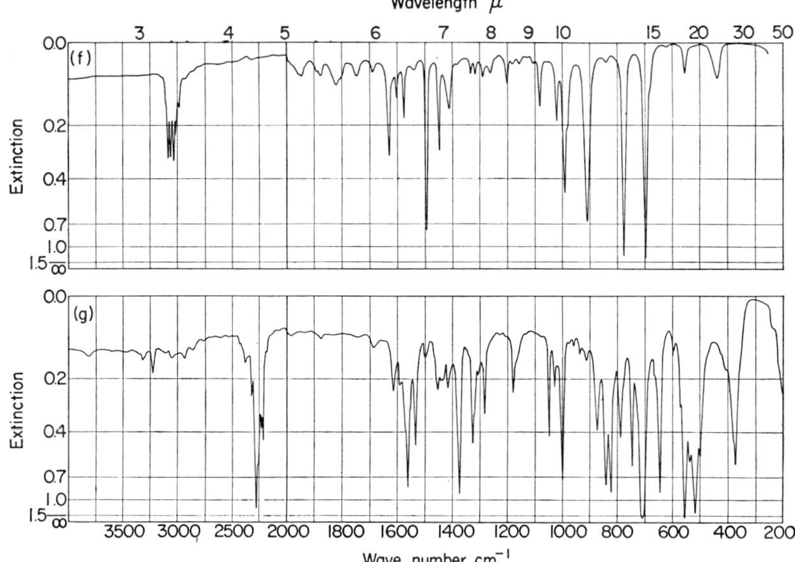

FIG. 124. IR spectra (for assignment of the bands, see footnote 6a). (These spectra were taken with the spectrophotometer model 225): (a) acetophenone; (b) acetophenone-d_5; (c) acetophenone-d_8; (d) carbinol; (e) carbinol-d_9 (containing a small amount of benzene); (f) styrene; (g) styrene-d_8.

with chloroformic acid ethyl ester in pyridine proceeds with 90% yield. Further, pyrolysis proceeds at 300°C rapidly and with a yield of almost 100%.

Procedure: For preparation of the ester, about 2 moles of previously distilled chloroformic acid ethyl ester was added to an ice-cooled solution of 1 mole carbinol in 400 ml pyridine. The mixture was allowed to stand 1 day at room temperature, and was then mixed with 500 ml dilute acetic acid. The ester was precipitated and separated. The aqueous phase was shaken three times with ether. The ester, together with the ethereal extract, was washed three times. Washing was with dilute acetic acid, saturated K_2CO_3 solution, and saturated KCl solution, followed by drying over dehydrated Na_2SO_4. After evaporating the ether, the ester was distilled (b.p. of ethyl ester 123°C at 11 Torr, of ethyl ester-d_9 127°C at 11 Torr).

Pyrolysis: The pyrolysis apparatus was so constructed that only the undecomposed ester refluxed, the temperature in the reflux condenser being regulated at 55°C with a thermostat. The reaction products were removed by freezing in a trap cooled with liquid nitrogen. The contents of the trap were thawed out slowly, during which process the CO_2 escaped. Then 0.1% *tert*-butyl catechol was added as inhibitor and the alcohol (C_2H_5OD) was evaporated at 200 Torr and 60°C, the reaction mixture was then distilled under a water pump vacuum. Despite cautious distillation, 3% diethyl carbonate, produced by rearrangement, was carried over, and could only be reduced to 0.5% by a further distillation (b.p. of styrene 38°C at 12 Torr, of styrene-d_8 52°C at 14 Torr).

Polymerization to perdeutero-polystyrene membranes: We proceeded exactly as in Section VI.1.1. However, it was necessary to bear in mind that styrene-d_8

polymerizes more slowly than styrene (on this point see footnote 15). We accordingly polymerized the mixture 4 days at 80°C, and then 10 additional days at 95°C. However, the membranes are less cross-linked than would be expected from the amount of employed divinylbenzene, as evidenced by the swelling observed after sulfonation. Undeuterated benzoyl peroxide was used as initiator, and undeuterated divinylbenzene as cross-linking agent. We cross-linked these membranes only 2%, to minimize contamination with hydrogen. A spectrum of a membrane thus prepared is shown in Fig. 1(g).

VI.2. Incorporation of $—SO_3^-$ Ions and $—SeO_2OH$, $—SeOOH$, $—P(H)(=O)OH$, and $—P(SH)(=O)OH$ Groups

VI.2.1. Incorporation of $—SO_3^-$ Ions

We have used chlorosulfonic acid for sulfonation[3] (see also footnotes 16–18). In this process, the $—SO_3^-$ ions are incorporated into the membrane in two stages. (For sulfonation of the perdeuterated membranes see Section VI.2.1.2.) First, $—SO_2Cl$ groups are incorporated; second, these groups are hydrolyzed and immediately hydrated.

We sulfonate mostly at 25°C, as described in the next section. By this procedure, membranes with between 0.5 and 1.5 $—SO_3^-$ ions per benzene ring may be obtained. If membranes with a degree of sulfonation less than 0.5 $—SO_3^-$ ion per benzene ring are required, we sulfonate at 0°C, to obtain a more or less uniform distribution of the incorporated ions in the membranes (Section V.2.1.3).

VI.2.1.1. Sulfonation at 25°C

First Stage

$$O_2S(Cl)(OH) + \text{[polystyrene unit: } CH_2\text{–}CH(C_6H_5)\text{–}CH_2\text{]} \longrightarrow \text{[}CH_2\text{–}CH(C_6H_4SO_2Cl)\text{–}CH_2\text{]} + H_2O$$

The membranes are placed in pure chlorosulfonic acid, as described below.

[15] A. Kirchner, *Makromol. Chem.* **96**, 179 (1966).

[16] F. D'Alelio, International General Electric Co., U.S. Patent No. 2,366,007, August 11, 1942.

[17] R. F. Boyer, Dow Chemical Co., U.S. Patent No. 2,500,149, February 21, 1947.

[18] A. Abbey, Dow Chemical Co., British Patent No. 670,145, November 18, 1949.

The sulfonation time at 25°C varies from 12 hours to 2 weeks, depending on the degree of cross-linking and the required degree of sulfonation.

The practical details of the procedure are as follows:

Before sulfonation, the polystyrene membrane is cut into pieces of the size required for investigation. The pieces are then thoroughly dried, i.e., for at least 3 days at about 10^{-1} Torr over P_2O_5. If the membrane is not absolutely dry when introduced into the chlorosulfonic acid, it rolls up and cannot be used.

Sulfonation is carried out in a weighing glass (preferably one with a ground glass lid fitting over the outside of the rim). The weighing glass is placed in a well-dried desiccator. To avoid a pressure buildup in the desiccator as a result of the evolution of HCl, it is connected by a tube to a vessel containing P_2O_5 which in turn communicates with the atmosphere through a capillary tube. During sulfonation, the desiccator is cautiously shaken once or twice each day.

Transfer of the membrane to the aqueous medium:

As much of the chlorosulfonic acid as possible is removed with a pipette. The membrane is then exposed to the atmosphere, so that the remainder of the chlorosulfonic acid can decompose by reacting with the moisture in the atmosphere.* The membrane is now placed in a desiccator over water for 1 hour at about 12 Torr. The sulfuric acid formed is thus so strongly diluted that the membrane may be placed directly in water without being destroyed. It is then washed in distilled water.

The IR spectrum of a membrane prepared in this way is reproduced in Fig. 1(b). Polystyrenesulfonyl chloride is formed by this procedure, as is to be expected (see footnote 19 and footnote 20, p. 530). This follows in the first place from the assignment of the bands (see Chapter II and Table A.1), and in the second place from a comparison of this spectrum with that of toluenesulfonyl chloride.[21]

Second Stage—Hydrolysis and Hydration

$$\text{(CH}_2\text{—CH(C}_6\text{H}_4\text{SO}_2\text{Cl)—CH}_2\text{)} + H_2O \longrightarrow \text{(CH}_2\text{—CH(C}_6\text{H}_4\text{SO}_2\text{OH)—CH}_2\text{)} + HCl$$

* The chlorosulfonic acid could be washed away with methylene or ethylene chloride, but there is the danger that some of the solvent will remain in the membrane and disturb the measurement by its strong IR absorption.

[19] E. Wertheim, "Textbook of Organic Chemistry," 2nd Ed. McGraw-Hill (Blakiston), New York, 1945.

[20] J. Houben and Th. Weyl, "Methoden der organischen Chemie," Vol. 9. Thieme, Stuttgart, 1955.

[21] G. Zundel, IR-Untersuchungen von Ionenaustauschern auf Polystyrolbasis, insbesondere im Hinblick auf die Ionenhydratation und Protonenbeweglichkeit. Ph.D. Thesis, University of Munich, 1960.

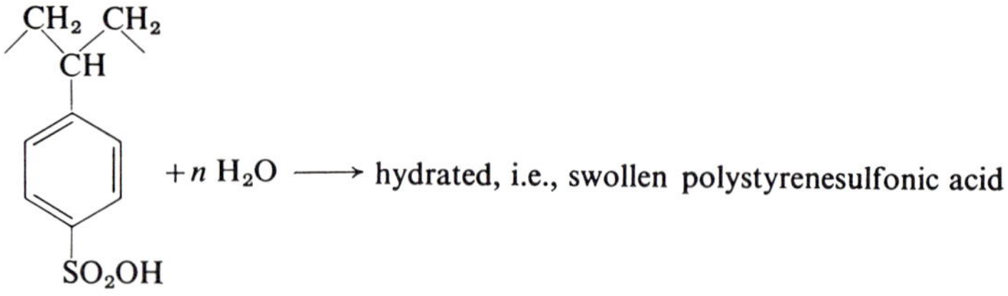

$$\text{[-CH}_2\text{-CH(-C}_6\text{H}_4\text{-SO}_2\text{OH)-CH}_2\text{-]} + n\,H_2O \longrightarrow \text{hydrated, i.e., swollen polystyrenesulfonic acid}$$

Polystyrenesulfonyl chloride hydrolyzes very slowly at 18°C in pure water. Hydrolysis might be expected to proceed more quickly in a basic medium.[20] In fact, hydrolysis of the polystyrenesulfonyl chloride membrane in 2 *N* NaOH at 18°C is complete in 2–10 hours, depending on the degree of cross-linking. In 0.1 *N* NaOH, complete hydrolysis at 18°C takes 1 to 3 days.

It is best to carry out the hydrolysis with 0.1 *N* NaOH, for the following reasons. First: if stronger NaOH is used, the hydrolysis proceeds so quickly that stresses arise in the membrane as a result of the rapid swelling produced. These stresses produce fine cracks, reducing the mechanical stability of the membrane. The high osmotic pressure, which momentarily arises as a result of the ensuing transference of the membrane from the 2 *N* solution into conductivity water, has a similar effect. Second: 0.1 *N* NaOH also appears to be a preferred concentration in the later preparation of the various salts (on this point see Section VI.3). On the one hand, if a concentration less than 0.1 *N* is used in exchangers with more than 5% cross-linking, practically no adsorption of neutral salt results,[22,23] so that in these cases subsequent washing of the membrane with conductivity water is unnecessary. On the other hand, ion exchange would proceed much more slowly if salt solutions with concentrations much less than 0.1 *N* were used, as the exchange rate is still rising with the concentration at values of the latter, in the region of 0.05 *N*, while it becomes independent of concentration at higher values of the latter.[24]

The membrane swells considerably during hydrolysis. This is because the hydrolized groups are hydrated at once. The membranes will shrink or roll up on hydration, if the sulfonation has not been uniform.

Are membranes of polystyrenesulfonic acid or its sodium salt produced by this method? IR spectra of membranes which we have prepared in this way, and which we have further prepared for IR investigation as described in Sections VI.3 and VI.4, are given in Fig. 1(c–f).

Membranes of polystyrenesulfonic acid or its sodium salt are in fact produced. This follows, first, from the assignment of the bands, especially

[22] H. P. Gregor, *J. Am. Chem. Soc.* **73**, 642 (1951).
[23] K. W. Pepper, D. Reichenberg, and D. K. Hale, *J. Chem. Soc.* (*London*) p. 3129 (1952).
[24] D. K. Hale and D. Reichenberg, *Discussions Faraday Soc.* **7**, 79 (1949).

from the assignment of the bands of the —SO_3^- ions in these spectra (on this point see Section II.2 and Table A.1), and second, very clearly, by direct comparison of these spectra with those of aqueous solutions of toluenesulfonic acid, Fig. 3 and Fig. 96.

A particularly important point is the demonstration that, aside from —SO_3^- ions, no —SO_2— groups have been incorporated to any degree. The suspicion arose that such incorporation might arise because in Houben-Weyl (see footnote 20, p. 531), in connection with the sulfonation of polystyrene with pure chlorosulfonic acid, the following statement is true: "... While on sulfonation of polystyrene strong decomposition, with sulfone formation ... occurs ...." If —SO_2— groups were present in the membranes, they would be indicated by the presence of a band in the region 1350–1300 cm^{-1} (footnotes 25–27). If such a band is found at all in the spectra of these substances, it is extremely weak [Fig. 1(e, f)]. Hence, only a very few —SO_2— groups are present.

Orientation of the substituents. In addition, the spectra yield information on the orientation of the substituents. We obtain this information from the changes of the out-of-plane bending vibration of the ⟩CH groups of the benzene ring on sulfonation (see Section II.2, and Table A.1). While the out-of-plane bending vibration occurs at 760 cm^{-1} in polystyrene [Fig. 1(a)], out-of-plane bending vibrations are found in sodium salts of polystyrenesulfonic acid at 836 and 776 cm^{-1}. The band at 836 cm^{-1} indicates the *p*-position, that at 776 cm^{-1} the *o*-position of the —SO_3^- ions [cf. Fig. 1(c–f) with Fig. 3]. The band at 836 cm^{-1} is also found when two —SO_3^- ions per benzene ring, one in the *o*- and the other in the *p*-position, are present. This shows that the pairs of ⟩CH groups vibrate almost completely independently of one another. It follows that the majority of the —SO_3^- ions in the membrane are in the *p*-position, and only a few in the *o*-position.

VI.2.1.2. Sulfonation of the Perdeuterated-Polystyrene Membranes

The sulfonation can be carried out, not only with chlorosulfonic acid, but also with concentrated sulfuric acid.[28] To establish whether this produced any difference, we sulfonated undeuterated polystyrene membranes, 6% cross-linked, with H_2SO_4 at 60°C, having added 0.1 mole% Ag_2SO_4 as catalyst. The spectrum of a membrane sulfonated in this way did not differ from that of one sulfonated with chlorosulfonic acid.

[25] L. J. Bellamy, "The Infra-red Spectra of Complex Molecules," 2nd Ed. Methuen, London, 1958.

[26] K. C. Schreiber, *Z. Anal. Chem.* **21**, 1168 (1949).

[27] D. Barnard, J. M. Fabian, and H. P. Koch, *J. Chem. Soc.* (*London*) p. 2442 (1949).

[28] K. W. Pepper, *J. Appl. Chem.* **1**, 124 (1951).

After this preliminary experiment, we sulfonated the perdeuterated membranes with pure deutero-sulfuric acid (prepared by adding D_2O to SO_3) to which 1 mole % Ag_2SO_4 had been added. The very weakly cross-linked membranes were treated for 2 days at 40°C with the sulfonation mixture. The latter was then poured off and the membrane placed for 1 day in a desiccator alongside a dish containing D_2O, so that the sulfuric acid remaining on the membrane was diluted. The membrane was then washed in pure H_2O. Spectra of membranes prepared in this way and further treated for IR-spectroscopic investigation, as described in Section VI.3 and VI.4, are shown in Fig. 1(h, i). Caution: If one dries these membranes, the deuterium atoms of the polymeric network are likely to exchange fairly rapidly with the acidic hydrogen atoms.

VI.2.1.3. Sulfonation at 0°C

The lower the degree of sulfonation, the less uniform is the distribution of the fixed ions in the membrane because the sulfonation time is so short that the chlorosulfonic acid cannot diffuse sufficiently into the interior of the membrane. A more uniform distribution of the substituents may be attained by making the diffusion rate in the membrane large compared with the incorporation rate of the substituents.

This ratio can be attained by allowing the polystyrene membranes to swell beforehand in tetrachloroethylene, toluene, or xylene according to the procedure of Boyer [17] and Abbey, [18] (as described also by Pepper [28]). However, all these agents show strong IR bands which can interfere with subsequent studies.

Alternatively, this ratio can be attained by dilution of the chlorosulfonic acid with sulfuric acid, a technique which appears at first sight less promising since incorporation of the groups and diffusion in such a system are assumed to be concentration dependent in the same way. However, the processes during the reaction are considerably more complicated. The rate of the sulfonation reaction does not decrease as much as was expected, particularly in very slightly cross-linked membranes because the membranes swell slightly even during sulfonation. Though the reaction rate is not reduced to the expected extent by dilution, the swelling still has a very favorable effect on the uniformity of distribution of the —SO_2Cl groups.

Membranes sulfonated in this way hydrolyze and swell much more quickly afterwards. This suggests that the swelling on sulfonation may be caused by partial solvolysis of the —SO_2Cl groups. This supposition is confirmed by the IR-spectroscopic study.

Finally we can reduce the incorporation rate relative to the diffusion rate by carrying out sulfonation at a lower temperature, e.g., at 0°C.

For this purpose we proceed as follows:
We prepared first a mixture of chlorosulfonic acid and sulfuric acid. In the series of experiments already discussed in Section V.11.2 the membranes were 5% cross-linked. For sulfonation, we used a mixture of 8% v/v chlorosulfonic acid and 92% v/v "water-free" sulfuric acid. We cooled this mixture to as low a temperature as possible and then poured it over the membrane, which was also cooled. It is particularly important that the membranes do not warm up on removal from the sulfonation mixture. When the mixture is poured off, the weighing glass is placed on ice and the chlorosulfonic acid is allowed to decompose by reaction with the water in the air.

VI.2.1.4. Kinetics of the Sulfonation Reaction

We investigated the kinetics of the sulfonation reaction[29] on a series of 5% cross-linked membranes which we sulfonated by the procedure given in the foregoing section. We titrated the membranes by the microtitration process of Sansoni.[30,31] We weighed the membranes with a quartz fiber balance of the type first described by Volmer[32] (also see footnotes 33 and 34). From a consideration of load capacity and sensitivity, a quartz fiber 30 cm long and 200 μ diameter was found to be the best.

The results are shown in Fig. 125. In Fig. 125(a) the number of $—SO_2OH$ groups per benzene ring is plotted and in Fig. 125(b) the reaction rate is plotted as a function of the period of sulfonation. We observe the following:

Result 133: The reaction rate is small at the start of the reaction; it then increases strongly, passes through a maximum, and finally decays exponentially. Interpretation: At first the chlorosulfonic acid–sulfuric acid mixture only wets the surface. It then slowly penetrates to the interior of the membrane, swelling the latter. The layer in which the reaction is proceeding thereby becomes thicker with advancing time. The rate of reaction should thereby increase, exactly as we have observed. The increase in the rate of reaction is opposed by the number of benzene rings still to be sulfonated becoming continually smaller. Thus the reaction rate passes through a maximum and then finally decreases.

Consequence: The course of the reaction shows us how unfavorable the conditions were, for a uniform distribution of the $—SO_2OH$ groups in the membrane at a low degree of sulfonation. A uniform distribution is attained only when the progress of the reaction is no longer determined by penetration of the sulfonation mixture. Likewise, under the favorable prevailing reaction

[29] G. Zundel and H. Metzger, *Z. Physik. Chem. (Leipzig)* **240**, 90 (1969).
[30] B. Sansoni, *Angew. Chem.* **75**, 164 (1963).
[31] B. Sansoni, Bundesforschungsanstalt für Strahlenschutz, Neuherberg near Munich, West Germany (private communication, 1963).
[32] M. Volmer, Photographische Umkehrungserscheinungen. Ph.D. Thesis, Leipzig University, 1910.
[33] T. N. Rhodin, *Advan. Catalysis* **5**, 39 (1953).
[34] A. Predwoditelew and A. Witt, *Z. Physik. Chem. (Leipzig)* **132**, 47 (1928).

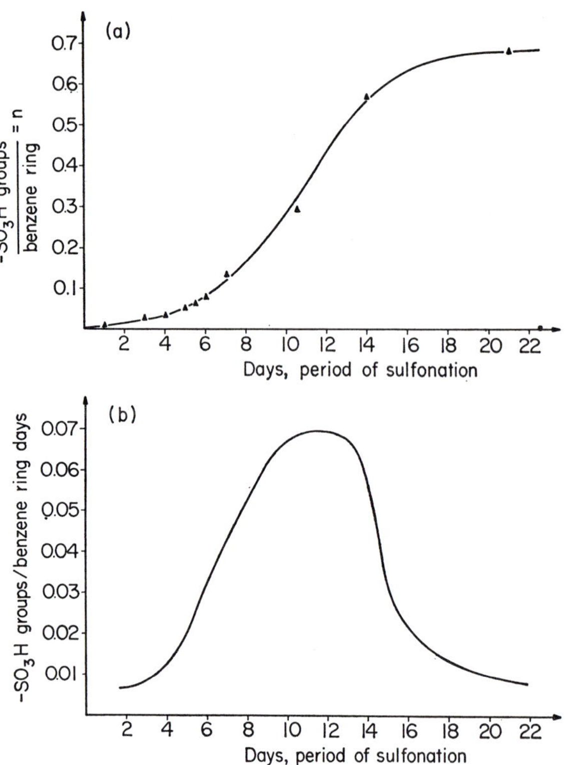

FIG. 125. Kinetics of the sulfonation reaction: (a) number of $—SO_3^-$ ions/benzene ring incorporated as a function of time; (b) reaction rate as a function of time.

conditions, we can only expect a fairly uniform distribution of the ions at a degree of sulfonation higher than 0.65 $—SO_2OH$ groups per benzene ring.

VI.2.2. INCORPORATION OF $—SeO_2OH$ GROUPS

We introduced the $—SeO_2OH$ groups into the polystyrene membranes by means of H_2SeO_4,[35] using Ag_2SeO_4 as catalyst. This is following the method of Boresch.[36]

The practical details of the process are as follows:

The membrane is prepared exactly as for sulfonation (Section VI.2.1.1). It is then placed in a weighing glass with H_2SeO_4 to which 3 mole % Ag_2SeO_4

[35] G. Zundel, *Z. Naturforsch.* **23b**, 119 (1968).

[36] C. Boresch, Farbenfabrik Bayer AG, Leverkusen-Bayerwerk, West Germany, West German Patent No. 942,624, May 3, 1956.

$$H_2SeO_4 + C_6H_5\text{–CH(CH}_2\text{–)(CH}_2\text{–)} \xrightarrow{Ag_2SeO_4} HOO_2Se\text{–}C_6H_4\text{–CH(CH}_2\text{–)(CH}_2\text{–)} + H_2O$$

has been added. This high concentration of catalyst is found to be advantageous, for the incorporation of the —SeO_2OH groups proceeds rapidly compared to the oxidation processes, caused by the H_2SeO_4, which destroy the polymer network. Under these experimental conditions, the oxidation processes do not interfere. The membrane is immersed, using a glass rod, until it is wetted all over. Thicker membranes float on the surface, since H_2SeO_4 has a high density; they therefore must be weighed down. The membranes must be shielded from light while they are in contact with the H_2SeO_4, since light promotes oxidative degradation of the membranes, as well as the formation of red and grey selenium deposits. These deposits, when they occur, interfere with IR examination by virtue of scattering. Such precautions are no longer necessary after the membranes have been removed from the H_2SeO_4.

According to Lenher,[37], the Ag_2SeO_4 is prepared by dissolving silver powder in selenic acid heated above 200°C. If—as described by Boresch[36]—the Ag_2SeO_4 is prepared from $AgNO_3$, the NO_3^- ions could by no means be quantitatively removed, even by thorough washing. Then —NO_2 groups were incorporated in the membranes, a process strongly preferred to the incorporation of the —SeO_2OH groups. If a large excess of the solution were used for the reaction, this led to only —NO_2 groups being found in the membrane.

The weighing glass, in a well-dried desiccator, is placed in a drying oven. Membranes with a medium degree of cross-linking (3–10%) are selenonated first at 60–65°C until it can be seen that the reaction has begun. This will be perceptible, after some hours at the latest, by a waviness of the membranes. The temperature is then reduced to 50–60°C and the membrane is left for a further 12 hours to 5 days, depending on the number of —SeO_2OH groups per benzene ring required. For membranes with a lower or a higher degree of cross-linking, the reaction is carried out at a lower or a slightly higher temperature, respectively.

The solution must now be removed from the membrane without precipitating the Ag_2SeO_4. The solution is first poured off, as nearly quantitatively as possible, and the membrane is washed several times with pure H_2SeO_4, being left each wash for a few minutes in the acid. This must be repeated until the H_2SeO_4 used for washing no longer becomes cloudy when mixed with water.

[37] V. Lenher, *J. Am. Chem. Soc.* **24**, 355 (1902).

The membrane is now placed for half an hour in a desiccator over water at 12 Torr. The H_2SeO_4 is thereby so strongly diluted that the membrane can be placed directly in water. It is then washed several times with distilled water.

Are membranes of polystyreneselenonic acid actually produced by this procedure? IR spectra of membranes produced in this way and further prepared for IR examination as described in Section VI.3 and VI.4 are shown in Fig. 4(a–d).

Polystyreneselenonic acid is, in fact, formed. This follows from the assignment of the bands (Section II.3 and Table A.2). In particular it can be seen that no pyro-selenonic acid ions are present, as only one band is observed in the region 625–425 cm^{-1}, the CSe stretching vibration (Section II.3, also see footnote 38).

However, in the case of the Na^+ salt for example [Fig. 4(c, d)], a weak band can be seen at 803 cm^{-1} when it is hydrated with H_2O, or two weak bands at 803 and 776 cm^{-1} when it is hydrated with D_2O. These are approximately the same for all membranes. They are caused by —SeO_2^- ions present in the membrane (on this point see Section VI.2.3). —SeOOH groups arise on incorporation of the —SeO_2OH groups by unwanted decomposition. Under our conditions, however, the number of such groups is extremely small.*

Note: We have seen in Section V.2.2 that —Se(=O)OH groups link up with one another, even when the membrane is hydrated, by very strong hydrogen bridges. If —Se(=O)OH groups are present with —SeO_3^- ions, this leads to a kind of additional cross-linking, which depends on the number of —Se(=O)OH groups present. Since, however, the selectivity of the exchanger depends on the degree of cross-linking, the selective properties of the exchangers with —SeO_3^- ions will, in this way, certainly be influenced by the number of —Se(=O)OH groups present.

Orientation of the substituents: The spectra in Fig. 4, together with the

* In Houben-Weyl's handbook[20] the following statement occurs on p. 1127: "The seleninic acid is oxidized with hot alkaline potassium permanganate solution,[39] or it is neutralized with ammonia and heated with the necessary quantity of aqueous potassium permanganate solution for a short time to boiling.[40]" It should then be possible to oxidize these —SeO_2^- ions, but we have not as yet tried it.

[38] A. Simon and R. Paetzold, *Z. Anorg. Allgem. Chem.* **303**, 39 (1960).

[39] F. L. Pyman, *J. Chem. Soc.* **115**, 170 (1919).

[40] P. L. N. Rao, *J. Indian Chem. Soc.* **18**, 1 (1941).

assignment given in Table A.2, also give information on this point. In this case we see that, in contrast to the phenomena observed with polystyrenesulfonic acid, only one band occurs which can be ascribed to the out-of-plane bending vibration of $>$CH groups on the benzene ring. It occurs at 831 cm^{-1} for the acid, or 823 cm^{-1} for the sodium salt. It is the out-of-plane bending vibration of two pairs of $>$CH groups on the benzene ring. From this follows that all the —SeO_2OH groups lie in the *p*-position. The same conclusion is naturally reached for the —SeOOH groups (see below).

In addition, we see from Fig. 4(c, d) and Fig. 5(a–d), that the bands at 760 and 698 cm^{-1} of polystyrene have almost vanished. Hence, under the given reaction conditions, almost all the benzene rings have received at least one substituent.

VI.2.3. Transformation of —SeO_2OH Groups to —SeOOH Groups

In Houben-Weyl's handbook[20] on p. 1128 we find: "Aromatic selenonic acids differ from the corresponding sulfonic acids in their pronounced oxidizing power, which is equal to that of H_2SeO_4. Just as the latter is reduced to seleninic acid by concentrated HCl, so are selenonic acids quantitatively reduced to seleninic acids by concentrated HCl, with evolution of chlorine.[41–45]"

We can accordingly easily convert our polystyreneselenonic acid into polystyreneseleninic acid.[35] For this, the membrane of polystyreneselenonic acid is kept in 12 *N* HCl for 1–2 days at 20°C. IR spectra of membranes produced in this way, and further treated for IR examination as described in Sections VI.3 and VI.4, are reproduced in Fig. 5(a, d).

We now compare the spectra in Fig. 5(a–d) with those in Fig. 4(a–d) in the region 900–800 cm^{-1}. In doing so, if we take into account the assignments (Sections II.3 and II.4, cf. Table A.2 with Table A.3), we see that the reaction

$$(\text{polystyrene unit})\text{–}C_6H_4\text{–}SeO_2OH + 2H^+ + 2Cl^- \longrightarrow (\text{polystyrene unit})\text{–}C_6H_4\text{–}SeOOH + H_2O + Cl_2$$

[41] M. Stoecker and F. Krafft, *Ber. Deut. Chem. Ges.* **39**, 2201 (1906).

[42] H. W. Doughty, *Am. Chem. J. Baltimore*, **41**, 329 (1909).

[43] R. Lesser and R. Weiss, *Ber. Deut. Chem. Ges.* **46**, 2644 (1913).

[44] A. Michaelis and P. Langenkamp, *Ann. Chem. Liebigs* **404**, 32 (1914).

[45] H. W. Doughty and F. R. Elder, *Intern. Congr. Appl. Chem., 8th, New York*, Washington Section IV Organic Chemistry **6**, 93 (1912).

has proceeded quantitatively under the given conditions, i.e., only $-Se(=O)OH$ groups are present in the membrane.

VI.2.4. Incorporation of $-P(H)(=O)OH$ Groups

We incorporate the $-P(H)(=O)OH$ groups into the membrane in two stages.[46] We carry this out following the procedure of Kressman and Tye,[47] as well as of McMaster and Glesner.[48] We first incorporate $-PCl_2$ groups and then hydrolyze .

First Stage—The Incorporation of the —PCl_2 Groups

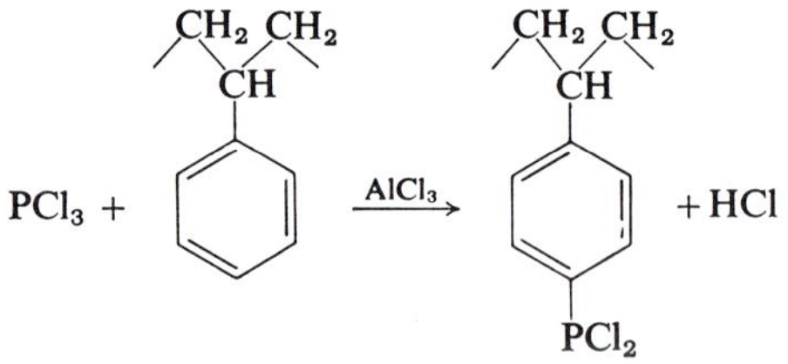

$$PCl_3 + \text{(polystyrene unit: } -CH_2-CH(C_6H_5)-CH_2-) \xrightarrow{AlCl_3} \text{(} -CH_2-CH(C_6H_4PCl_2)-CH_2-) + HCl$$

The membranes are prepared exactly as they were for sulfonation (Section VI.2.1.1). They are then introduced into a weighing glass containing 10 ml distilled PCl_3 (b.p. 75.5°C). The polystyrene membranes swell very rapidly in PCl_3; the less strongly they are cross-linked, the more they swell. So that the membranes shall not be destroyed by swelling, it is necessary to take care that they are wetted all over as quickly as possible with PCl_3. (Membranes with less than 3% cross-linking swell so rapidly in PCl_3 at 20°C that they are destroyed.) The weighing glass is placed in a desiccator, which must be well greased with Teflon grease. The membranes then are allowed to swell at room temperature in PCl_3, those which are weakly cross-linked for some minutes, those more strongly cross-linked for some hours.

[46] G. Zundel, *Kunststoff-Rundschau* **15**, 166 (1968).

[47] R. E. Kressman and F. L. Tye, Permutit Co., Ltd., London, British Patent No. 726,918, March 23, 1955.

[48] E. L. McMaster and W. K. Glesner, Dow Chemical Co., U.S. Patent No. 2,764,563, September 25, 1956.

Now 1 to 2 gm $AlCl_3$ are cautiously added. The $AlCl_3$ had been allowed to stand for about 15 minutes exposed to the air, with repeated stirring. A part of the $AlCl_3$ reacts with water from the air, which promotes the subsequent reaction considerably[10] (see p. 236). The reaction is carried out at 60–65°C. For this, the desiccator is placed in a drying oven. In order that no excess pressure arises in the desiccator, its stopcock is briefly opened several times. The duration of the reaction is 1–4 days, dependent on the degree of phosphination required. Some PCl_3 always evaporates off. For this reason it would have been better to place the weighing glass in an acid-resistant autoclave. However, such equipment was not available. The emission of vapors from the drying oven can largely be prevented by placing a dish of concentrated NaOH in the oven.

Transference of the membrane to the aqueous medium: The weighing glass with the membrane is cooled to 0°C, the PCl_3 is poured off as completely as possible, and acetone* is added cautiously, drop by drop. CCl_4 is recommended for this purpose,[47,48] but we avoided using CCl_4, as we were not certain of being able to remove it quantitatively from the membrane afterwards. The acetone dissolves the remaining $AlCl_3$ and can be poured off. The same procedure is repeated at room temperature, so that the $AlCl_3$ is completely removed. After pouring off the acetone, water may now be added. The membrane is then carefully washed in water.

Second Stage—Hydrolysis and Hydration

$$[-CH_2-CH(C_6H_4PCl_2)-CH_2-] + 2H_2O \xrightarrow{NaOH} [-CH_2-CH(C_6H_4P(=O)(H)OH)-CH_2-] + 2HCl$$

$$[-CH_2-CH(C_6H_4P(=O)(H)OH)-CH_2-] + n\,H_2O \longrightarrow \text{hydrated polystyrenephosphinic acid}$$

* Caution: If the $AlCl_3$ solution is allowed to stand at room temperature, the $AlCl_3$ reacts very violently with the acetone after a short time.

The $—PCl_2$ group hydrolyzes extremely slowly at 18°C in water, whether or not it is pure or acidified. The hydrolysis, however, proceeds very quickly in 0.1 *N* NaOH at 18°C. Accordingly the membranes are first treated with water to which only a little NaOH has been added, so that the membrane shall not be destroyed by the swelling associated with rapid hydrolysis. The NaOH concentration is so chosen that the membrane swells up in about 15 to 20 minutes. In strongly phosphinated membranes, especially if only weakly cross-linked, this appears as very considerable extension of the membrane. Afterwards, the membrane is placed for some hours in 0.1 *N* NaOH, to make sure that all the groups have been hydrolyzed. Highly cross-linked or weakly phosphinated membranes can be placed directly in 0.1 *N* NaOH.

Are membranes of the Na^+ salt of polystyrenephosphinic acid produced in this way? IR spectra of membranes produced in this way and further prepared for IR examination as described in Sections VI.3 and VI.4 are shown in Fig. 6(a–d).

Membranes of polystyrenephosphinic acid, or its sodium salt, are actually produced. This follows, in the first place, from the assignment of the bands, principally those of the $—P(H)(=O)(OH)$ groups (Section II.4 and Table A.4), and, in the second place, very clearly by direct comparison of the spectra in Fig. 6(a, b) with those of aqueous solutions of benzenephosphinic acid [Fig. 9(a)].

Orientation of the substituents: In polystyrenephosphinic acid one band is found at 830 cm^{-1}. This band is the out-of-plane bending vibration of two pairs of $\gtrdot CH$ groups on the benzene ring. It follows that in polystyrenephosphinic acid all the fixed ions are in the *p*-position.

If unsubstituted benzene rings were still present in the membrane whose spectra are reproduced in Fig. 6, bands would have to occur at 760 and 698 cm^{-1}. Because these bands are not observed to any extent in the spectra in Fig. 6, it follows that, by proceeding as above, with 7% cross-linked membranes, after 3 days, every benzene ring carries at least one $—P(H)(=O)(OH)$ group.

When trying to incorporate $—P(=O)(OH)(OH)$ groups in the same way, it is found that $POCl_3$ does not react with polystyrene even at its boiling point (105.3°C).

Membranes with $—P(=O)(OH)(OH)$ groups have been made from membranes of polystyrenephosphinic acid by oxidation with H_2O_2. The spectra of such membranes reproduced in Fig. 8 still indicate a few unoxidized —P(H)(=O)(OH) groups together with the —P(=O)(OH)(OH) groups. This is shown by the bands of PH stretching vibration at 2310 cm^{-1} (for the salt) and 2365 cm^{-1} (for the acid) which have almost, but not completely vanished. This process of preparation, however, cannot well be used, as the polymer network in the membrane is too strongly attacked.

VI.2.5. Incorporation of —P(SH)(=O)(OH) Groups

Membranes of polystyrenethiophosphonic acid can be made by a procedure corresponding to that for polystyrenephosphinic acid,[46] again following the instructions of McMaster and Glesner.[49] Hence only certain differences between the procedures will be pointed out:

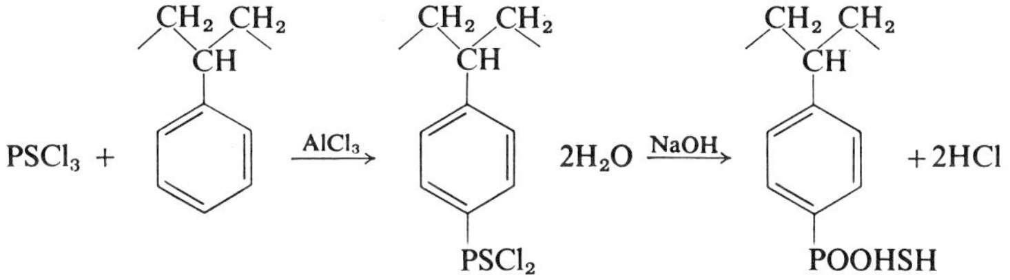

$$PSCl_3 + C_6H_5\text{-}CH(CH_2)(CH_2) \xrightarrow{AlCl_3} Cl_2PS\text{-}C_6H_4\text{-}CH(CH_2)(CH_2) \quad 2H_2O \xrightarrow{NaOH} HSOOP\text{-}C_6H_4\text{-}CH(CH_2)(CH_2) + 2HCl$$

Distilled $PSCl_3$ (b.p. 125°C) is used. The membranes swell much less rapidly in $PSCl_3$ than in PCl_3, so that even slightly cross-linked membranes are not destroyed by this swelling. The membranes must therefore be allowed to swell a little longer. Then 10 ml $PSCl_3$ and 2 gm $AlCl_3$ are added and the reaction is carried out at 80 to 90°C.

The hydrolysis also proceeds much more slowly. Membranes with more than 8% cross-linking can be hydrolyzed directly in 2 *N* NaOH. Slightly cross-linked membranes are first treated with a less concentrated NaOH solution.

Are membranes of the sodium salt of polystyrenethiophosphonic acid actually produced by this process? IR spectra of membranes produced in this way and further prepared for IR investigation as described in Sections VI.3 and VI.4 are reproduced in Fig. 7(a–d).

[49] E. L. McMaster and W. K. Glesner, Dow Chemical Co., U.S. Patent No. 2,764,564, September 25, 1956.

Membranes with $-\mathrm{P}(=\mathrm{O})(\mathrm{SH})(\mathrm{OH})$ groups are, in fact, formed. However, they also contain $-\mathrm{P}(=\mathrm{S})(\mathrm{OH})(\mathrm{OH})$ groups in tautomeric equilibrium. This follows from the assignment of the bands of these spectra (Section II.4 and Table A.5).

The band observed with the sodium salt at 1435 cm^{-1} and the shoulder at 1060 cm^{-1} are often much more intense than in the spectra shown here. This is the case when the incorporation reaction is carried out at a rather higher temperature and also when the $PSCl_3$ has not been distilled before the reaction. The cause of this is not known at present. On the one hand, it may be supposed that, under these conditions, two $-\mathrm{P}(=\mathrm{O})(\mathrm{SH})(\mathrm{OH})$ groups per benzene ring are present, for in these cases a weak band is also found at 880 cm^{-1} which is characteristic for the out-of-plane bending vibration of a single $\rangle$CH group in the benzene ring, i.e., for 1:2:4 substitution. On the other hand, it may be supposed that in these cases the polymer network has been attacked, for a weak shoulder is also observed on the band of the $-CH_2-$ stretching vibration at 2926 cm^{-1}, on the side toward higher wave numbers. These results will have to be clarified by future studies.

Orientation of the substituents: If the deviations discussed in the foregoing paragraph are ignored, a single band for an out-of-plane bending vibration is always found at 830 cm^{-1} (acid) or 835 cm^{-1} (salt). This is the out-of-plane bending vibration of two pairs of $\rangle$CH groups in the benzene ring. From this it follows that *p*-substitution normally also occurs in polystyrenethiophosphonic acid. As in the previous cases, it is found that at least one acid group is present as substituent on almost every benzene ring under the given reaction conditions, since the bands at 760 and 698 cm^{-1} have almost completely vanished.

VI.3. Preparation of Acids and Salts

The procedure varies slightly according to whether the acid is strong or weak.

With polystyrenesulfonic acid, whose true degree of dissociation is nearly one, we exchanged the cations with 0.1 *N* salt solutions. For the reasons for chosing this particular concentration, see Section VI.2.1.1. We used chloride solutions whenever possible,* since traces of halogen anions remaining in the

* This was impossible only in the case of Pb^{2+} ions, where we used the nitrate. It was then necessary to wash the membranes free of Cl^- ions with aqueous HNO_3 solution before exchanging cations. Otherwise $PbCl_2$ deposits would have formed, and they would have disturbed the observations by scattering.

membranes have no IR bands. With some of these salts, such as $AlCl_3 \cdot 6H_2O$ more or less strong hydrolysis occurs. We therefore acidified these, taking care at the same time that the hydrogen ion concentration remained considerably below that of the ions to be incorporated.

With weaker acids, the cations are exchanged by the neutralization reaction. In the polystyrenethiophosphonic, polystyreneseleninic, and polystyrenephosphinic acids it is necessary to use not 0.1 *N* but 2 *N* alkali solutions. The second OH group of polystyrenethiophosphonic acid is very weakly acid. In the spectrum of the Na^+ salt of this acid, there are indications that some of these groups are not neutralized.

We carried out the exchange as follows: The solutions were prepared with conductivity water in polyethylene bottles. The membranes are first changed into the acids (the reason for this is given below). They are then treated eight times with the appropriate solutions for periods of 15 minutes to 4 hours in each case, according to the size of the ions and the degree of cross-linking. This is carried out in polyethylene Petri dishes. For membranes with 5% cross-linking or more, if 0.1 *N* solutions are used, subsequent washing with conductivity water is unnecessary, for at these concentrations only an insignificant amount of neutral salt adsorption takes place.[22,23]

Are the ions quantitatively exchanged in this treatment? The IR spectra of the acids are different from those of the salt forms in one respect. They show either a powerful continuous absorption (Section V.11) or broad bands in the region 3100–2100 cm^{-1} (Section V.2). Hence, in order to obtain a criterion for the quantitative exchange of the cations, it is always necessary to prepare the salts from the acids. If the salts are prepared according to the procedure described above, either the continuous absorption or these bands vanish completely. It follows from this that in this treatment the cations are always* almost completely exchanged for the hydrogen ions. In the cases of those salts which protolytically decompose water, it is always necessary to consider the spectra of the thoroughly dried membranes, as under these conditions no protolytic decomposition of water molecules occurs (Section IV.19); otherwise the hydrolysis protons give the impression of having undergone an incomplete exchange.

Preparation of membranes containing different cations together: For this purpose the membranes are treated as in the preceding section, but with solutions containing the required cations in the proper proportions. These proportions are determined, first, by the ratio in which the ions are to be present in the membrane, and second, by selective behavior. We have prepared membranes containing Na^+ and H^+ ions together (Section V.11.2).

* Among the exchanger forms which we studied, only the Tl^{3+} salt of polystyrenesulfonic acid exhibited special features in the observed exchange (Section IV.20).

The necessary data on the selective behavior were obtained from a paper by Myers and Boyd.[50] We determined the ratio of Na^+ to H^+ ions in the membranes by Sansoni's very accurate coulometric microtitration method.[30,31]

VI.4. Preparation of Membranes for IR Spectroscopy

For IR measurement, the membranes are removed from the solution. This may be done in various ways, according to the nature of the membrane:

Procedure 1: All membranes with $—SO_3^-$, $—SeO_3^-$, and SeO_2^- ions and the corresponding acids as well as the acids with $—P(H)(=O)(OH)$ and $—P(SH)(=O)(OH)$ groups may be prepared in this manner. First the membrane is fished out with chromatography paper. For this procedure, it is very important that the paper used should have a short staple length and exert a slow capillary action. The membrane is placed on the chromatography paper and allowed to dry out slightly; it is then raised at the edge with a spatula and carefully removed from the paper with tweezers. As the membrane tears very easily in this operation, some practice is necessary. It is essential that the membrane be slightly dried before it is lifted from the paper, but it must not be too dry. In this operation there is no danger of contamination.

Procedure 2: Membranes of the salts of polystyrenethiophosphonic and polystyrenephosphinic acids must be prepared as follows, because they contract too strongly in drying on paper. We first extract water from the membrane by placing it in another liquid miscible with water, e.g., acetone. If the membrane contracts too rapidly, this stage is split up into several steps, using mixtures of the appropriate liquid with water and reducing the water concentration at each stage. In the sodium salt of polystyrenephosphinic acid we used acetone; for the sodium salt of polystyrenethiophosphonic acid we used ether dried over sodium. It is important that the liquid used should be easily removed from the membrane.

The membrane can now be placed in the sample holder and, in this, in a special cell for IR investigation.

Caution: Traces of NH_3 in the laboratory atmosphere can very quickly change the acid membranes into NH_4^+ salts. The bending vibration band of the NH_4^+ ion lies at 1440 cm^{-1}. The NH_4^+ ion also exchanges in the transition from H_2O to D_2O hydration. If a band at about 1440 cm^{-1} is that of the NH_4^+ ion, it vanishes on the transition from H_2O to D_2O hydration. The corresponding band of the ND_4^+ ion at 1070 cm^{-1} is then found. If such an error should have occurred, these bands bring it to light.

[50] G. E. Myers and G. E. Boyd, *J. Phys. Chem.* **60**, 521 (1956).

CHAPTER VII

IR INVESTIGATION METHOD

VII.1. Carrying Out of Measurements

The membranes must be examined at a specified degree of hydration and at a specified temperature.

The specified degrees of hydration were obtained by using reproducible humidities of the air in contact with the membranes. We could produce these reproducible relative humidities in the cell by means of saturated salt solutions—water vapor pressures are given (footnotes 1–3) (see Table 18)—or by means of sulfuric acid solutions of various concentrations.[4] We dried the membranes at 10^{-3} Torr, inserting a liquid nitrogen trap between the cell and the pump. The limit of dryness approached in this way at 25°C we term "thorough drying." The procedure in the D_2O hydration is described on p. 261. A corresponding procedure is followed in the HDO hydration. Such hydration is obtained with a mixture of H_2O and D_2O (see p. 31). Since the OH or OD bands of HDO under these conditions are very weak, thicker membranes must be used. It is helpful to place several membranes one on top of the other, but then reflection losses become considerable. Furthermore, interferences by superposition can have a disturbing effect.

If the membranes are to be investigated at temperatures near room temperature, the cell described in Section VII.2.1.2 is used. For investigations at lower temperatures (down to 85°K), the IR low-temperature cell described in Section VII.3.1 is suitable.

At investigations near room temperature, the procedure is as follows. The membrane is placed in a sample holder as described in Section VII.2.2. The holder is then inserted in the cell. The solution used to produce the required water vapor pressure has already been placed in a small vessel

[1] R. H. Stokes and R. A. Robinson, *Ind. Eng. Chem.* **41**, 2013 (1949).

[2] A. Wexler and S. Hasegawa, *J. Res. Natl. Bur. Std.* **53**, 19 (1954).

[3] H. Pfennig, *Naturwissenschaften* **49**, 81 (1962).

[4] E. Glueckauf and G. P. Kitt, *Trans. Faraday Soc.* **52**, 1074 (1956).

TABLE 18

RELATIVE ATMOSPHERIC HUMIDITY OVER SATURATED AQUEOUS SALT SOLUTIONS AT 25°C, AFTER STOKES AND ROBINSON[1]

Salt	Relative atmospheric humidity (%)
$K_2Cr_2O_7$	98.00
KNO_3	92.48
$BaCl_2 \cdot 2H_2O$	90.19
KCl	84.26
KBr	80.71
NaCl	75.28
$NaNO_3$	73.79
$SrCl_2 \cdot 6H_2O$	70.83
$NaBr \cdot 2H_2O$	57.7
$Mg(NO_3)_2 \cdot 6H_2O$	52.86
$LiNO_3 \cdot 3H_2O$	47.06
$K_2CO_3 \cdot 2H_2O$	42.76
$MgCl_2 \cdot 6H_2O$	33.00
$K(C_2H_3O_2) \cdot 1.5H_2O$	22.43
LiCl	11.05
NaOH	7.03

attached to the cell (see Section VII.2.1.2). The rate with which the water vapor pressure equilibrium between the salt solution and the membrane is attained depends on the degree of cross-linking, the concentration of ions, and the nature of the ions in the membranes; it also depends on the required moisture content of the membrane, i.e., the equilibrium itself which is to be attained. The attainment of the equilibrium takes between six hours and some days in an evacuated cell, according to conditions. For this period the cell is placed in a thermostatically controlled oven. The vacuum right-angle stopcock (Fig. 127) is then closed, the cell is placed in the spectrophotometer, and the spectrum is recorded. The IR investigation of a membrane at low temperatures is described in Section VII.3.3.

Our spectra have been recorded with an IR spectrophotometer made by Perkin-Elmer GmbH, Ueberlingen-Bodensee, West Germany, model 221 with a prism grating interchange unit. Some of the spectra were taken with other spectrophotometers, as indicated. In the following text, however, not only the auxiliary equipment but also its properties are described exactly. This makes it possible to build corresponding auxiliary equipment for other spectrophotometers easily. To obtain measurements of the accuracy required for the studies described in Section IV, the thermostatic control of the spectrophotometer must be better than provided normally. Present instruments no longer suffer from these shortcomings.

If weak bands of H_2O and CO_2, as present in the atmosphere, are superimposed on the spectra, the fault lies in unsymmetrical conditions in the sample and reference beams. This is termed lack of compensation. This disturbance can be eliminated if an evacuated cell of the same length, in which the same water vapor pressure is produced as in the sample cell, is placed in the path of the reference beam. It is advantageous if the cell in the reference beam has windows of the same material as those of the cell containing the sample, since the reflection losses at the windows of the latter cell are thus compensated. For the cell in the reference beam we use a normal gas cell, with such windows, to which a small detachable vessel containing the solution for producing the required vapor pressure is attached.

VII.2. IR Cell and Sample Holders

The cell and sample holders here described are suitable for studies at temperatures close to room temperature.[5]

VII.2.1. IR Cell

VII.2.1.1. Properties of the Cell

(a) The membranes can be examined in the cell at a definite vapor pressure, in particular at a definite water vapor pressure.

(b) The cell can be evacuated to attain vapor pressure equilibrium rapidly. The liquid used to produce the vapor pressure can be temporarily frozen, so as not to pump it away.

(c) It is possible to separate the space containing the sample from that containing the liquid used to produce the vapor pressure. This has three advantages: First, the sample space can be separated from the glass space (Fig. 127) during the measurement; thus a disturbance of the vapor pressure equilibrium can be avoided. Second, when the liquid used to produce the vapor pressure is changed, the vapor pressure over the membrane does not change in an ill-defined manner. This is particularly important when it becomes necessary to approach vapor pressure equilibrium from the same direction, i.e., in adsorption or desorption. Third, this is of importance in hydration with D_2O or HDO (see p. 261).

(d) The cell has a large thermal capacity (water jacket) and is thermally well isolated. Hence the cell can be removed from the thermostatically controlled oven in which it is kept and can be placed in the spectrophotometer

[5] G. Zundel, A. Murr, and G.-M. Schwab, *Z. Naturforsch.* **17a**, 1027 (1962).

for recording the spectrum without the necessity of providing further thermostatic control.

(e) The IR-transparent cell windows consist of humidity-resistant material.

(f) The membrane in holders can rapidly be placed in the cells.

(g) The cell can easily be placed in the spectrophotometer.

VII.2.1.2. Construction of the Cell

The cell consists essentially of metal and glass parts.

Figure 126 shows the different cross sections of the metal parts. The larger constructional elements 3, 4, and 6 are indicated by slightly larger figures.

The IR beam enters the cell at 1 and leaves at 2.

Part 3, made of corrosion-resistant metal, is a part of a normal liquid cell, which insures rapid fitting of the cell in the spectrophotometer (g).* Part 3 must fit well in the cell holder slot of the spectrophotometer, i.e., it must be an exact fit, so that the position of the cell in the spectrophotometer is always the same for all measurements (g). Part 3 is fixed to the housing (4) with four screws.

The housing (4) is closed on one side by the cell window (5), which is cemented in place with synthetic resin cement. Part 6 closes the opening through which the sample holder is introduced (f). Part 6 is pressed against the housing by means of part 8. Part 7 is an O-ring seal of rubber, 10 is a washer.

Part 6 consists of a rotation-symmetrical aluminum part, with an internal screw thread (11) to hold the cell window (13) fast by means of the thread ring (12). Part 14 indicates the rubber sealing rings, 15 is a washer.

The housing (4) is made of brass. To provide a good vacuum seal (b) between the interior of the cell and the jacket containing the liquid used for thermostatic control, the following procedure is used. Before the cell is used, the jacket is filled with vacuum-sealing lacquer and the cell interior is evacuated. After some hours the vacuum is released, the lacquer is poured off, and the coating of lacquer remaining on the inner walls of the jacket is allowed to dry. The liquid used for thermostatic control (d) can be introduced either through the nozzles (19) or through the opening (20) provided for a thermometer. The jacket can be filled with water, which is recommended because of the large thermal capacity (d). However, the thermal capacity of paraffin oil is also sufficient to maintain the cell at a constant temperature during measurement. Paraffin oil has the advantage of not disturbing the measurements if small quantities of it penetrate into the interior of the cell through a soldered seam. The nozzles (19) make it possible to circulate the thermostatic liquid. This, however, is unnecessary for measurements at 25°C. In this case the nozzles can simply be closed with rubber caps, or omitted from the construction of the cell. Part 17 is the connection for the glass parts described below. The tube (18) permits evacuation of the cell interior (b). A polyvinyl chloride tube connects the nozzle (18) with a high-vacuum stopcock (Schiff's type) (not shown). The cell (4), together with the glass part inserted in it, is finally surrounded by foamed plastic (not shown) (d).

The sample holder is clamped in the springs (16) (f).

The cell windows (5) and (13) consist of KRS 5 (thallium–iodide–bromide, composition: 44% TlBr, 56% TlI) (e).

Figure 127 shows the glass parts. They are connected to the metal housing by the

* The letters in parentheses refer to the properties of the cell listed in the preceding section.

ground glass cone (23) and socket (17). If the cell is not vacuum-tight, the leak is usually to be found in this joint. To prevent leakage, the cone is cemented into the socket with a solvent-soluble cement.

The glass parts serve to maintain the vapor pressure in the cell (a). The liquid used in each case is introduced into the container (24). This can be cooled to freeze the liquid temporarily (b). Part 26 is connected to the container (24) by the spherical joint (25).

Parts 26 and 27 are essentially a vacuum right-angle stopcock. This makes it possible to separate the space over the liquid used to produce the vapor pressure from the sample space (c) as required.

The high-vacuum stopcock (Schiff's type) fitted at the top of part 27 makes it possible to evacuate the cell or the glass part (b) alone, when the vacuum right-angle stopcock is closed.

Procedure for Hydration with D_2O or HDO

For this purpose D_2O or the corresponding mixture of H_2O and D_2O is introduced into the container (24). The cell is now carefully evacuated through the stopcock fitted at (27). If the liquid begins to boil, the stopcock is closed. After some minutes, the connection between the glass part and the sample space is interrupted by means of the right-angle stopcock. Then the sample is dried by pumping out through the stopcock at tube (18). Afterwards the sample space is again connected with the glass part and the sample is hydrated again. If this is repeated once or twice, the water in the membrane is exchanged quantitatively. Stepwise drying of the D_2O-hydrated membrane is carried out by first closing the stopcock, then pumping until the degree of hydration required is reached, closing the stopcock, and waiting until equilibrium in the sample chamber has been attained. The required vapor pressure may be produced with saturated salt solutions, exactly as in the case of H_2O hydration. The vapor pressures over such salt solutions are unfortunately not known. Such vapor pressures do not differ greatly from those of solutions of the same salts in H_2O.

VII.2.2. Sample Holder

We have used sample holders of various types, according to the type of experiment.[5] The first sample holder (Section VII.2.2.1) is suitable for experiments in which the humidity over the sample is not very different from that in the atmosphere of the laboratory. The second holder (Section VII. 2.2.2) is for use when the sample is to be dried. This type is necessary because membranes tightly stretched during drying will rupture as they contract slightly in this process. In this case, however, it must be borne in mind in the subsequent evaluation of the spectra that the membrane contracts slightly with increasing drying (Fig. 74).

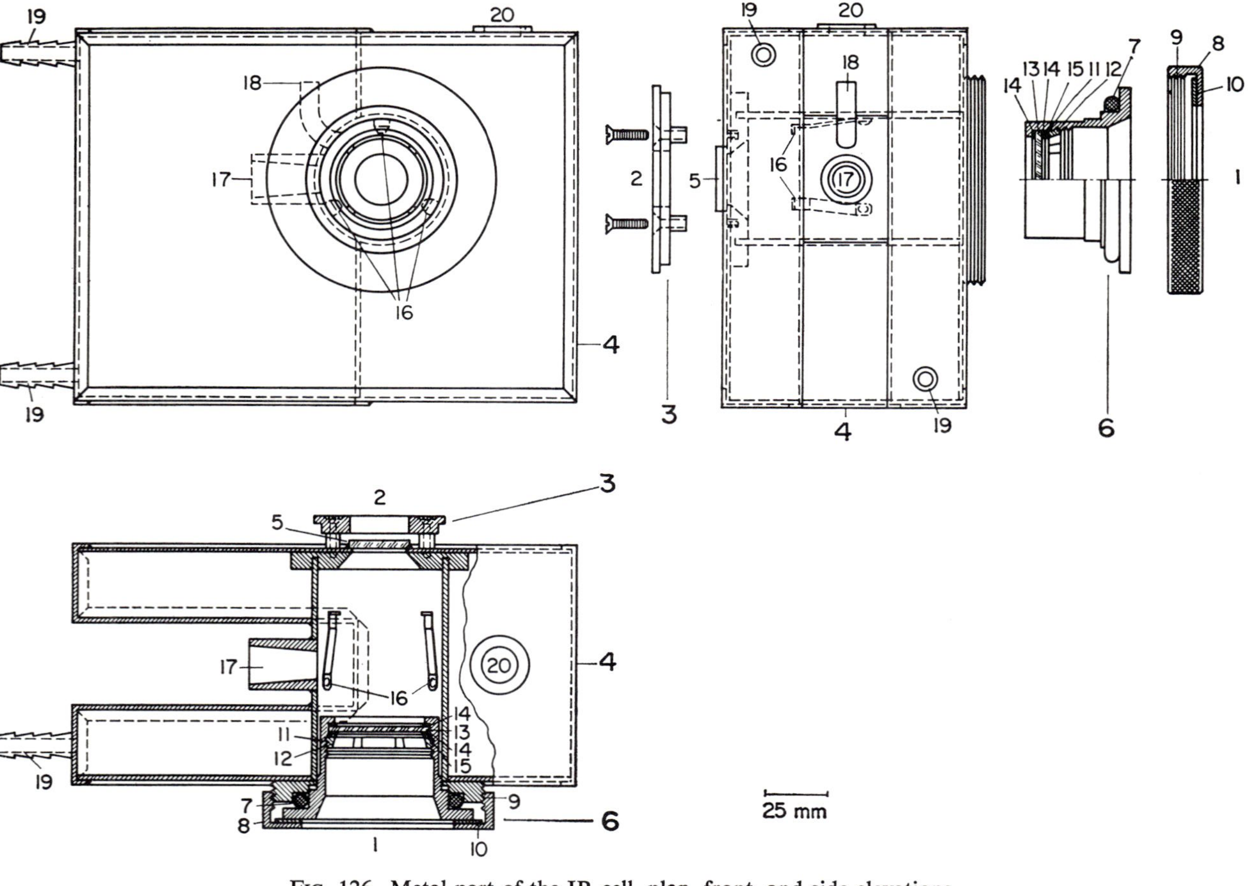

FIG. 126. Metal part of the IR cell, plan, front, and side elevations.

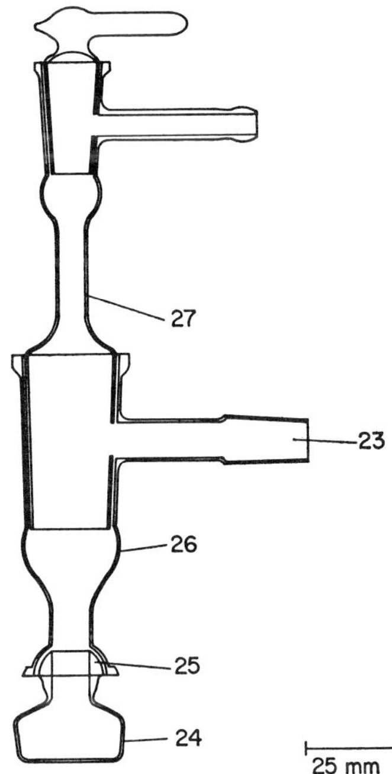

FIG. 127. Glass parts of the IR cell.

VII.2.2.1. Construction of the First Sample Holder

In Fig. 128 there is shown a sample holder in which the membrane is held in place magnetically (a).*

Parts 2 and 3 are made in the magnetic alloy Magnetoflex 35 (52% Co, 13% V, 35% Fe; Supplier: W. C. Heraeus GmbH, Hanau, West Germany). They are longitudinally magnetized and lie side by side.

Part 2 is cemented to part 1 in such a way that part 3, on part 2, is located by the glass ring (1). For this purpose, the glass ring (1) must extend 0.2 to 0.5 mm above part 2. The membrane to be examined is clamped between parts 2 and 3 by the force of magnetic attraction between them (a). So that part 3 may be moved onto or away from part 2, when inserting or removing the membrane, a blunt brass nail is cemented to part 3.

The metal parts are coated with a thin layer of polyethylene at their points of contact with the membrane (b). The slits (6) make it possible to grasp the sample holder and place it in the IR cell with tweezers (c).

VII.2.2.2. Construction of the Second Sample Holder

Figure 129 shows this type of sample holder.

The assembly of this holder is as follows. The glass ring (1) is first filled with synthetic resin cement, which is cured. The "window" (2) and the slit (3) are machined out, and pieces of platinum foil (4) are cemented to the synthetic resin parts.

* The letters in parentheses refer to the properties listed in Section VII.2.2.3.

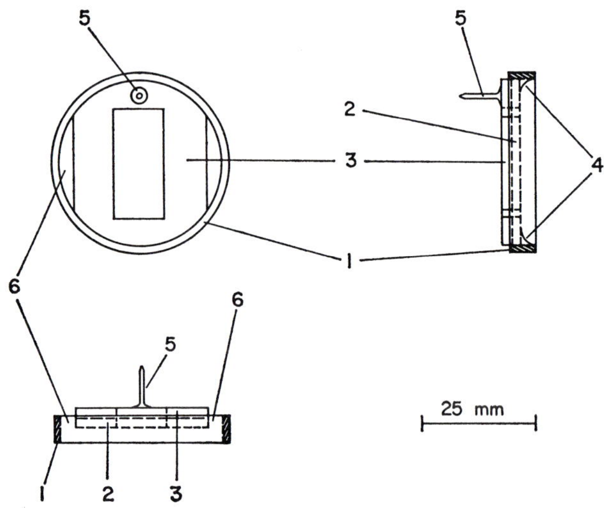

FIG. 128. First sample holder.

The membrane is clamped between the platinum foil (b) and the synthetic resin (a). The slit (3) makes it possible to grasp the sample holder and place it in the IR cell with tweezers (c).

VII.2.2.3. Properties of the Sample Holders

(a) The membrane can be quickly placed in the holder and removed from it again without damage. This is possible in the first holder through the use of the magnetic clamping arrangement. In the second holder, the membrane is only clamped between the platinum foil and the plastic part. The membrane can, in this case, be placed quickly in the holder, if the platinum foil is bent slightly upwards.

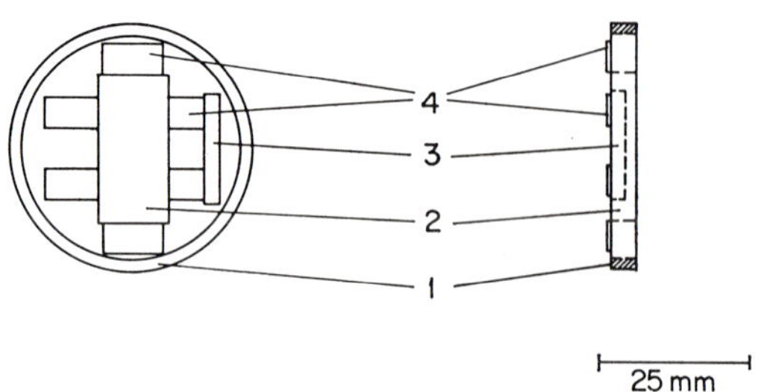

FIG. 129. Second sample holder.

(b) Some of the membranes—particularly the acid membranes—attack metals. The holders of the first type must therefore be thinly coated with polyethylene (rhodium plating has been found inadequate).

(c) The holders can easily and quickly be fitted into the cell.

VII.3. IR Investigations at Low Temperatures

A reproducible degree of hydration is first produced in the membrane at 25°C, as previously described. The membrane is then cooled to a low temperature instantaneously. Cooling must proceed so rapidly that the degree of hydration remains unaltered. The description of the construction of a cell which has been shown to be suitable for investigations at temperatures down to 85°K is given in Section VII.3.1 and that of the corresponding sample holder in Section VII.3.2. The procedure for carrying out an experiment with this cell is described in Section VII.3.3.

VII.3.1. Construction of a Low-Temperature Cell[6]

Figure 130(a, b) shows the cell in two perpendicular sections. Boundary edges behind the plane of section are omitted for clarity. Arabic numerals in the text give the numbers of the parts described. Indices a and b indicate the figure in which the part specified can be seen more clearly.

The body of the cell proper is a cylinder (3_b). This is surrounded by a U-shaped vessel (4_a), containing the cooling medium. Two caps (1_b) and (2_b), which can be evacuated, close the ends of the cylinder. These are provided with IR-transparent windows (19_b) and (20_b). The light enters the cell at (5_b) and exits at (6_b). The sample holder, in which the membrane is fixed, is clamped in the cylinder (3_b) by means of the springs (31_b). Manipulations can be carried out on the sample in its holder, when the cell is closed and cooled, by means of the fitting (7_a).

The Dewar caps (1) and (2): Each one of the Dewar caps consists of a rotatable part (25_b) in V4A steel (wall thickness 3 mm) to which a pressed spherical (24_b) 1 mm thick V4A steel sheet is welded. All welding of V4A steel was carried out by the argon arc process. Steel is preferred over other metals, although it is more difficult to machine, not only because of its lower thermal conductivity, but also because of its specific heat which decreases much less rapidly with decreasing temperature than is the case with other metals.

Tombac tubes (11_b) are cemented to the caps at (12_b) with synthetic resin cement. These tubes provide the connection to a vacuum. They are 500 mm long and have an internal diameter of 10 mm. Their arrangement is flexible enough so that the cell can be removed from, or replaced in, the spectrophotometer during use. The caps are lacquered, particularly at the welded seams, to prevent leakage. A vacuum of 10^{-3} Torr is sufficient to almost completely eliminate heat exchange by thermal conduction through the air. In addition, the caps are silvered inside and outside, to minimize heat exchange by radiation.

[6] G. Zundel, *Chem.-Ingr.-Tech.* **35**, 306 (1963).

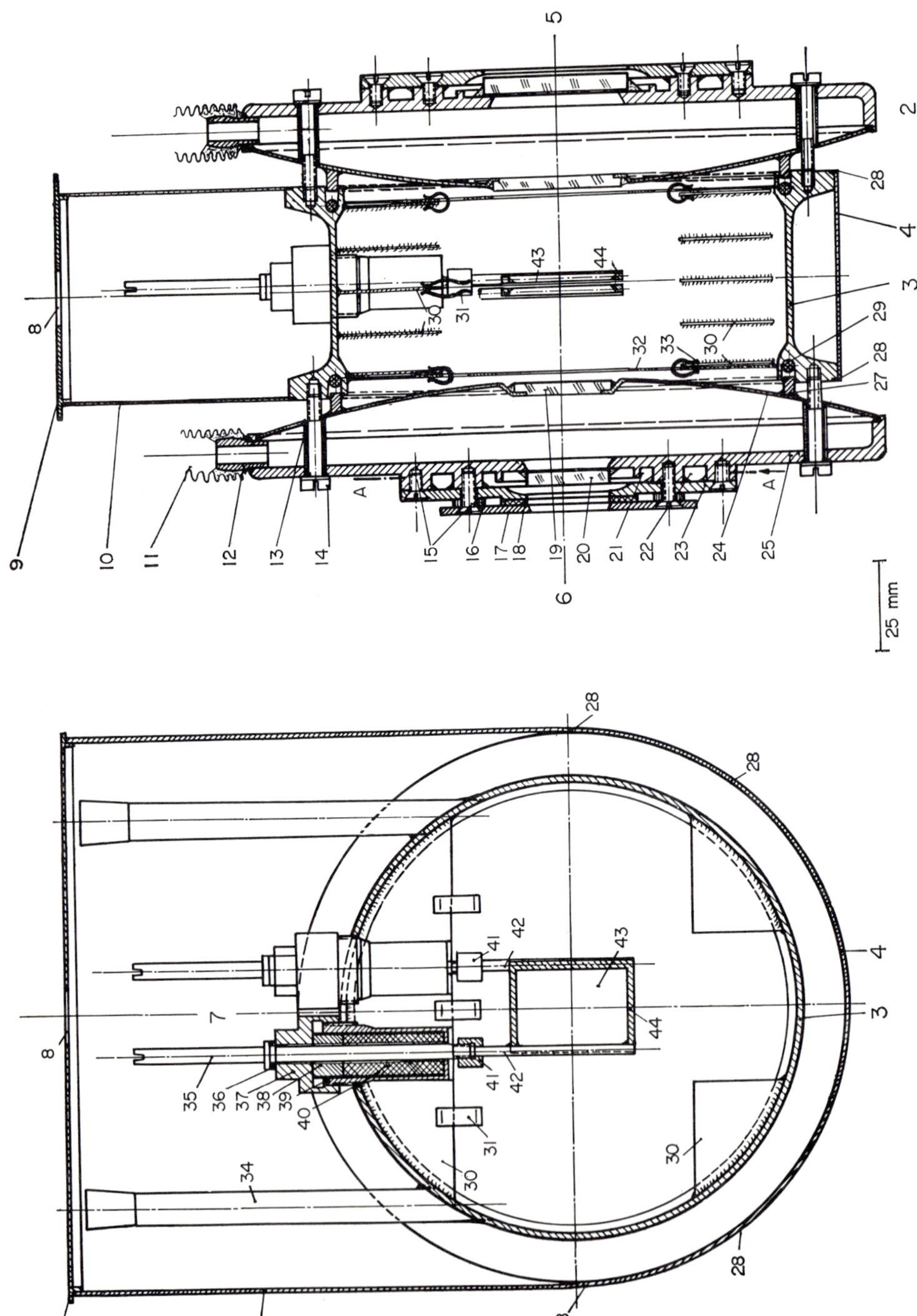

2
3
4
5
6
7
8
9
10
11
12
13
14
15
16
17
18
19
20
21
22
23
24
25
27
28
29
30
31
32
33
34
35
36
37
38
39
40
41
42
43
44
A
25 mm

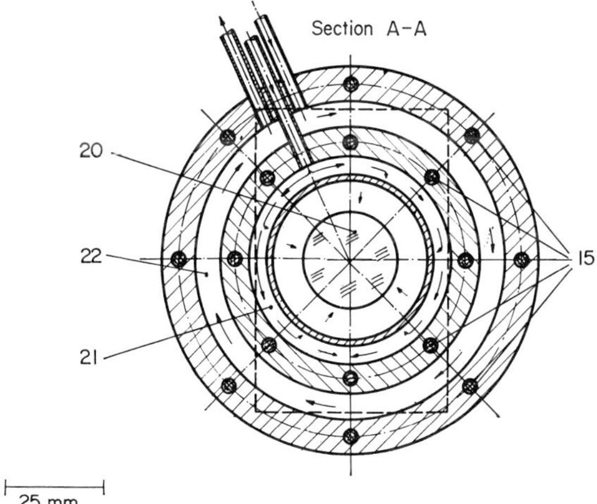

FIG. 131. Section through the window heating arrangement of the low-temperature cell.

The caps have windows of NaCl (20_b) on the "outside" and of germanium (19_b) (n-doped, resistance at least 50 Ω cm) on the "inside." The windows are cemented in place with synthetic resin cement. Nearly 80% of the intensity of the radiation is lost in passing through these four windows, but the remaining energy is still sufficient to provide perfect spectra. The inside windows are subjected to considerable fluctuations of temperature. Germanium and steel have different coefficients of thermal expansion. To allow for this, the steel sheet is bent, as shown, in the neighborhood of the cemented regions.[7] In addition, less hardener was used in the synthetic resin cement at these points, so that it could retain some elasticity.

The outside windows of the cell must be kept warm enough to prevent moisture condensation; thus they are surrounded by two ring channels (21_b) and (22_b). Water at about 30°C circulates in the outer ring channel (22_b), the inner (21_b) serves to flush the windows with gaseous nitrogen. This is shown in a sectional drawing in Fig. 131. This is a section, in the direction A–A, perpendicular to the plane of the drawing in Fig. 130(b). The cover plates (23_b), which close the ring channels, are held in place with screws (15_b) and also cemented to seal off the water channels.

The plate (17_b) serves to support the cell in the spectrophotometer. The parts (16_b) are spacers. They provide the space between the plates (17) and the cap (1), which is necessary for insertion of the cell in the spectrophotometer. Part (18_b) is a rubber sealing ring. It insures that the dry nitrogen in the spectrophotometer will flow out.

Tubes (13_b) of V4A steel are welded in the caps (1) and (2) at six places. They accommodate the screws (14_b) which press the caps against the cell body (3). Teflon O-rings

[7] V. Roberts, *Rev. Sci. Instr.* **31**, 251 (1954).

FIG. 130. Low-temperature cell: (a) section perpendicular to the direction of the beam; (b) section in the direction of the beam, side view.

(29_b), lying in grooves in part (3), provide a seal. If the screws (14) are tightened, the ring (27_b) of V4A steel, which is welded to the cap, presses on the Teflon rings (29_b).

Cell body and cooling medium container (3) and (4): The cell body (3) consists of a V4A steel turned part. The wall of the cylindrical part of the cell body is 1.2 mm thick. The cell body (3) is welded at (28_b) to a U-shaped piece of V4A steel sheet, 1 mm thick. In Fig. 130(a), these welded joints are in front of and behind the plane of the drawing. The side walls (10_b) of the U-shaped vessel thus formed are of V4A steel sheet, 0.5 mm thick. Liquid nitrogen can be poured in through the opening (8_b) in the lid (9_b). The cooling fins (30) are welded to the cell body (3). To keep out the heat which tends to flow along the caps into the interior of the cell, two additional polished radiation shields of copper (32_b) are clamped to the outer cooling fins with springs (33_b).

The U-shaped vessel and the lid (9) are isolated with layers of styrofoam (not shown). In addition, the outer walls of this vessel are polished on both sides to reduce heat exchange by radiation.

The tubes (34_a) make it possible to flush the cell with dry gaseous nitrogen and, if necessary, to evacuate it. Finally, a low-temperature thermometer or a thermocouple can be introduced into the cell through one of the tubes. In Fig. 130(a) these tubes lie in front of and behind the plane of the drawing. In the cell which we used, three such tubes are fitted. They are made of V4A steel, have each a soldered standard ground conical brass socket, and are welded to the cylinder (3). These tubes function as a valve if, at any time, liquid nitrogen should penetrate into the interior of the cell and evaporate when the cell warms up.

Attaching the sample holder to the cell: The sample holder (Section VII.3.2) can be inserted into the springs (31_b) after the lid (2) has been removed.

Fitting (7_a) for carrying out manipulations on the sample: The two manipulation fittings (7) are shown in Fig. 130(a), one in section, the other in elevation. In this figure the position of these two fittings relative to one another is shown (42_b–44_b).

The two V4A steel sockets (39_a) are welded into the cylinder jacket. It would probably be preferable to set each of the sockets (39_a) higher, i.e., with only the lower end in the turned part. This would make it possible to give the turned part a smaller radius, so that the whole cell would be easier to handle. In each of these sockets is a V-sleeve packing of Teflon (40_a) which is pressed together by the brass screw cap (37_a). The pressure is transmitted through the parts (38_a) to the V-sleeve packing. The parts (38) are hard polyvinyl chloride. The brass shafts (35_a) (4 mm in diameter) are passed through the V-sleeve packings. The height of these shafts, and hence of the manipulation fittings, can be adjusted by inserting washers at (36). These brass shafts can be turned comparatively easily even at 85°K, so as to actuate the manipulation fittings.

A round brass-rod (42_a) (3 mm in diameter) is clamped to each of the brass shafts (35_a) by the part (41_a). A brass frame (44_a) is soft-soldered to each of these brass rods, and a piece of 10 μ platinum foil (43_b) [shown in Fig. 130(b) only as a line] is cemented to the frame with synthetic resin cement. By this means the membranes can be covered for instantaneous quenching (Section VII.3.3). This special fitting (41–44) can be removed from the cell by unscrewing the parts (41).

VII.3.2. Membrane Support for Low-Temperature Study

Two important facts must be considered in designing a holder. First: We must be able to quench the membrane instantaneously. Hence the thermal capacity of the holder must be as small as possible. Second: The membrane must not tear on quenching. Hence the membrane must not be stretched too tightly in the holder. A sample holder designed with these considerations in mind is shown in Fig. 132. The sample holder consists of two parts (45), made of sheet brass, pressed together with the screws (46). These parts can be inserted in the springs (31) provided for this purposc in the cell [Fig. 130(b)]. A piece of platinum foil (47) is clamped between the brass sheets (45). The cut out of the platinum foil (48) is shown in Fig. 132; it is turned over at the edge. The membrane (50) is laid in the fold (49) and the sample holder with the membrane is then placed in the cell.

VII.3.3. Investigation of a Membrane at 85°K

The sample holder—described in the above section—with the membrane in position, is placed in the cell. The membrane is now hydrated to a specified degree by placing a vessel containing an aqueous salt or sulfuric acid solution in the cell, closing the latter with the Dewar cap (2). When vapor pressure equilibrium is attained in the cells, the polystyrenesulfonic acid membrane is covered with the platinum foil covers (43_b), by means of the manipulation fittings (7). The cell is then opened, the salt solution used to produce the vapor pressure is removed, and the sample is quickly immersed in liquid nitrogen and quenched instantaneously. The water-heating and nitrogen-flushing of the cell windows is then set in operation, liquid nitrogen is poured into the U-shaped vessel, and the cell is cooled. The vessel containing liquid nitrogen, in which the membrane has been quenched, is then removed from the cell, the second radiation shield is placed in position, and the cell is then closed again with its Dewar cap. After about 15 minutes, the membrane in the

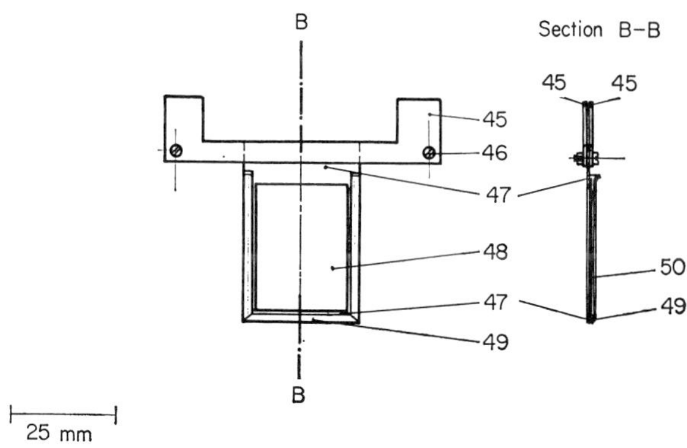

Fig. 132. Sample holder for low-temperature studies.

cell has reached a temperature of 85°K. The cell is now placed in the spectrophotometer, the membrane is uncovered by means of the manipulation fitting, and the spectrum is recorded. For each measurement, i.e., cooling the cell and 1 hour's observations, about 5 l of liquid nitrogen are needed.

VII.4. Evaluation of the Spectra

The Calibration Curve of the Spectrophotometer

To obtain values of wave numbers as exactly as possible, we correct the values obtained from the spectra with the help of a calibration curve. This is particularly important for the investigations described in Chapter IV.

We determine the calibration curve of the spectrophotometer with the help of the calibration bands given in Table 19. These values of wave numbers are obtained from footnote 8, or, if not given there, from footnote 9. Some wave number values which were not to be found in either of these sources are taken from footnote 10. The wave number of the ammonia band at 3333.5 cm^{-1} is taken from a paper by Plyler *et al.*[11]

TABLE 19

CALIBRATION BANDS EMPLOYED

Band	Wave number (cm^{-1})	Band	Wave number (cm^{-1})
air	3904.2 ± 0.1		
air	3852.1 ± 0.1	air	1395.8 ± 0.1
air	3567	air	1336.6 ± 0.1
NH_3	3333.5	polystyrene	1154.3 ± 0.3
polystyrene	3083	polystyrene	1028.0 ± 0.3
polystyrene	2850.7 ± 0.3	NH_3	908.2 ± 0.1
air[a]	2349.9	air	721
polystyrene	1601.4 ± 0.3	air	667.3
air	1436.7 ± 0.1		

[a] Sharp minimum between the strong CO_2 bands.

[8] "Tables of Wavenumbers for Calibration of Infrared Spectrometers." Butterworths, London and Washington, D.C., 1961.

[9] Instruction Manual Model 221 Prism-Grating Interchange. Instrument Div., Perkin-Elmer Co., Norwalk, Connecticut, 1960.

[10] Instruction Manual Ultrared-Spectrophotometer 221 UR–G–105/61. Bodenseewerk Perkin-Elmer and Co., Ueberlingen/Bodensee, West Germany, 1961.

[11] K. Plyler, L. R. Blaine, and M. Nowak, *J. Res. Natl. Bur. Std.* **58**, 195 (1957).

Wave Numbers of the Maxima of Broad Bands

In many cases the bands we have studied are very broad. To determine the exact wave number of the band maximum, we draw the locus of the midpoint of a chord parallel to the wave number axis. The wave number of the point of intersection of these loci with the spectral curve is then the exact wave number of the band maximum. A condition for the success of this procedure is that the band does not overlap another band too strongly and that too powerful interferences are not superimposed.

Overlapping of Broad Bands

If two bands in the spectrum lie so close as to overlap, this produces a shift of the band maxima so that they approach one another slightly. Since the half width, and hence the overlapping of the bands, also depends to some extent on the resolving power of the spectrophotometer, it is important, for precise comparisons, to give the conditions under which the spectra have been recorded. Our spectra were taken with spectrophotometer model 221, with the prism-grating interchange unit, made by Perkin-Elmer GmbH, Ueberlingen/Bodensee, West-Germany. Unless otherwise specified, we use the slit program 960 for investigations with the cell described in Section VII.2.1.2, and the slit program 980 for investigations with the low-temperature cell. If a spectrum has been recorded on another spectrophotometer, this is specified.

The shape of the individual bands can be represented by functions, as summarized by Jones and Sandorfy.[12] If such shapes are assumed for the individual bands of a band complex, and if, in addition, something is known about the individual bands, a band complex can be resolved into its components, as has been done by Busing and Hornig[13] and Schultz and Hornig[14] (see also footnotes 15–18). However, we have not followed this procedure.

[12] R. N. Jones and C. Sandorfy, *in* "Chemical Applications of Spectroscopy" (W. West, ed.), p. 279 ff. Wiley (Interscience), New York, 1956.

[13] W. R. Busing and D. F. Hornig, *J. Phys. Chem.* **65**, 284 (1961).

[14] J. W. Schultz and D. F. Hornig, *J. Phys. Chem.* **65**, 2131 (1961).

[15] J. Pitha and R. N. Jones, *Can. J. Chem.* **44**, 3031 (1966).

[16] R. N. Jones, R. Venkataraghavan, and J. W. Hopkins, *Spectrochim. Acta* **23A**, 925 (1967).

[17] R. N. Jones, R. Venkataraghavan, and J. W. Hopkins, *Spectrochim. Acta* **23A**, 941 (1967).

[18] J. Pitha and R. N. Jones, *Can. J. Chem.* **45**, 2347 (1967).

APPENDIX

(Tables A.1–A.5)

TABLE A.1

THE ASSIGNMENT OF THE BANDS OF POLYSTYRENE, POLYSTYRENESULFONYL CHLORIDE, AND POLYSTYRENESULFONIC ACID AND ITS Na^+ SALT

Polystyrene and perdeutero-polystyrene [Fig. 1(a, g)]	Polystyrenesulfonyl chloride [Fig. 1(b)]	Polystyrenesulfonic acid and perdeutero-polystyrenesulfonic acid [Fig. 1(c, d, h)]	Na^+ salts of polystyrenesulfonic acid and of perdeutero-polystyrene-sulfonic acid [Fig. 1(e, f, i)]
		3700–2500 cm^{-1} OH stretching vibrations of H_2O molecules followed by a continuous absorption toward smaller wave numbers further details in Chapter V	3700–3000 cm^{-1} OH stretching vibrations of H_2O molecules further details in Chapter IV
CH[a]: 3083 cm^{-1} w, 3061 cm^{-1} m, 3027 cm^{-1} ms CD: 2287 cm^{-1} w, 2273 cm^{-1} ms	CH: 3088 cm^{-1} w, 3065 cm^{-1} w, 3030 cm^{-1} vw [a]	CH[a]: 3065 cm^{-1} w, 3030 cm^{-1} vw CD: 2273 cm^{-1} w	CH[a]: 3066 cm^{-1} w, 3030 cm^{-1} vw CD: 2279 cm^{-1} w, 2258 cm^{-1} vw
stretching vibrations of the >CH (>CD) groups of the benzene ring			
2924 and 2851 cm^{-1} —CH_2— 2193 and 2100 cm^{-1} —CD_2—	2930 and 2859 cm^{-1}	2928 and 2856 cm^{-1} —CH_2— 2193 and 2100 cm^{-1} —CD_2—	2927 and 2855 cm^{-1} —CH_2— 2193 and 2100 cm^{-1} —CD_2—
stretching vibrations of the —CH_2— (—CD_2—) group, antisymmetric and symmetric			
		OH: 2950 and 2405 cm^{-1} [b] OD: 2240 and 1805 cm^{-1} [b] Fig. 60 2950 (OH) 2240 (OD) stretching vibration in the hydrogen bridges of the —S(=O)(O····XO)…(OX····O)(O=)S— group	

		2405 (OH) 1805 (OD) first overtone of the OH (OD) bending vibration of these groups	
		2750–1850 cm^{-1} OD stretching vibrations of D_2O molecules followed by a continuous absorption toward smaller wave numbers further details; Chapter V	2750–2200 cm^{-1} OD stretching vibration of D_2O molecules further details, Chapter IV
		at about 1690 cm^{-1}	at about 1641 cm^{-1} at 11% relative atmospheric humidity surrounding the membrane
		scissor vibration of the H_2O molecule	
2000–1700 cm^{-1} four weak bands combination vibrations or overtones of the out-of-plane bending vibrations of the >CH groups of the benzene ring			
1601, 1583, 1494, and 1453 cm^{-1} (H) 1568, 1540, 1376, and 1323 cm^{-1} (D,d_8)[c]	1594, 1487, and 1412 cm^{-1}	1601, 1495, and 1413 cm^{-1} (H) 1568, 1544, and 1320 cm^{-1} (D,d_7)[c]	1602, 1496, and 1411 cm^{-1} (H) 1568, 1544, and 1316 cm^{-1}(D,d_7)[c]
	skeleton stretching vibrations of the benzene ring		
masked 1051 (D,d_8)	1452 cm^{-1}	1451 cm^{-1} —CH_2— 1040 cm^{-1} —CD_2—	1451 cm^{-1} —CH_2— 1044 cm^{-1} —CD_2—
	scissor vibration of the —CH_2— (—CD_2—) group, termed in-plane bending vibration		

Polystyrene and perdeutero-polystyrene [Fig. 1(a, g)]	Polystyrenesulfonyl chloride [Fig. 1(b)]	Polystyrenesulfonic acid and perdeutero-polystyrenesulfonic acid [Fig. 1(c, d, h)]	Na^+ salts of polystyrenesulfonic acid and of perdeutero-polystyrene-sulfonic acid [Fig. 1(e, f, i)]
	1374 cm^{-1} antisymmetric and 1172 cm^{-1} symmetric SO stretching vibration of the —$S(=O)_2$—Cl group	1350 cm^{-1} [b]-antisymmetric and 1172 cm^{-1} symmetric stretching vibration of the SO bonds with double-bond character in the —S(=O)(=O)—OH group[d]	
		at about 1200 cm^{-1} doublet	at about 1200 cm^{-1} doublet
		antisymmetric stretching vibration of the —SO_3^- ion; degeneracy removed, therefore split into two bands for different modes of vibration, further details in Sections IV.1 and IV.2.	
		1034 cm^{-1} (H) 1015 cm^{-1} (D,d_7)	1040 cm^{-1} (H) 1016 cm^{-1} (D,d_7)
		symmetric stretching vibration of the —SO_3^- ion	
1069 cm^{-1}	1081 cm^{-1}	1128 cm^{-1} (H) (1090 cm^{-1} D,d_7) vanishes on drying 1097 cm^{-1} (H) (1056 cm^{-1}D,d_7) appears on drying[b]	1128 cm^{-1} (H) and (1090 cm^{-1} D,d_7)

in-plane skeleton vibrations of the benzene ring with strong participation of the substituents (pp. 12 and 123)			
1181, 1154, 1069, and 1028 cm^{-1}(H) 866, 839, 820 and 786 cm^{-1} (D,d_8)		1011 → 1001 cm^{-1} (H) shifted with progressive hydration. Fig. 2 826 cm^{-1} (D,d_7)	1009 cm^{-1} (H) 826 cm^{-1} (D,d_7)
in-plane bending vibrations of the ⟩CH (⟩CD) groups of the benzene ring; further details, Section V.1.3			
907 and 841 cm^{-1} (H) 759 and 655 cm^{-1} (D,d_8) out-of-plane bending vibrations of the ⟩CH (⟩CD) groups in the benzene ring			
		907^b cm^{-1} stretching vibration of the SO bond with single-bond character in the —S(=O)(=O)—OH group	
760 cm^{-1} (H) 675 cm^{-1} (D,d_8)	836 cm^{-1}	838 cm^{-1} (H) 732 cm^{-1} (D,d_7)	836 cm^{-1} (H) 732 cm^{-1} (D,d_7)
out-of-plane bending vibration of the five ⟩CH (⟩CD) groups in the benzene ring characteristic of monosubstituted benzene ring	out-of-plane bending vibration of the pairs of ⟩CH (⟩CD) groups in the benzene ring		

Polystyrene and perdeutero-polystyrene [Fig. 1(a, g)]	Polystyrenesulfonyl chloride [Fig. 1(b)]	Polystyrenesulfonic acid and perdeutero-polystyrenesulfonic acid [Fig. 1(c, d, h)]	Na^+ salts of polystyrenesulfonic acid and of perdeutero-polystyrene-sulfonic acid [Fig. 1(e, f, i)]
	773 cm^{-1}	777 cm^{-1} (H) 684 cm^{-1} (D,d_7)	776 cm^{-1} (H) 684 cm^{-1} (D,d_7)
	out-of-plane bending vibration of four >CH (>CD) groups in the benzene ring		
698 cm^{-1} (H) 549 cm^{-1} (D,d_8)			
out-of-plane skeleton bending vibration characteristic of monosubstituted benzene ring			
	at 688 cm^{-1} shoulder	671 cm^{-1} (H) 649 cm^{-1} (D,d_7) is shifted by about 10 cm^{-1} toward smaller wave numbers on thorough drying	676 cm^{-1} (H) 649 cm^{-1} (D,d_7)
	ring vibration with strong participation of the substituents (essentially CS stretching vibration)		

[a] The position of these bands was determined precisely by means of extension of ordinate, always using thoroughly dried membranes.
[b] Thoroughly dried membrane studied.
[c] Vibration of polystyrene d_8 or polystyrenesulfonic acid d_7 respectively.
[d] In the case of the —S(=O)(=O)—OD group, this band lies, not at 1350 but at 1340 cm^{-1}.

TABLE A.2

THE ASSIGNMENT OF THE BANDS OF POLYSTYRENESELENONIC ACID AND ITS NA^{+} SALT

Polystyreneselenonic acid [Fig. 4(a, b)]	Na^{+} salt of polystyreneselenonic acid [Fig. 4(c, d)]
3700–2500 cm^{-1} OH stretching vibrations of H_2O molecules followed by continuous absorption toward smaller wave numbers further details, Chapter V	3700–3000 cm^{-1} OH stretching vibration of H_2O molecules further details, Chapter IV
3083 cm^{-1} vw 3055 cm^{-1} m at 3026 cm^{-1} vw shoulder	at 3087 cm^{-1} w shoulder 3047 cm^{-1} w at 3017 cm^{-1} vw shoulder
Stretching vibrations of the $\rangle$CH groups of the benzene ring[a]	
2926 and 2855 cm^{-1}	2924 and 2853 cm^{-1}
stretching vibrations of the —CH_2— group, antisymmetric and symmetric	
OH: 2880 and 2385 cm^{-1} [b]; OD: 2160 cm^{-1} [b] weak with strong degree of hydration, increases on drying 2880 (OH) 2160 (OD) stretching vibration in the hydrogen bridges of the O····XO —Se=O O=Se— OX····O 2385 (OH) first overtone of the OH bending vibration of these groups Fig. 113; further details, Sections V.2. and V.3	

Polystyreneselenonic acid [Fig. 4(a, b)]	Na$^+$ salt of polystyreneselenonic acid [Fig. 4(c, d)]
2750–1850 cm^{-1}	2750–2200 cm^{-1}
OD stretching vibrations of D_2O molecules followed by continuous absorption toward smaller wave numbers further details, Chapter V	OD stretching vibrations of D_2O molecules further details, Chapter IV
1766 and 1722 $cm^{-1\,b}$	
see text, p. 18	
at about 1650 cm^{-1}	1642 cm^{-1} at 11% relative atmospheric humidity surrounding the membrane
scissor vibration of the H_2O molecule	
1589, 1485, and 1412 cm^{-1}	1591, 1488, and 1411 cm^{-1}
skeleton stretching vibrations of the benzene ring	
1449 cm^{-1}	1448 cm^{-1}
scissor vibration of the —CH_2— group	
at about 1260 cm^{-1} broad OH bending vibration in the hydrogen bridges of the —Se(=O)(–OH···O=)···(O···HO–)Se(=O)— group [cyclic dimer: —Se=O / O···HO / OH···O / O=Se—]	
1200 cm^{-1}	1206 cm^{-1}
scissor vibration of the D_2O molecule	
1187^b and 1011 cm^{-1}	1184^b and 1012 cm^{-1}
in-plane bending vibrations of the >CH groups of the benzene ring	
1072 cm^{-1}	1074 cm^{-1}
in-plane skeleton vibration of the benzene ring with strong participation of the substituents	

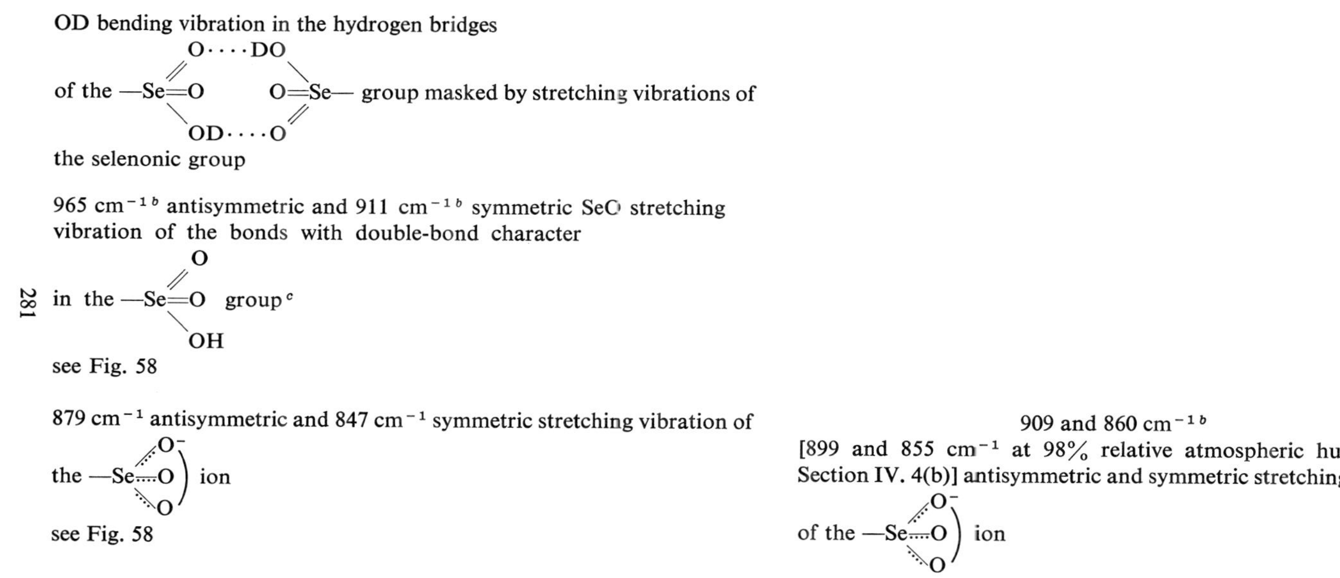

OD bending vibration in the hydrogen bridges

of the —Se(=O)(—O····DO—)(—OD····O=)Se— group masked by stretching vibrations of the selenonic group

965 cm^{-1} [b] antisymmetric and 911 cm^{-1} [b] symmetric SeO stretching vibration of the bonds with double-bond character in the —Se(=O)(=O)—OH group [c]

see Fig. 58

879 cm^{-1} antisymmetric and 847 cm^{-1} symmetric stretching vibration of the —Se(O)(O)(O)$^{-}$ ion

see Fig. 58

909 and 860 cm^{-1} [b]

[899 and 855 cm^{-1} at 98% relative atmospheric humidity, see Section IV. 4(b)] antisymmetric and symmetric stretching vibrations of the —Se(O)(O)(O)$^{-}$ ion

Polystyreneselenonic acid [Fig. 4(a, b)]	Na^+ salt of polystyreneselenonic acid [Fig. 4(c, d)]
831 cm^{-1} [b]	823 cm^{-1}
out-of-plane bending vibration of the pairs of the >CH groups	
the band at 831 cm^{-1} is relatively intense, since a small number of $—SeO_2H$ groups is present, whose SeO stretching vibration lies approximately at this wave number see Table A.3	810–780 cm^{-1} shoulder or weak band, caused by the stretching vibration of a small number of $—SeO_2^-$ ions see Table A.3
725 cm^{-1} [b] SeO stretching vibration of the bond with single-bond character in the $—Se(=O)(=O)OH$ group Fig. 58	
547 cm^{-1}	547 cm^{-1}
ring vibration with strong participation of the substituents (CSe stretching vibration)	

[a] The position of these bands was determined precisely by means of extension of ordinate, always with thoroughly dried membranes.
[b] Thoroughly dried membrane studied.
[c] These two bands are found to be shifted by a few cm^{-1} toward larger wave numbers when a D_2O-hydrated membrane is dried.

TABLE A.3

THE ASSIGNMENT OF THE BANDS OF POLYSTYRENESELENINIC ACID AND ITS Na^+ SALT

Polystyreneseleninic acid [Fig. 5(a, b)]	Na^+ salt of polystyreneseleninic acid [Fig. 5(c, d)]
3700–3000 cm^{-1}	3700–3000 cm^{-1}
OH stretching vibrations of H_2O molecules, further details, Chapter IV and V	
3042 cm^{-1} m 3015 cm^{-1} vw	3037 cm^{-1} w 3016 cm^{-1} w shoulder
stretching vibrations of the >CH groups of the benzene ring[a]	
2924 and 2852 cm^{-1}	2926 and 2854 cm^{-1}
stretching vibrations of the $—CH_2—$ group, antisymmetric and symmetric	
OH: 2960 and 2420 cm^{-1} [b]; OD: 2195 cm^{-1} [b] intensity independent of the degree of hydration 2960 (OH) 2195 (OD) stretching vibration in the hydrogen bridges of the 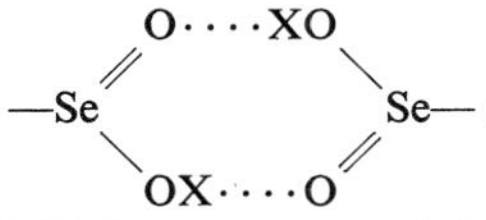 —Se(=O···XO)(—OX···O=)Se— group 2420 (OH) first overtone of the OH bending vibration of these groups further details, Section V.2 and V.3	
2750–2200 cm^{-1}	2750–2200 cm^{-1}
OD stretching vibrations of D_2O molecules, further details, Chapters IV and V	
1765 and 1705 cm^{-1} [b] see text, p. 18	

<table>
<tr><th>Polystyreneseleninic acid
[Fig. 5(a, b)]</th><th>Na^{+} salt of polystyreneseleninic acid
[Fig. 5(c, d)]</th></tr>
<tr><td>at about 1640 cm^{-1}</td><td>at about 1650 cm^{-1} [c]
at 11% relative atmospheric humidity surrounding the membrane</td></tr>
<tr><td colspan="2">scissor vibration of the H_2O molecule</td></tr>
<tr><td>1591, 1485, and 1409 cm^{-1}</td><td>1592, 1487, and 1408 cm^{-1}</td></tr>
<tr><td colspan="2">skeleton stretching vibrations of the benzene ring</td></tr>
<tr><td>1449 cm^{-1}</td><td>1449 cm^{-1}</td></tr>
<tr><td colspan="2">scissor vibration of the —CH_2— group</td></tr>
<tr><td>at about 1260 cm^{-1} broad OH bending vibration in the hydrogen bridges of the —Se(=O)(OH) ⋯ Se— group (O····HO, OH····O)</td><td></td></tr>
<tr><td>1205 cm^{-1}</td><td>1209 cm^{-1}</td></tr>
<tr><td colspan="2">scissor vibration of the D_2O molecule</td></tr>
<tr><td>1187 [b] and 1014 cm^{-1}</td><td>1187 cm^{-1} w [b] and 1015 cm^{-1}</td></tr>
<tr><td colspan="2">in-plane bending vibrations of the >CH groups of the benzene ring</td></tr>
<tr><td>1069 cm^{-1}</td><td>1069 cm^{-1}</td></tr>
<tr><td colspan="2">in-plane skeleton vibration of the benzene ring with strong participation of the substituents</td></tr>
<tr><td>shoulder at about 940 cm^{-1} OD bending vibration in the hydrogen bridges of the —Se(=O)(OD) ⋯ Se— group (O····DO, OD····O)</td><td></td></tr>
</table>

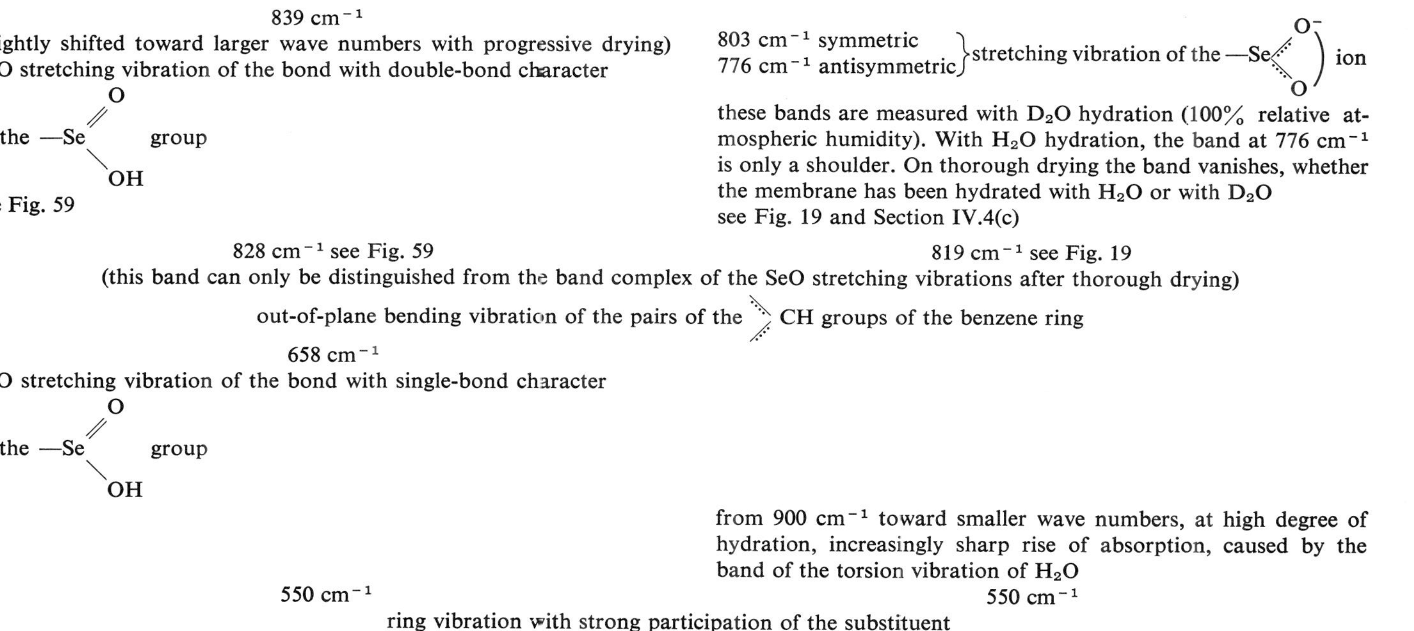

$839\ cm^{-1}$ (slightly shifted toward larger wave numbers with progressive drying) SeO stretching vibration of the bond with double-bond character in the —Se(=O)(OH) group see Fig. 59	$803\ cm^{-1}$ symmetric, $776\ cm^{-1}$ antisymmetric: stretching vibration of the —Se(O)(O⁻) ion these bands are measured with D_2O hydration (100% relative atmospheric humidity). With H_2O hydration, the band at $776\ cm^{-1}$ is only a shoulder. On thorough drying the band vanishes, whether the membrane has been hydrated with H_2O or with D_2O see Fig. 19 and Section IV.4(c)
$828\ cm^{-1}$ see Fig. 59	$819\ cm^{-1}$ see Fig. 19
(this band can only be distinguished from the band complex of the SeO stretching vibrations after thorough drying)	
out-of-plane bending vibration of the pairs of the CH groups of the benzene ring	
$658\ cm^{-1}$ SeO stretching vibration of the bond with single-bond character in the —Se(=O)(OH) group	from $900\ cm^{-1}$ toward smaller wave numbers, at high degree of hydration, increasingly sharp rise of absorption, caused by the band of the torsion vibration of H_2O
$550\ cm^{-1}$	$550\ cm^{-1}$
ring vibration with strong participation of the substituent (essentially CSe stretching vibration)	

[a] The position of these bands was determined precisely by means of extension of ordinate, always with thoroughly dried membranes.
[b] Thoroughly dried membrane studied.
[c] As this band has a very low intensity at 11% relative atmospheric humidity; the wave number given is only approximate.

TABLE A.4

THE ASSIGNMENT OF THE BANDS OF POLYSTYRENEPHOSPHINIC ACID AND ITS NA^+ SALT

Polystyrenephosphinic acid [Fig. 6(a, b)]	Na^+ salt of polystyrenephosphinic acid [Fig. 6(c, d)]
3700–3000 cm^{-1}	3700–3000 cm^{-1}
OH stretching vibrations of H_2O molecules, further details, Chapters IV and V	
3056 cm^{-1} vw 3039 cm^{-1} vw 3025 cm^{-1} w, sharp band	at about 3060 cm^{-1} vw shoulder 3030 cm^{-1} w, broad 3016 cm^{-1} w, sharp
stretching vibrations of the >CH groups of the benzene ring[a]	
2927 and 2854 cm^{-1}	2926 and 2856 cm^{-1}
stretching vibrations of the —CH_2— group, antisymmetric and symmetric	
OH: 2750 cm^{-1} [b] broad and vst; OD: 2060 cm^{-1} [b] broad and st 2750 (OH) 2060 (OD) stretching vibration in the hydrogen bridges of the X–P(=O····XO)–OX····O=P(–X) group (—P=O····XO / OX····O=P group) further details, Sections V.2 and V.3	
2180 cm^{-1} [b] broad and vw first overtone of the OH bending vibration of these groups (but see also Section V.3)	

2750–2200 cm^{-1}	2750–2200 cm^{-1}
OD stretching vibrations of D_2O molecules, further details, Chapters IV and V	
2365 cm^{-1}	2304 cm^{-1} this band is continuously shifted toward larger wave numbers with increasing degree of hydration with 98% relative atmospheric humidity surrounding the membrane, it lies at 2312 cm^{-1}
PH stretching vibration	
1722 cm^{-1} PD stretching vibration of the —P(=O)(D)(OD) group	this H is not exchanged for D at 25°C with D_2O in the case of the Na^+ salt
1655 cm^{-1} [b] see text, p. 27	
at about 1650 cm^{-1} still very weak with 98% relative atmospheric humidity surrounding the membrane. Only clearly recognizable by its decrease with increasing drying (Fig. 116)	1645 cm^{-1} at 11% relative atmospheric humidity surrounding the membrane
scissor vibration of the H_2O molecule	
1602, 1495 (vw), and 1407 cm^{-1}	1602, 1495 (vw), and 1405 cm^{-1}
skeleton stretching vibrations of the benzene ring	
1450 cm^{-1}	1450 cm^{-1}
scissor vibration of the $—CH_2—$ group	

Polystyrenephosphinic acid [Fig. 6(a, b)]	Na^+ salt of polystyrenephosphinic acid [Fig. 6(c, d)]
at about 1170 cm^{-1} a broad, intense doublet PO stretching vibration of the bond with double-bond character in the —P(H)(=O)(OH) group	1183 cm^{-1} [b] 1056 cm^{-1} [b] the first band is shifted toward 1162 cm^{-1} with increasing degree of hydration and the second toward 1045 cm^{-1}. These values are reached at 98% relative atmospheric humidity surrounding the membrane. Further details Section IV.4.b antisymmetric and symmetric PO stretching vibration of the —P(H)(O)(O)$^-$ ion
masked by the broad band at 1170 cm^{-1}	shoulder at 1200 cm^{-1} only clearly recognizable with 100% relative atmospheric humidity surrounding the membrane
scissor vibration of the D_2O molecules	
1131 cm^{-1}	1131 cm^{-1} 1080 cm^{-1} shoulder vanishes on drying
in-plane skeleton vibrations of the benzene ring with strong participation of the substituents	
masked	1016 cm^{-1}
in-plane bending vibrations of the >CH groups of the benzene ring	
see text, p. 26 ff.	985 cm^{-1}
characteristic for the P-phenyl group	
977 cm^{-1}	969 cm^{-1}

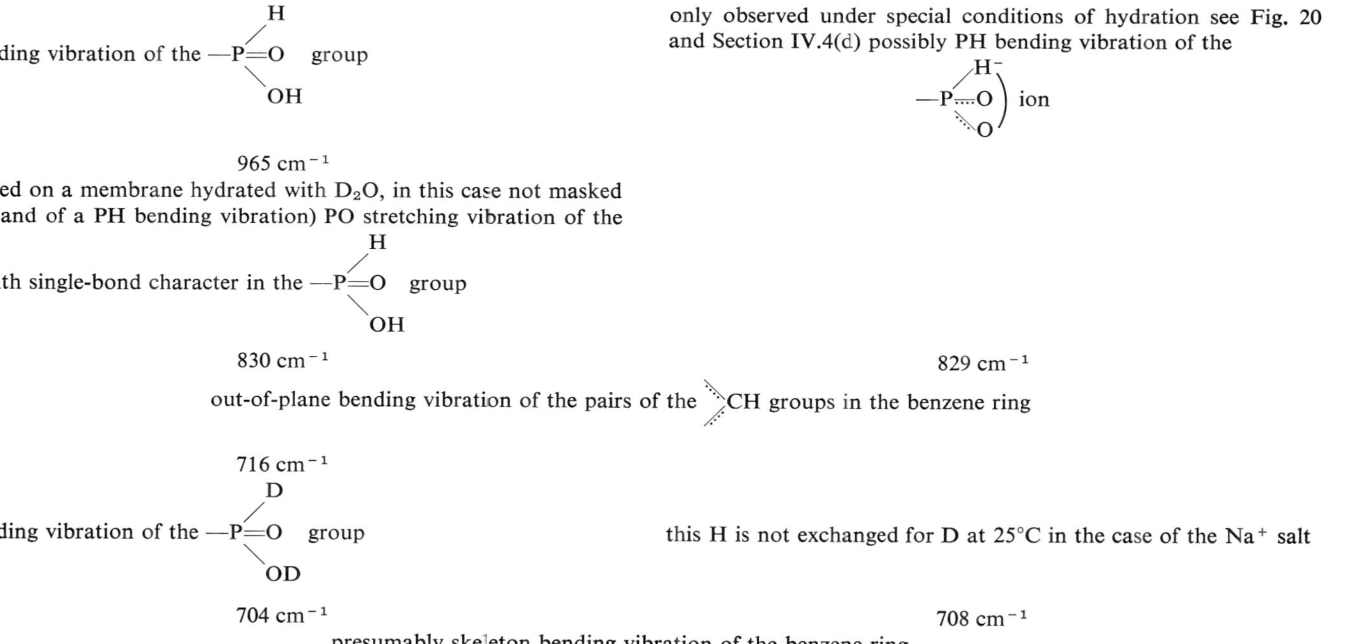

PH bending vibration of the —P(H)(=O)(OH) group	only observed under special conditions of hydration see Fig. 20 and Section IV.4(d) possibly PH bending vibration of the (—P(H)(O)(O))⁻ ion
965 cm^{-1} (measured on a membrane hydrated with D_2O, in this case not masked by the band of a PH bending vibration) PO stretching vibration of the bond with single-bond character in the —P(H)(=O)(OH) group	
830 cm^{-1}	829 cm^{-1}
out-of-plane bending vibration of the pairs of the >CH groups in the benzene ring	
716 cm^{-1} PD bending vibration of the —P(D)(=O)(OD) group	this H is not exchanged for D at 25°C in the case of the Na^+ salt
704 cm^{-1}	708 cm^{-1}
presumably skeleton bending vibration of the benzene ring	
	at high degree of hydration, from 900 cm^{-1} toward smaller wave numbers, increasingly sharp rise of absorption, caused by the band of the torsion vibration of H_2O

[a] The position of these bands was determined precisely by means of extension of ordinate, always with thoroughly dried membranes.
[b] Thoroughly dried membranes studied.

TABLE A.5

THE ASSIGNMENT OF THE BANDS OF POLYSTYRENETHIOPHOSPHONIC ACID AND ITS NA^+ SALT

Polystyrenethiophosphonic acid [Fig. 7(a, b)]	Na^+ salt of polystyrenethiophosphonic acid [Fig. 7(c, d)]
3700–3000 cm^{-1} OH stretching vibrations of H_2O molecules followed by a weak continuous absorption toward smaller wave numbers further details, Chapter V	3700–3000 cm^{-1} OH stretching vibrations of H_2O molecules further details, Chapter IV
3045 cm^{-1} broad, vw 3025 cm^{-1} sharp, vw	3045 cm^{-1} broad and vw
stretching vibrations of the >CH groups of the benzene ring[a]	
2926 and 2858 cm^{-1}	2926 and 2858 cm^{-1}
stretching vibrations of the $—CH_2—$ group, antisymmetric and symmetric	
OH: 2855 cm^{-1} [b,c] broad and vst; OD: 2155 cm^{-1} [b,c] broad and st 2855 (OH) 2155 (OD) stretching vibration in the hydrogen bridges of the $—P(SX)(=O\cdots XO)—OX\cdots O=P(XS)—$ group further details, Section V. 2 and V.3	
2290 cm^{-1} [b,c] broad and vw first overtone of the OH bending vibration of these groups (but see also Section V.3)	

<table>
<tr><td>2750–2200 cm^{-1}
OD stretching vibrations of D_2O molecules followed by weak continuous absorption toward smaller wave numbers
further details, Chapter V</td><td>2750–2200 cm^{-1}
OD stretching vibrations of D_2O molecules

further details, Chapter IV</td></tr>
<tr><td>1690 cm^{-1} [b]</td><td>1690 cm^{-1} [b]
weak shoulder</td></tr>
<tr><td colspan="2">see text, p. 27</td></tr>
<tr><td>at about 1650 cm^{-1} only clearly recognisable by its decrease with progressive drying Fig. 120</td><td>at about 1640 cm^{-1} but not precisely measurable because masked by other bands</td></tr>
<tr><td colspan="2">scissor vibration of the H_2O molecule</td></tr>
<tr><td>1603 and 1412 cm^{-1}</td><td>at about 1590 and 1412 cm^{-1}</td></tr>
<tr><td colspan="2">skeleton stretching vibrations of the benzene ring</td></tr>
<tr><td>1449 cm^{-1}</td><td>1450 cm^{-1}
often almost completely masked by the band at 1435 cm^{-1}</td></tr>
<tr><td colspan="2">scissor vibration of the —CH_2— group</td></tr>
<tr><td></td><td>at about 1435 cm^{-1}
very broad and sometimes very intense, presumably the band found in many P-phenyl compounds (?)</td></tr>
<tr><td>at about 1170 cm^{-1} very intense and broad PO stretching vibration of the bond with double-bond character in the —P(=O)(SH)(OH) group</td><td>1218 cm^{-1} [b]
1037 cm^{-1} [b]
the band at 1218 cm^{-1} becomes a doublet on thorough drying with increasing degree of hydration, all these bands are shifted by about 5 cm^{-1} towards smaller wave numbers (Fig. 21) see Section IV.4.b antisymmetric and symmetric PO stretching vibration of the —P(S⁻)(O)(O) ion</td></tr>
</table>

Polystyrenethiophosphonic acid [Fig. 7(a, b)]	Na⁺ salt of polystyrenethiophosphonic acid [Fig. 7(c, d)]
the scissor vibration of the D_2O molecule is masked by the broad band at about 1170 or 1218 cm^{-1}, respectively	
at about 1130 cm^{-1}	1123 cm^{-1} arises on drying at about 1070 cm^{-1} a shoulder arises on strong degree of hydration (Fig. 21)
in-plane skeleton vibration of the benzene ring with strong participation of the substituents (see p. 52)	
996 cm^{-1} [b] and shoulder at 940 cm^{-1} [b]	979 cm^{-1} [b] is shifted with increasing degree of hydration towards 970 cm^{-1} (value with 98% relative atmospheric humidity surrounding the membrane) characteristic for the P-phenyl group
antisymmetric and symmetric stretching vibration of the $>P(=)(OH)_2$ group in the $—P(=S)(OH)—OH$ group (see text, p. 26 and p. 220 ff.)	
	890 cm^{-1} weak broad
830 cm^{-1}	at about 835 cm^{-1}
out-of-plane bending vibration of the pairs of the $>CH$ groups in the benzene ring	
	from 900 cm^{-1} toward smaller wave numbers with increasing degree of hydration, increasingly sharp rise of absorption, caused by the band of the torsion vibration of H_2O see Fig. 41

[a] The position of these bands was determined precisely by means of extension of ordinate, always with thoroughly dried membranes.
[b] Thoroughly dried membrane studied.
[c] The position of these bands depends on the degree of thiophosphonation. They were measured on the membrane shown in Fig. 7(a, b).

AUTHOR INDEX

Numbers in parentheses are reference numbers and indicate that an author's work is referred to although his name is not cited in the text.

SUBJECT INDEX

B

C

R

S

T